P9-DVR-636

ABOUT ISLAND PRESS

Island Press, a nonprofit organization, publishes, markets, and distributes the most advanced thinking on the conservation of our natural resources—books about soil, land, water, forests, wildlife, and hazardous and toxic wastes. These books are practical tools used by public officials, business and industry leaders, natural resource managers, and concerned citizens working to solve both local and global resource problems.

Founded in 1978, Island Press reorganized in 1984 to meet the increasing demand for substantive books on all resource-related issues. Island Press publishes and distributes under its own imprint and offers these services to other nonprofit organizations.

Support for Island Press is provided by Apple Computers, Inc., The Mary Reynolds Babcock Foundation, The Educational Foundation of America, The Charles Engelhard Foundation, The Ford Foundation, The George Gund Foundation, The William and Flora Hewlett Foundation, The Joyce Foundation, The J. M. Kaplan Fund, The John D. and Catherine T. MacArthur Foundation, The Andrew W. Mellon Foundation, The Joyce Mertz-Gilmore Foundation, Northwest Area Foundation, The Jessie Smith Noyes Foundation, The J. N. Pew, Jr., Charitable Trust, The Rockefeller Brothers Fund, The Florence and John Schumann Foundation, The Tides Foundation, and individual donors.

WAR
ON
WASTE

Can America Win Its Battle with Garbage?

LOUIS BLUMBERG
ROBERT GOTTLIEB

Foreword by Jim Hightower

Washington, D.C. □ Covelo, California

Text design by Irving Perkins Associates
Cover design by Ben Santora

Library of Congress Cataloging-in-Publication Data

Blumberg, Louis, 1948–
War on waste : can America win its battle with garbage? / Louis Blumberg, Robert Gottlieb.
p. cm.
Includes index.
ISBN 0-933280-92-0. — ISBN 0-933280-91-2 (pbk.)
1. Refuse and refuse disposal—United States. 2. Refuse and refuse disposal—Political aspects—United States. I. Gottlieb, Robert. II. Title.
TD788.B58 1989
363.72'8'0973—dc20 89-11169
CIP

Printed on recycled, acid-free paper

Manufactured in the United States of America

10 9 8 7 6 5 4 3

To Ellen and Marge
and our families

ACKNOWLEDGMENTS

The origins of this book stemmed from the Comprehensive Project Report produced in the spring of 1987 by ten graduate students (including one of the authors, Louis Blumberg) connected with the Urban Planning Program at the University of California at Los Angeles. The UCLA LANCER Report, as it was called at the time, was also supervised by Robert Gottlieb, who was ably assisted by Bev Pitman, a doctoral student who was the project's teaching assistant.

As soon as it was released, the UCLA LANCER Report had a direct impact on the issue at hand: the construction of the 1,600 tons per day mass burn incinerator that was to serve as a model facility for the city of Los Angeles as well as municipalities across the country. Its analysis and critique of the rush to incineration and the LANCER decision-making process paralleled the language and position put forward by Los Angeles mayor Tom Bradley on announcing his decision to reverse his prior support and urge cancellation of the project. On the other hand, several of LANCER's supporters, including consultants and attorneys, were unhappy with the Report and communicated their displeasure with the dean of the Graduate School of Architecture and Urban Planning, the chancellor's office, and even the Board of Regents. The Report, however, was also praised by a number of public officials, publications, and community activists, and ultimately received a 1989 national award from the American Planning Association.

In the midst of such strong feedback, the two of us decided to transform the Report into a book, expanding its focus and tackling the overall dimension of the complex and fascinating politics and policies of the world of garbage. In doing so, we were able to rely on the continuing help of two of the members of the Report team, Rubell Helgeson, who daily confronts such issues in her capacity as chief planning aide for Los Angeles City

Council member Ruth Galanter, and Terry Bills, currently with the Los Angeles County Department of Regional Planning.

Many other people helped enormously in various stages of the creation of this book. April Smith did substantial legwork and offered insights with her research skills and public interest passion. We had steadfast support and help from several at the UCLA Urban Planning Program, including Marsha Brown, Erika Roos, Debbie Jolly, Tony Miranda, Donna Mukai, Kimberly Maxwell, Anne Hartmere (the Planning school librarian) and Dean Richard Weinstein. While one of us (Robert Gottlieb) was a visiting faculty at the University of Wisconsin at Madison, he also received support and help from Irene Kringle and Judith Patt, among others.

Several people reviewed the manuscript and offered meaningful advice and comments. For their help, we would like to thank Terry Bills, Rubell Helgeson, Brian Lipsett, Martin Melosi, and Tom Webster. We also would like to thank Mary Burns, Tania Lipschutz, Will Collette, Ellen Connett, Paul Connett, Martha Gildart, Dan Knapp, Peter Montague, Penny Newman, Ariel Parkinson, Ann Peterson, Michael Picker, Drummond Pike, Judith Silver, Portia Sinott, Peter Sinsheimer, Linda Sunnen and Deborra Samuels at the EPA Region 9 library, David Tam, and Mary Lou Van Deventer, as well as the many others who shared their experiences and insights with us.

We also have our friends and family who heard and talked garbage while we piled through the issue; we want to especially thank Stanley and Margery Blumberg, whose support and encouragement were so important, Casey and Andy, for their joy and inspiration, Twyla and Gypsy for their nesting while we worked, and Ellen Friedman and Marge Pearson for their love and patience as deadlines absorbed our lives. And we want to thank once again the UCLA LANCER Comprehensive Project group—Terry Bills, Rick Fisher, Rubell Helgeson, Ted Lopez, Gero Leson, Jim Mather, Adam Relin, Robin Toma, and Mark Wald—for starting the process and producing such a terrific Report.

CONTENTS

FOREWORD

During the Great Depression and World War II, the United States had serious, innovative, and successful conservation and recycling programs. We taught a conservation ethic. In so doing, we saved money, we saved energy, we saved jobs, we saved resources. Just two generations later, we have been taught to be wasteful. Today, our durable goods are anything but durable, designed as they are for planned obsolescence, and nearly all our nondurable goods are sold in throwaway packaging. We produce enormous quantities of waste, then try to bury it or burn it and forget it. But it cannot be forgotten, of course—it washes up on our beaches, it reappears as air pollution, it leaches into our water supply, it comes back to haunt us. A throwaway society is not a sustainable society.

How did it happen that we went from learning to conserve to learning to be wasteful? During the economic and international crises of the 1930s and 1940s, policy-makers saw no choice but to teach conservation and to provide the infrastructure to make it work, practically and economically. But when the end of the war brought the possibility of yet another depression, policy-makers essentially panicked. They no longer wanted us to conserve; they wanted us to buy, buy, buy. They decided that rampant consumerism was needed to keep the economy going. Hence the shutdown of recycling industries, the acceptance of planned obsolescence, the exponential growth of food and fiber processing, and the hyping of products through saturation advertising. Our production-oriented society changed to a consumption-oriented society. When it comes to resource policy, we have lost the conservation and recycling worldview that used to be part of everyday life.

Conservation is nothing new; it is a commonsense approach to which we must return, with ample public policy support. We need to look at our waste as a great economic and environmental opportunity. Every year we toss into city dumps 50 million tons of paper, 41 million tons of food and yard wastes, 13 million tons of metals, 12 million tons of glass, and 10 million tons of plastic. The Environmental Defense Fund has calculated that, as a nation, we throw away enough iron and steel to supply domestic auto-makers continuously; enough glass to fill the twin towers of New York City's World Trade Center every two weeks; enough aluminum to rebuild our commercial air fleet every three months; and enough office and writing paper to build a 12-foot Great Wall coast to coast every year. In one year, a large American city dumps more aluminum than can be mined from a small bauxite mine and more copper than can be mined from a medium-sized copper mine. The opportunity we face is, of course, that we can accomplish more sustainable practices by tapping these resources through reuse and recycling rather than going back to the mines and forests and depleting our store of resources.

Today, our actions are often ahead of our thinking on resource issues, forcing us to fight battles of our own creation. We resort to new technologies as a substitute for comprehensive ideas to deal with the way we use our resources. With solid waste, we can no longer rely on putting all our garbage in our soil, so we tinker with ways to put it in our air through incineration. So attractive is the idea of "making it all go away" that we play a shell game, moving our waste from land to air to water to land. Nature can absorb a certain amount of waste; indeed, nature needs replenishing of nutrients. But large concentrations of garbage in dumps or pollution-belching incinerators and their ash do not accomplish this goal—they merely rob some areas and poison others.

In numerous arenas—from energy production and use to chemically dependent agricultural practices to urban planning and toxic waste management—the shortsighted abuse of nature, resource depletion, and a belief in salvation through technology are the hallmarks of public and private policy. Yet, maintaining sustainable ecological systems requires understanding of natural processes and achieving a harmonious working relationship with nature, none of which can be substituted with technology.

In agriculture, the chemical war on pests has burned out organic material in our soils, created hundreds of resistant strains of agricultural pests, and contaminated our groundwater. The modern marvels of increased production through chemistry—combined with farm policies hostile to family farmers—have bankrupted farmers. As a result, responsibility for the food we eat is placed in the hands of multinational corporations whose primary concern is not whether there will be food, water, and land for tomorrow's children, but whether there will be hefty profits in their next quarterly report.

As part of the consumerism blitz of the last four decades, the food industry has converted from raw or simply processed foods to highly processed and packaged foods. While people can't be forced to eat more, they can be made to pay more through processing and packaging gimmicks. Certainly, some of the move to convenience in the food market is good, but it has gotten completely out of hand. Farmers receive a small percentage of the retail price spent by consumers for food products; the money goes for packaging and processing. Today farmers are paid less for the food they produce than is spent on the package it is sold in! The food is eaten, the package dumped. It shouldn't take too much brainpower to realize that if we value packaging more than we value food, we will continue to get newfangled high-quality packaging, while we get lower-quality food and drive farmers out of business. And still the public must find a way to dispose of the paper, plastic, metal, and glass packaging.

A sustainable society will produce its goods in ways that inure to the benefit of the people producing them and assure replenishment of resources needed for continued production and use. The irony in our system is that we send raw materials from the countryside to the cities, but what the cities send back is unusable waste. As landfill space becomes increasingly scarce in urban areas, more prime agricultural lands are converted into waste dumps, which then leak and render neighboring lands unusable. When we finally run out of land and out of money for the landfill battles, we move on to the next technological fix—incineration.

A voice for environmental sanity is now rising from among those who have to live with the trash, the landfills, the air pollution. These people are housewives, working men and women, minorities, the poor—in short, those who have traditionally been denied voice in the policy-making process. In community after community, people are pressing for commonsense waste management. Their demands confront policy-makers and managers who want to simply and quickly empty the truck or the incinerator so they can fill it again. While we ask that others do what we must do for ourselves, the leaders from the grass roots proclaim what for so long we have denied: we can and must be more responsible for our waste. Neighborhoods are organizing recycling efforts and getting cities to adopt them. When such options are put properly to people, they are proving their willingness to contribute. There is a new *esprit de corps,* with people realizing that they can contribute to saner policies.

In the midst of this reevaluation of our waste policies and practices, Louis Blumberg and Robert Gottlieb remind us of the lessons of history. *War on Waste* provides a history and sociology of solid waste management over the last century. For those who cannot, or will not, remember yesterday, this book chronicles how intentional, if not conscious, choices in solid waste management policy have come together to create the crisis we face today, and it advises us that the problems of solid waste disposal are neither new nor unique. Blumberg and Gottlieb remind us that artificial stimulation of consumption in order to fuel production creates a society founded upon trash and waste, and such a society is unsustainable.

We are that society. We must begin to look for those practices that are environmentally and economically sustainable, not for the next quarterly report, but for generations. Those practices that are not sustainable must be abandoned. We must think out the long-range consequences of our actions and adjust our public policies accordingly.

A long-term, sustainable solution to reducing our mountains of garbage will require us to conserve instead of waste, to recycle instead of dump, to compost instead of bury, to fix instead of buy. It will require a profound readjustment in our economy, and it will create new kinds of jobs to replace old ones. Americans know the right thing to do, if given the chance to do it. When people are a part of the policy debates and the policy-making process, rather than being treated simply as receptors at the end of a production-advertising-consumption scheme, they will act in responsible and innovative ways. We do have the ability to act on our common sense; what we need is an economy, a philosophy, and a public policy that support people in their efforts to act on that common sense.

War on Waste offers options for the future. We can no longer just send our growing garbage pile to new technological shrines; we must, instead, finally accept responsibility for the garbage crisis, broaden our vision, and let our actions follow that vision. The litany of options—waste reduction, reuse, and recycling—is not new, and the authors make no claims as to their popularity. But given the growing mountains of waste threatening to bury us before we bury them, we will soon have no choice.

Jim Hightower
Commissioner of Agriculture,
State of Texas

INTRODUCTION

It is 5:00 P.M., the evening of December 31, 1999. The trash shuttle *Refuse 1* has just landed at Vandenberg Air Force Base near Los Angeles from its daily trip to the moon. Along with dozens of other shuttles that make similar runs, these space ships are charged with disposing of an increasing percentage of the country's solid waste. The space disposal program, which began in 1995, was established in response to the continuing closure of landfills and difficulties facing other disposal technologies.

As recently as ten years ago, when the program was first proposed, the idea of hauling trash to the moon had been dismissed as absurd. "It would be prohibitively expensive," critics warned. But with the sharp escalation of the costs of earth landfilling and incineration, the two favored options before the trash shuttle, moon disposal costs have become quite favorable in comparison.

Recycling, of course, is still trotted out prominently as a component of the moon-based waste management strategy. Waste planners argue that both approaches are compatible, though it is clear that the trash shuttle program also depends on certain economies of scale, requiring that a substantial volume of waste be guaranteed. Reducing the level of waste at its source, and thereby restructuring the production process, remains unpopular with the designers of the solid waste moon disposal program. Their slogan, "Less Waste Means Fewer Jobs," has been an effective rejoinder to the growing community opposition to moonfilling.

Since the Division of Wastes was established as a division within NASA, the trash-to-moon project has actually begun to turn a profit. Prospects for future moonfilling appear bright, with most industrialized nations competing for shuttle space. Scientific studies show that moonfill capacity will not be exhausted for 300 years. Furthermore, federal officials at the NASA Division of Wastes have also devised plans for new hazardous waste moon sites. Critics worry about potential problems with the moon environment,

> but the moonfill proponents argue that a waste site, after all, is nothing more than the environmental equivalent of a moonscape.

An absurd fantasy? Perhaps. But today local, regional, and federal governmental agencies are scrambling in search of a successful way to get rid of the country's growing volume of solid wastes. Policy-makers are eager for quick solutions. They are predisposed toward technical fixes and "one-stop" resolutions to the disposal dilemma. But when these methods fail, as they have in the past and continue to do today, the search becomes more frantic, more desperate. We are left then with such images as trash barges sailing the high seas in search of dumping grounds—modern-day scenarios as incongruous as the "trash shuttle" itself.

This increasingly intractable problem of dealing with the trash is not simply a present-day issue. The same concerns preoccupied municipalities for more than 100 years, gradually becoming more complex as they increased in magnitude.

The solutions for what to do with the trash have, over the years, also turned out to be problematic. Whether wastes were incinerated, separated, cooked, reduced, dumped on land or on sea, or fed to the hogs, the techniques became too costly, too harmful, too controversial. All these methods finally would be supplanted in the 1950s by a cheaper, more proficient technology. The era of the landfill had arrived, fueled by cheap land and expanding cities, and utilized by industries and municipalities that generated new and larger volumes of wastes.

But landfills, made attractive in part by their disguised economics (costs were determined by what it took to bury the trash, not to clean up the resulting pollution), had a relatively short reign. New problems emerged, changing both the political and economic balance sheet. Regulations were developed, community opposition became an explosive factor, and the question of environmental hazards rose to the fore. By the 1980s, landfill space was running short, and municipalities, on the front lines of the waste issue, began to seek alternatives.

Then, almost as suddenly, a new way out of the dilemma presented itself: burn the trash rather than bury it, but this time recover energy in the process. With barely time to design their Requests for Proposals (RFPs), public officials eagerly jumped onto a waste-to-energy bandwagon, placing considerable hopes on what they perceived to be a comprehensive, low-risk, cost-efficient, high-tech solution. These waste managers were particularly attracted to the technology of *mass burn incineration,* plants that burned unsorted municipal solid waste in high-temperature incinerators, and also generated electricity. The remaining ash residue would then be hauled away to local landfills. By the mid-1980s, incinerators had come to be widely touted—by EPA officials, waste industry executives, politicians, and sanitation engineers—as the best way to reduce the amount of trash for landfilling.

This incineration-based strategy was developed in major metropolitan areas such as New York City, Baltimore, Chicago, and Detroit; mid-sized cities such as Bellingham, Washington, Fresno, California, and Tulsa, Oklahoma; and even such smaller communities as Key West, Florida, Portsmouth, New Hampshire, and Long Beach, New York.

Between 1980 and 1987, when this option peaked, more than 100 mass burn incinerator plants came on line, burning upwards of 45,000 tons per day of garbage. These facilities disposed of a little less than 5 percent of the municipal waste stream, up from 1 percent at the end of the previous decade. More impressively, 210 *additional* plants were being proposed, of which 72 had reached an advanced stage of construction by January 1988. These plants would handle, according to U.S. Environmental Protection Agency (EPA) estimates, 190,000 tons of garbage per day by 1990, about 20 percent of the waste stream. Some public officials were predicting that incineration could become, as early as the year 2000, *the* reigning method of disposal, even surpassing landfills in the amount of trash disposed.

Los Angeles was one of those cities that decided to make the shift from landfills to incineration. In the wake of political controversies in the late 1970s surrounding the area's declining landfill space, the city's Bureau of Sanitation began to reevaluate a new waste management strategy. Within a few years, the choice had been made. The city would build, over the next couple of decades, three huge mass burn incineration facilities, each capable of burning 1,600 tons per day of the city's growing waste stream.

These three plants, known collectively as LANCER (Los Angeles City Energy Recovery Project), were to handle from 55 to 70 percent of the city's municipal waste stream. The reliance on incineration underscored the skepticism of city officials toward other, more environmentally benign waste management alternatives such as recycling, composting, and waste reduction.

Between 1977, when the Bureau of Sanitation's search first began, and 1987, when a final decision on the fate of LANCER was announced by Los Angeles mayor Tom Bradley and ratified by the City Council, a sharp, protracted, and contentious debate over incineration took place, parallel to similar debates taking place throughout the country. This debate provided a revealing glimpse into the politics of garbage and the search for disposal strategies among municipalities and regional, state, and federal agencies, which have come to regard solid waste problems as one of the most demanding, controversial, and potentially expensive items on their agenda.

Intrigued by these issues, the two of us—and nine others—participated in an intense one-year investigation of the LANCER project and its relation to the key issues of solid waste management. One of us (Robert Gottlieb), as a member of the faculty of the UCLA Graduate School of Architecture and Urban Planning, supervised the project, including the research and writing of a 450-page report that was released in June 1987. The other (Louis Blumberg) joined the nine other master's and doctoral students from the schools of Urban Planning, Environmental Science and Engineering, and Law in researching and producing the report.

This report was undertaken as part of the UCLA Urban Planning Program's Comprehensive Research Project, a group effort produced on behalf of a specific client to explore, analyze, and make recommendations on a topic of major policy significance. California state senator Art Torres (D—Los Angeles), who monitored both the incineration and solid waste issues in his capacity as chair of the Senate Committee on Toxics and Public Safety Management, became the group's primary client, with California Assembly member Tom Hayden (D—Santa Monica) also serving as project client. At the time, the LANCER debate was at its fiercest and the UCLA report itself became a subject of interest and debate. After the report's release, events moved quickly toward a climactic decision on LANCER's fate—an outcome that would have national significance.

This book is an outgrowth of that report and subsequent research into the continuing controversies surrounding the politics of garbage. The material from the report has been greatly expanded and restructured, with the focus shifting to the national scene and the key issues and different players involved. The book explores the history of the waste issue and analyzes how contemporary policy debates represent basic views on our system of production and consumption. It is a portrait of a complex and changing arena of public life, one which touches directly on our daily lives and our surrounding communities.

The waste issue first became prominent in the late nineteenth century, when sharp debate emerged over the appropriate means of collection and disposal; a debate that would influence the direction of solid waste policy well into the twentieth century. Fifty years later, the waste issue refocused on dramatic and widespread changes in the patterns of production and consumption that fundamentally altered the nature of the waste stream. These changes coincided with the rise of the "sanitary landfill." But landfills raised concerns about contamination of the land, the water, and the air, concerns that intensified in the late 1960s and 1970s. A new era of uncertainty had arrived, compounded by the rapid rise and eclipse of incineration and the uneven search for other alternatives.

This book is about our contemporary politics of garbage, its historical background, and its future direction. It focuses on the decline of landfills and the reemergence of incineration as a catch-all disposal strategy as well as the growing social movements that have emerged to oppose and, at times, undermine a successful implementation of this strategy. It explores the economic, environmental, and health questions that play a key role in evaluating the different waste disposal options. It looks at the rapidly expanding waste industry and the public officials and agencies who confront these solid waste dilemmas. And it provides a view of the conflicts and turmoil associated with the development of alternative strategies and their challenge to existing arrangements and policies.

This book, ultimately, is about how policy is framed, whether at the level of government decision-making, industry activity, or community action. It is a book that raises questions about the changes that have been wrought by our industrial and urban order and the trends that are still unfolding. And with the talk of impending crisis on the lips of policy-makers, the fanciful image of *Refuse 1* transporting its solid waste load to the moon can also be construed, in an age of unresolved industrial and urban direction, as one possible metaphor of the times.

PART I

THE CONTEXT: SEEKING THE FIX

1

THE GROWTH OF THE WASTE STREAM: THE NARROWING OF OPTIONS

EXPORTING GARBAGE

A garbage crisis, we are told once again, is at hand. As a nation, we have begun to worry that the growing mounds of wastes will only continue to increase as the means of disposal become further restricted. "We're filling up our lives with garbage," a singer laments, as government agencies and public officials urgently cast about for a solution. The waste dilemma—the squeeze on options even while we produce more trash—has become the centerpiece of the politics of garbage.

This mood of crisis manifests itself in countless ways, including extraordinary attempts to export the problem, here or abroad. Numerous municipalities, counties, and states, particularly those with heavier concentrations of industry and greater urban density, have attempted to send their wastes to less dense, often poorer areas. This has created what Ohio congressman Thomas A. Luken, chairman of the Subcommittee on Transportation, Tourism, and Hazardous Materials, calls "a garbage war between the states." California seeks to dispose in Arizona, New York looks to Vermont, Minnesota makes a move on Iowa.[1] New Jersey, especially, has been an active exporter, probing the possibilities of dumping its wastes in Pennsylvania, Kentucky, Ohio, and West Virginia. These states, though constrained by the commerce clause of the Constitution, have nevertheless sought to pass legislation to halt New Jersey's aggressive export policy.[2]

But it is the city of Philadelphia and the saga of its ash barge that provides perhaps the most striking example of this form of garbage imperialism. During the 1980s, Philadelphia sought to rely on incineration to reduce the amount of its municipal trash earmarked for distant landfills. As a consequence, local officials were stuck with a new, and more difficult, problem: how to dispose of the city's incinerator-generated ash, particularly after residents sued to compel the city to remove thousands of tons of ash residue piled up near two of the city's main incinerators. In an ironic twist, the ash had been hauled to an overextended and actively exporting New Jersey, until that state intervened. Some Philadelphia ash was also shipped to Virginia, until local residents near the landfill successfully stopped it. When Philadelphia contracted with a private landfill in Ohio, nearby residents formed a human chain around the dump when the ash arrived, effectively blocking the trucks carrying the unwanted ash. This contract, like all the others, was also withdrawn.

With its land-based disposal options under attack, the city finally arranged with two private companies to ship the ash abroad. One of the two cargo ships, a seventeen-year-old freighter named the *Khian Sea,* immediately took to the seas to find a disposal spot outside our country's borders. Loaded with more than 14,000 tons of Philadelphia's ash, the tankers wandered through the Caribbean, first stopping in the Bahamas, where the cargo was rejected. Next stop was Panama, where a hastily arranged deal collapsed after the international environmental group Greenpeace released EPA memos detailing the heavy metals and toxic dioxins present in the ash.

The *Khian Sea* then left Panama, seeking a new haven for its cargo. Strife-torn Haiti was the next port of call, where another deal was arranged, this time with Haitian military leaders who agreed to dump the ash on a beach near the port of Gonaïves. To minimize trouble, local dockworkers were told that the cargo consisted of nothing more than fertilizer and did not present a health hazard. But when that country's political opposition—alerted once again by Greenpeace—found out about the dumping and brought it to the attention of the public, the action was brought to a halt, although an undetermined amount of the ash, possibly as much as several thousand tons, remained behind.

The *Khian Sea,* subsequently renamed twice, hopefully to lessen attention to its mission, once again took to the oceans, leaving the Atlantic for other continents and destinations. Through 1987 and 1988, the ship sailed in vain, traveling to Africa, then Sri Lanka, and finally Singapore, where its cargo mysteriously disappeared. The freighter's last owner, an obscure, shadowy company named Romo Shipping, sent a message, ultimately forwarded to Philadelphia officials, claiming the ash had indeed been discharged but declining to say where. "Owners will not advise location," the message read, "because world press and waste monopolies would probably have instigated usual false propaganda against shipping cargo, creating frustration of discharge." Court documents later suggested that the ash barge, a kind of stateless and homeless by-product of the garbage crisis, had actually dumped the ash in the Indian Ocean.[3]

The Philadelphia experience, it turns out, has become the rule rather than the exception in the costly and sometimes bizarre search to dump the trash. Exporting scandals—in which incinerator ash or other wastes are either unloaded illegally or under questionable circumstances—have taken place in a number of African and Latin American countries, such as Nigeria and Guinea-Bissau, and even in England, which has become a "haven for garbage" because of its relatively lax standards.[4] In the United States, an increasing number of cities are now confronting the kinds of problems Philadelphia has experienced. In the process, community has been pitted against community, county against county, state against state, and even country against country.

The solid waste dilemma is not limited to the issue of garbage export. It ultimately raises questions about the source, volume, and nature of the wastes that are being generated. Policy-makers have placed a special emphasis on disposal technologies as they seek a solution they hope will be sufficiently risk-free and cost-effective. This end-of-the-pipe approach is just one more example of the prevalent strategy of pollution control that pervades environmental policy in this country. Rather than seeking a strategy to prevent pollution, with its demands on the system of production and consumption, high-tech disposal strategies are predicated on avoiding both political fallout and public input. They are designed as crisis management techniques, presumably capable of offering immediate answers and comprehensive solutions.

This, however, is not the first time crisis talk has been associated with the politics of garbage. The history of wastes, particularly municipal solid wastes, can be traced to the growth of the modern city and the consolidation of the first industrial revolution. While the problem has grown in magnitude and some crucial components of the waste stream have changed qualitatively as well as quantitatively, the issues of garbage and its disposal are deeply rooted in contemporary urban and industrial life.

GARBAGE AS GOLD

The modern conception of solid waste management first emerged in the 1890s under the influence of the sanitary reform movement and the municipal reforms associated with the Progressive Era.[5] Responding to the sanitary problems generated by the rapid industrialization and urbanization of the second half of the nineteenth century, reformers in many urban centers began to lobby for municipal control over the collection of urban wastes. Prior to that point, waste had been viewed largely as an individual responsibility, involving a rather limited governmental role. As late as 1880, according to census figures, only 43 percent of all cities provided some minimum form of collection. Just 24 percent of the cities maintained municipally operated garbage collection systems, and an additional 19 percent contracted out for the service.[6] This lack of public intervention exacerbated the steady accumulation of garbage in the streets. Compared with Europe, where disposal technologies had been more substantially developed, there was a higher per capita generation of waste in the United States. These wastes, moreover, were growing at a faster rate.[7]

Increasingly, public health reformers linked sanitation practices to the incidence of disease and health risks in urban areas. In an era of "sewer socialism" and progressive reform, municipal governments quickly responded, partly to preempt state intervention and to maintain a measure of "home rule." By the turn of the century, a growing number of cities provided at least a rudimentary level of solid waste collection and disposal. By 1915, 50 percent of the major cities had some form of municipally controlled collection service, while an additional 39 percent provided the service under private contract. By the late 1930s, the figure for services offered or contracted by cities with populations over 100,000 had jumped to 100 percent.[8] Still unresolved was the related question of what methods to use for collection and disposal.

From the outset of the Progressive Era, the problem of how to approach the collection of wastes preoccupied municipal officials as much as any other metropolitan issue. The relatively disorganized and limited forms of collection available in the nineteenth century were compounded by a diverse and rapidly growing waste stream, from horse manure to an increasing array of industrial wastes and the mounds of garbage that rotted on sidewalks and in the streets and alleyways. Following the example of New York City, a group of sanitary engineers in the 1890s argued for a system of primary or *source* separation of waste at the household level, which would allow for the recovery of the marketable elements of the waste. These could then be sold to help defray the city's overall costs. Further, source separation advocates argued that such a waste stream would be more sanitary, with the perishable component of the wastes collected and disposed of on a more frequent basis.[9]

Another faction of engineers advocated a combined system of collection, in which all of the elements of household waste were collected simultaneously. This approach, it was claimed, would work easier at the household level, while reducing costs in the long run with a simpler method of collection.[10]

The division between source separation and combined collection preoccupied engineering societies and became a frequent subject of debate at meetings of the American Public Health Association. Different cities ultimately adopted one or the other method, determined in part by their disposal strategy.[11]

Similar to the collection debate, the issue of refuse disposal became an area of contention. By the turn of the century, a variety of waste disposal practices had come to be adopted by municipalities, ranging from land and water disposal (including ocean dumping) to incineration, reduction, or some combination of methods (see table 1.1). Water and ocean dumping came under the greatest attack with the rise of the "public sanitarian" movement. Dumping wastes in surface waters was seen as merely shifting one community's waste to another, while the public health impact assumed only secondary importance. The practice of ocean dumping was finally ruled illegal in 1933 for municipal solid wastes, although industrial and commercial wastes continued to be exempted. These economic interests justified their largely unregulated discharge practices by identifying streams and other water sources as "nature's sewers," a position largely unchallenged by municipal officials and engineers.[12]

Table 1–1
WASTE DISPOSAL PRACTICES

METHOD	1899	1902	1913
	%	%	%
Dumped on land	70	46.5	61
Dumped in water	3	2.5	3
Incineration	16	29.5	7
Sanitary fill	—	—	7
Combination of methods	—	1.5	11
No systematic method	11	.5	—
No data	—	19.5	11

SOURCE: *Garbage in the Cities,* Martin Melosi, p. 164.

From the 1880s to the 1930s, land dumping remained the most prevalent method of waste disposal, though hardly favored by public health officials. Already by the 1890s, concerns were being raised about the health risks posed by large open dumps. At the same time, it was quickly becoming apparent that the sanitary engineers, who themselves were beginning to dominate waste management decisions, also frowned upon land disposal methods. They preferred instead either of the two methods—incineration or reduction—most directly linked to the debates over collection methods.

In those cities where wastes were incinerated, combined collection emerged as the preferred method. Cities that relied on some form of reduction activity were more likely to encourage collection based on household separation. By World War I, neither method had fully displaced the other, and even within the same city, different methods prevailed on a neighborhood-by-neighborhood level.[13]

In the case of incineration, sanitary engineers borrowed heavily from the methods then being developed in Europe, including mobile and stationary systems. The introduction of incinerators in the United States, however, involved a series of bad choices and eventual failures. Problems of faulty design and construction, in addition to inadequate preliminary studies, contributed to a widespread malfunctioning of these systems. Often the U.S. incinerators were used only to burn wet garbage without the inorganic materials that the European incinerators relied upon to maintain combustion. As a result, of the 180 incinerators built in the United States between 1885 and 1908, 102 had been abandoned by 1909.[14]

Two years later, however, engineers were proclaiming the arrival of a new generation of incinerators. In the decade after 1910, incineration came into widespread use. By the 1930s, at its peak, between 600 and 700 cities had built incineration plants. Avoiding some of the design problems of the earlier technology, incineration from a stationary source became a significant method of disposal of municipal wastes in the wake of the Progressive reforms.[15]

Beyond straight incineration as a waste reduction system, the British and Germans led the way with technology to produce energy from incineration. The first plant to produce electricity from incineration was developed in Great Britain in the mid-1890s. By 1912, 76 such plants in Great Britain produced energy, as did 17 more on the Continent.

Although a pilot project was built in New York City in 1905, "waste-to-energy," as the technology became known, failed to develop in the United States for another 60 years. The cost of producing energy significantly increased total costs, which contrasted with other, relatively cheap energy sources. Further, the emergence of the regulated monopoly utility (another reform of the Progressive Era) reduced the possibility of establishing a waste-to-energy technology as an alternative energy source. Despite the appeal of this progressive-sounding, "multiple use" process, the more limited (non-energy-generating) form of incineration predominated.[16]

The other major innovative method of the 1890s involved the technology of reduction. This essentially entailed "cooking" the garbage to extract a variety of marketable by-products, including grease and tankage, the dried animal solids sold as fertilizer. The history of reduction paralleled that of the early incineration projects. Also based on European technologies, the first-generation reduction plants (most of which were built in the Midwest) proved to be failures. By 1914, only 22 of the 45 plants built in the United States were still in operation.[17]

During World War I, the reduction concept was given additional backing by government officials concerned about war-related materials scarcity. Key ingredients for military products such as glycerine and nitric acid could be derived as garbage by-products, and some municipalities—New York City, for example—actively pursued them as income-generating.[18] After the war, however, interest in garbage by-products declined.

The reduction method was limited by its inability to dispose of the inorganic elements of the waste stream. As a result, only 10 to 30 percent of the total solid waste stream in this period could be disposed of through reduction. Moreover, given the need to allow organic wastes to putrefy, odor problems became significant, and lawsuits against the plants were common. As a result, few reduction plants remained in operation after World War I.[19]

The reduction process nevertheless reflected a new willingness among sanitation engineers to pursue recycling strategies. These approaches were based less on environmental concerns such as reducing the volume of wastes than on their efficiency and cost features. The perception of waste as a sanitary and aesthetic problem was complemented increasingly by the concept of "waste as wealth," which was tied to the overall conservationist or utilitarian approach regarding resource policy in the early 1900s. Articles with such titles as "Profits from Garbage" and "Revenue from Municipal Waste" appeared in the public health and sanitation engineering publications of that period.[20]

In addition to the reduction method, there were several other processes that sought to utilize solid waste by-products. One process, patented in Texas in 1915, transformed solid waste into fuel bricks called "oakcoal." A British technology from around the same period, known as "coalesine," produced a fuel briquette from pulverized waste. There were also a number of attempts to utilize the ash from incinerators for building materials. The most successful of these was adopted in New York, where ash and street sweepings were used as landfill along the shoreline.

As the problems of solid waste disposal continued to preoccupy municipalities, some older, less "scientific" methods were reintroduced. A relatively widespread method that came into prominence around the turn of the century was swine feeding; that is, directing the garbage component of wastes to hogs and other animals as feed. While it never developed in many of the larger cities, especially those along the Atlantic seaboard, it became a significant factor in the solid waste disposal activities of several smaller and medium-sized cities, including Los Angeles. Swine feeding was actively encouraged during World War I as a means of increasing food production; by 1917, 35 percent of all cities monitored in one survey actively utilized this method. That figure increased to 44 percent in 1925, and then leveled off at 39 percent by 1930.[21] When a series of swine epidemics occurred in the 1950s and several operations were shut down, including one involving the U.S. Army, public health regulations were issued to prevent the feeding of raw garbage to pigs. While "cooking" the garbage solved this particular set of problems, the additional cost proved to be too much, and by 1970, only 4 percent of all collected food waste was being used as swine food.[22]

"THE LOOK IS YOU"

In 1960, a prolific and popular writer named Vance Packard produced yet another in his string of exhortative best-sellers criticizing the system of values and choices that had come to characterize post–World War II American society. Entitled *The Waste Makers,* Packard's book opened with a poignant quote from novelist Dorothy Sayers, who wrote: "A society in which consumption has to be artificially stimulated in order to keep production going is a society founded on trash and waste, and such a society is a house built upon sand." Elaborating on this theme, Packard proceeded to discuss such key marketing and production-related practices as planned obsolescence and "throwaway" products. As a result of these changes in product design and packaging, Packard warned, a waste crisis was in the making.[23]

Packard's analysis, though substantially integrated into the popular discourse about waste, tended to be dismissed by both policy-makers and academics as an overly dramatic popularization of themes and a misstatement of issues. The rapid changes in the industrial base and the patterns of consumption that had developed in the post–World War II era were still being celebrated as contributing to America's national prosperity and dominance in international markets. The notion of waste, Vance Packard's best-selling books notwithstanding, was rejected in favor of the concept of "productivity" to describe a system based on producing more goods and expanding consumer markets. The related concept of "consumerism" highlighted, for both marketing experts and economists, the presumption that the system could meet "human needs" by offering a wide range of differentiated products.[24]

The period from the Progressive Era to the 1960s that Packard focused on can be seen as a prologue to the contemporary period of crisis and contention around solid waste issues. During this period, the per capita generation of wastes increased significantly while the composition of the waste stream changed as well. New, potentially hazardous consumer and industrial products and processes entered the marketplace, coinciding with a major revolution in packaging and materials use. These changes were tied to the rise of the petrochemical industry, which in turn substantially influenced this transformation of the waste stream (see table 1.2).

Table 1–2
MATERIALS DISCARDED INTO THE MUNICIPAL SOLID WASTE STREAM
(IN MILLIONS OF TONS AND PERCENTS)

	1970		*1984*		*2000*	
MATERIALS	TONS	%	TONS	%	TONS	%
Paper and paperboard	36.5	33.1	49.4	37.1	65.1	41.0
Glass	12.5	11.3	12.9	9.7	12.1	7.6
Metals	13.5	12.2	12.8	9.6	14.3	9.0
Plastics	3.0	2.7	9.6	7.2	15.5	9.8
Rubber and leather	3.0	2.7	3.3	2.5	3.8	2.4
Textiles	2.2	2.0	2.8	2.1	3.5	2.2
Wood	4.0	3.6	5.1	3.8	6.1	3.8
Other	—	0.1	0.1	0.1	0.1	0.1
Food wastes	12.7	11.5	10.8	8.1	10.8	6.8
Yard wastes	21.0	19.0	23.8	17.9	24.4	15.3
Miscellaneous inorganics	1.8	1.6	2.4	1.8	3.1	2.0
Totals	110.3	100.0	133.0	100.0	158.8	100.0

SOURCE: "Characterization of Municipal Solid Waste in the United States, 1960–2000," Franklin Associates, EPA, Office of Solid Waste, PB-178323, 1986.

The growth of advertising with its impact on production decisions and consumer habits also played a critical role in restructuring the waste issue. Advertising became responsible, in part, for the system's tendency toward overproduction and the elevation of waste generation into an essential by-product of economic activity. With the advent of television, advertising's "perfect medium," advertising mushroomed into a billion-dollar industry and a central facet of production.[25]

The expanded level of advertising and marketing activities was most pronounced in influencing the role of packaging in the production system. Previously, packaging had been basically a subset of the manufacturing process. There were no trade associations, no consulting firms nor discrete spheres of activity such as packaging design. Nor were there specific organizational or financial operations distinct from the manufacturer itself. Packaging was functional, not a strategic element in a successful sales campaign, as it eventually evolved.[26]

The shift in packaging's role during the post–World War II era, when both advertising and petrochemicals became dominant factors in the production system, was thorough and profound. The concentration of production and distribution in the food and consumer products industries meant that small firms serving local markets were supplanted by a few companies serving national and international markets. These trends were encouraged by the changes in packaging, while at the same time they helped precipitate the shift in packaging's role. Packaging, for example, had direct bearing on the rise of supermarkets and the demise of the general store. And, with the elimination of the salesclerk, the package became, as Walter Stern, a packaging industry figure, pointed out, the producer's "sole representative at the sales decision point." Relying on advertising messages and product claims, the package, Stern continued, conveys "the message he wants to communicate. Look. This product is different. This product will satisfy, benefit, totally respond to your need. It will really do what it says it will do. You won't like it only now. You'll like it in your home. Even weeks from now."[27]

By the 1960s, packaging had been elevated to equal status with manufacturing itself, and was considered an essential component in the movement of goods from producer to consumer. The era of the "growing package," as one *Time* magazine article put it, had emerged, associated with such production innovations as "disposable," "one-way," or single-use products and "convenience living."[28] Packaging now vied with advertising for a lion's share in the costs of production, and the two activities became so fully intertwined that most of the technical and instrumental capabilities of the package, such as its ability to extend product life, had become secondary to its marketing role. "Today," one industry speaker pointed out at a 1969 conference on packaging wastes, "packaging is recognized as too important to be assigned anything less than a specific responsibility and strategy in the goals and management procedures of most companies. . . . Packaging is too closely related to the company's profit and loss picture," the speaker concluded, "and it is too definitely connected to a firm's growth potential to be left to chance."[29]

This 1969 conference billed itself as the first of its kind and was largely dominated by industry interests such as package users, designers, and producers, as well as the plastics, glass, paper, and ferrous metal companies and trade associations. The meeting was instructive for both its representations of the problem and its recommendations for possible solutions. Industry-related groups with such names as the California Anti-Litter League ("People, Not Materials, Create Litter" was its motto),[30] the Foundation for Conservation of our Environment, the Package Design Council, Package Engineering, and Flexible Plastics Corporation, as well as major producers such as Dow Chemical, Crown Zellerbach, Continental Can, and Monsanto, all argued that while waste disposal had indeed become a problem, in part because of packaging growth, government intervention was not the answer.

The growth of packaging wastes, in fact, at least in terms of measurement by weight, had peaked by this period. Though increases during the 1940s and 1950s of the per capita generation of container and package discards had been substantial, there developed a veritable explosion of packaging wastes in the 1960s, coinciding in part with the dramatic rise in the use of petrochemical-related materials, especially plastics, for packaging materials. According to a 1986 EPA waste stream analysis, packaging wastes jumped from 0.73 pounds per capita per day (pcd) in 1960 to 1.05 pounds pcd in 1970. Today, packaging represents more than one-third of the entire waste stream.[31]

In the same vein, the petrochemical industry in general and plastics in particular were transformed from nearly nonexistent industries in the 1920s to multi-billion-dollar operations by the 1970s, with most of that growth taking place in the late 1940s, 1950s, and 1960s. The use of plastic, in both packaging and a wide range of consumer products, became particularly attractive in this period because of its versatility and price. The market for plastic products first developed during the World War II years, when government funding and subsidies for the potential military application of plastics also laid the groundwork for its future use. While government policy focused on the use of plastic as a substitute product in order to save other raw materials, a natural market was established among veterans, who, for example, became familiar with the use of a broad range of plastic materials, including the plastic can. The development of the industry, moreover, was strengthened during the Korean War with its increases in new products and industry subsidies.[32]

The market for plastic products took off most dramatically during the late 1950s and 1960s when the prices for plastic compounds such as polyethylene began to drop significantly as production levels increased. The plastics industry moved aggressively to expand its markets in several different directions, especially in substituting for other materials. Plastic materials were lighter and tended to cost less during this boom period for chemical and petroleum products, and this versatility allowed them to be made into an unending variety of forms and textures. As a result, the proportion of plastic materials in a wide range of manufactured products increased dramatically in a relatively short period of time. In 1960, for example, less than 1 percent by weight of an average passenger car consisted of plastic materials; by 1984, that figure had increased to 6.4 percent, which translated into about 200 pounds of plastic per automobile, or 1 million tons of plastic wastes in the waste stream! This amount of plastic waste from automobiles, moreover, was expected to triple by the year 2000. Diaper backings, made from linear low-density polyethylene

(LLDPE)—also used in the production of plastic bags, sacks, and a variety of other plastic packaging products—was another example of a product transformed in part after 1960 by the introduction of plastics as a key raw material.[33]

This industry growth, in turn, had direct bearing on the waste stream. "Solid wastes are produced at essentially every step in the manufacture of plastics," two EPA consultants wrote in 1981, and that was particularly true regarding its end use or "postconsumer segment." Furthermore, the largest single end use that had evolved in the 30 years since plastics began to saturate the market was in packaging: plastic bags, plastic wrapping, foam containers, plastic bottles, and so forth. From literally no impact on the waste stream prior to World War II, plastics had come to represent, by the 1970s, its fastest-growing segment, with predictions of future growth at the same accelerated rates.[34]

The advent of plastic products in turn helped to reshape waste disposal issues. The glass container industry, for example, responding partly to the pressure of a competitive product, helped engineer the decline of the returnable bottle, long a fixture in the marketplace. As late as 1947, refillable bottles maintained a nearly 100 percent market share of the glass container business. But with the advent of the plastic and glass "one-way" nonrefillable bottles, as well as the metal beverage can, the share of the market for refillable bottles rapidly declined over the next 30 years. Meanwhile, single-use disposable products captured literally all of the extensive overall growth in the beverage container industry that occurred in this period. Not only did its market share decline, but there also was an actual decline in the total production of refillables, while disposables were making enormous gains.[35]

Plastic bottles especially turned out to be the prototypical single-use, nonrecyclable product. As late as 1977, one EPA study estimated that there was still no calculable measure of recycling and recovery of the various kinds of plastic containers on the market. And more than ten years after that, despite protracted battles over container recycling and reuse in more than a dozen states, more than 99 percent of all plastic containers continued to be discarded as one-way products.[36]

The decline of the recyclable product was by no means limited to either plastics or glass containers. The paper products industry, facing strong competition from a range of new petrochemical-based product alternatives, successfully marketed a range of its own "convenience" items, including paper diapers and other chemically treated paper goods. One paper goods manufacturer that witnessed rapid increases in sales and income during the late 1950s began to describe itself as a "totally integrated" producer of packaged goods with its emphasis on disposable paper products.[37]

By the 1950s and 1960s, the proliferation of new consumer products, product design modifications that focused on style changes while reducing a product's use life, and the continuing drive to develop and secure new markets for such products were all encouraged by the rapidly expanding share of advertising and promotion as a cost of production. The rise of television especially contributed to the prominent role of advertising in the decision-making structure of production. Within less than a decade after its introduction on a large scale, television had achieved a level of consumer penetration that "no other advertising medium has ever approached," as one marketing analyst noted. Given its ability to reach the entire family and establish "captive audiences," television became advertising's most successful means of "indoctrinating" and creating new sets of preferences, or what some analysts called "invented" or "artificial" needs.[38] To consume, according to this television/advertising environment, became an essential form of self-expression. "The Look," as two critics of the 1960s scathingly wrote, had become "You."[39]

This consumer goods revolution of the post–World War II era was fueled by the expansion of markets both here and abroad, and the increase in income levels that lasted for an unprecedented two decades. Advertising helped transform the American family and the American public into captive audiences to be delivered to the buyer of advertising time. Marketing and advertising personnel spoke of this creation of artificially stimulated needs as a form of "forced consumption," converting the purchase of goods into a set of rituals and a form of induced satisfaction. This revolution in consumption, furthermore, framed the changes in the generation of solid wastes. "We need things consumed, burned up, worn out, replaced, and discarded at an ever increasing pace" in order to alleviate the pressures of overproduction, an article in the *Journal of Retailing* proclaimed in the spring of 1955.[40] Yet even with the severe recessions of the 1970s and early 1980s, which slowed down both consumer spending and waste generation levels, the changes wrought by petrochemicals and other new products, as well as advertising with its emphasis on continually expanding markets and packaging's role in differentiating products, a major permanent niche for disposable products had been secured in the contemporary industrial order. The evolving patterns of production and consumption had set the stage for the contemporary solid waste dilemma.

THE RISE OF THE SANITARY LANDFILLS

Despite these enormous industrial changes and the resulting growth and change in the waste stream, the garbage issue, as reviewed by sanitary engineers, municipal officials, and other policy-makers, appeared, through the 1950s and much of the 1960s, to be under control. Solid waste management continued to be an exercise in developing new, more comprehensive disposal technologies, primarily land-based.

Land disposal offered a number of advantages. The availability of cheap real estate at the outskirts of the growing cities and suburbs of the 1940s and 1950s allowed such methods to be developed at a relatively low cost. This abundant undeveloped land also meant, it was assumed, that landfills could easily handle any increase in volume brought about by a rapidly changing waste stream. By expanding landfill capacity, public officials assumed they could ignore waste generation issues. Even if garbage was not identified as a revenue source, as Progressive Era officials had done, it did offer, at least indirectly, a measure of a flourishing economy. Expanding landfills became yet another yardstick of productivity.[41]

To be sure, most landfills in this period were little more than open pits, the breeding grounds for rats, flies, vermin, and windswept fires.[42] Local opposition to landfills was already significant by the 1950s, when policymakers began to promote a more refined, and presumably environmentally benign, method of disposal, the much-heralded "sanitary landfill." Though more complex and costly than the open pit, the economics of the sanitary landfill, influenced primarily by land-use patterns and the inexpensive price for land, remained an attractive means to offset public discontent about the hazards of land disposal. If the landfill had triumphed as the dominant method of getting rid of trash, the sanitary landfill became in turn the symbol of the most technically advanced form of that method.

Yet the meaning of "sanitary landfill" varied widely. In practice, many sanitary landfills were similar to the open dumping that remained by far the most prevalent method of land disposal well into the 1970s. For most waste managers, the sanitary landfill distinguished itself from the open dump primarily by its practice of covering the waste "with a layer of earth at the conclusion of each day's operation or at such more frequent intervals as may be necessary," as described by the American Society of Civil Engineers. Sanitary engineers in fact emphasized during the 1950s and 1960s that the sanitary landfill was a method based on "the principles of engineering" capable of eliminating any "nuisances" (odor problems, for example) "or hazards to public health or safety," as opposed to simply burying the wastes without any additional intervention.[43]

The "sanitary" method was first developed in England during and immediately following World War I. Application in this country varied, with critics suggesting that "sanitary" dumps were "neither sanitary in original conception or in final management."[44] The first actual sanitary landfills in the United States were established in New York City and in Fresno, California, in the 1930s. By 1940, the Army began research and experimentation in this area. Five years later, nearly 100 municipalities had either developed or initiated plans for a sanitary landfill. These were already perceived as preferable to other forms of disposal for their ability to minimize the health concerns associated with open dumping and incineration, while utilizing engineering expertise in an era when technology was presumed capable of solving any number of social problems.[45]

The rise of sanitary landfills also roughly coincided with the decline in incineration that began during the 1940s and 1950s. One 1959 survey estimated that 1,400 cities were using landfill cover or related techniques, and a number of articles in trade publications at that time argued that the sanitary landfill had become by far the preferred method of disposal for municipal officials and sanitary engineers. Furthermore, these officials were strongly attracted to the economics of landfills based on the availability of cheap and easily obtainable landfill sites, with average tipping fees (the fee paid for the waste disposed) comparing favorably with other disposal methods.[46]

By the mid-1960s, two parallel factors regarding the solid waste issue—the hazards of both land-based and incineration disposal methods and the increasing reliance on technology—converged. On the one hand, increased generation of wastes had finally been recognized as creating a "refuse disposal problem that far outstrips the waste handling resources and facilities of virtually every community in the nation," according to a report by the Bureau of Solid Waste Management, the federal agency then housed in the Department of Health, Education, and Welfare, and most directly involved in the solid waste issue. Despite the appeal of the sanitary landfill, health and environmental problems continued to preoccupy the waste disposal business. According to the bureau, 94 percent of all land disposal operations and 75 percent of incinerator facilities were, in terms of air and water pollution, insect and rodent problems, and physical appearance, clearly inadequate. The problems of waste disposal had become "obvious and appalling—billowing clouds of smoke drifting from hundreds of thousands of antiquated and overburdened incinerators, open fires at city dumps, wholesale on-site burning of demolition refuse . . . acres of abandoned automobiles that blight the outskirts of our greatest cities."[47] Open dumping and incineration were of particular concern. This was reflected in a series of episodes, including an incident at Washington, D.C.'s Kenilworth dump, which threatened to unravel the system relied upon by nearly every municipality in the country.

Since World War II, the city had disposed of much of its wastes at the Kenilworth dump located near a black residential neighborhood not far from the Capitol. Over the years, garbage had been reduced at Kenilworth by open burning, a practice similar to that at landfill sites across the country. Like many of these "nonsanitary" landfills, problems of odor, public health concerns (the dump had become a breeding ground for rats and was also considered a major source of air pollution), and community opposition had forced officials to scale back the size of the dump and to convert some of the land into a park. The city's primary disposal system, four large antiquated incinerators located throughout the city, had also become increasingly overburdened in handling the area's growing solid waste. This caused city officials to keep Kenilworth open until a new trash disposal contract was arranged, despite objections from residents.[48]

These delays, however, led to a tragic episode that brought the issue of open dumps to the fore. While it was still operating, Kenilworth continued to rely on fires lit on a daily basis. These were not well monitored and in the late afternoon tended to shift direction rapidly. Then, one day in February 1968, three boys from the neighborhood were playing at the site, a common occurrence. Suddenly, one of the uncontrolled fires changed direction and overtook the children. One of the boys, a seven-year-old black child named Kelvin Mock, was burned to death. The ensuing controversy not only shut down Kenilworth but also became a key factor in the evolving legislative debates over open dumps and other waste disposal practices.

The passage of the Air Quality Act in 1967 further intensified this debate. Despite the professed emphasis on safer, "sanitary" methods of disposal, almost all landfill sites had problems similar to Kenilworth's. Most incinerators, moreover, presented a range of health and environmental problems, particularly air emissions. The new air quality legislation exposed these problems while increasing the economic costs of this disposal option by forcing the adoption of new pollution control technologies to meet the act's emissions requirements. Around the time of the Kenilworth fire, there were more than 350 municipal waste incinerators in operation, many of which would become too costly and obsolete due to the new regulatory environment and the changing economics of disposal.[49]

While facing the concerns that had developed over open dumps and the hazards of the older incineration technologies, public officials at both the local and federal levels had become increasingly attracted to the notion that newer forms of technology and engineering would provide the answer. By the late 1960s, sanitary landfills and a new generation of incinerators were cautiously touted as the means to break the emerging solid waste management dilemma within narrowing disposal options. Federal legislation passed in 1965 and amended in 1970 with new provisions gave the federal government a clearly defined role in the solid waste issue for the first time. Most noteworthy, the federal government was able to intervene directly in the research and development of the new technologies, primarily through financial support of demonstration projects.[50]

This reliance on technology for solutions, still embraced by public officials and sanitary engineers, no longer enjoyed, by the late 1960s, unchallenged public support. Systematic criticism of sanitary landfills and the new incineration technologies had yet to emerge, but a number of new social movements had begun to address the waste issue from a different perspective than that of government and industry. Their concerns were more systemic than technical. The problem of "forced consumption" and acquisitive lifestyles were seen as a form of "domestic imperialism," as described in one 1960s essay, with environmental degradation, especially air, water, and land pollution, viewed as consequences of the social structure and actions by elites rather than of inadequate technology. Some groups, many of them locally based, became mistrustful of a technological fix and began to advocate—and practice, in some communities—voluntary action as a means to reduce litter and other waste. The community-based recycling programs of the late 1960s and early 1970s, for example—nearly all of which were organized by volunteers rather than public officials—were initiated not just as a way to establish a solid waste management alternative, but as a kind of cultural rejection of the "throwaway" market.[51]

Partly as a result of growing community pressure, some limited local and state measures, such as packaging review ordinances, were developed to address the extraordinary growth of the waste stream. Nearly all of these efforts were defeated in the legislative or judicial arenas or substantially modified and weakened, mostly through a strong counterattack by the packaging, petrochemical, and beverage container industries.[52]

Both industry and government officials, furthermore, were able to deflect criticism about production choices by placing responsibility on the individual consumer. "Packages don't litter, people do," proclaimed one executive from the American Can Company. Another American Can executive, William May, also suggested at an industry-led conference that though "convenience packaging" was "rapidly becoming problem packaging," industry officials had to be absolved from blame since they were "caught in the middle." "How do we continue to satisfy consumer demands for easy-open, readily disposable, adequately protective packages," May declared, "with built-in assurance that irresponsible disposal by the user will not result in advancing the litter and pollution problem?"[53]

This concept of "blame the victim" was immeasurably strengthened when leading advocates of environmental legislation, such as Edmund Muskie (D—Maine), a prominent Senate environmental leader during the late 1960s, and a number of traditional conservationist groups, endorsed this industry-promoted position of individual responsibility. Muskie, who helped steer the 1970 solid waste legislation, the Resource Recovery Act, through Congress, wrote that it was "easy to blame pollution only on the large economic interests, but pollution is a by-product of our consumption-oriented society. Each of us must bear his share of the blame."[54] For the traditional conservationist groups, and even many of the newly emerging environmental organizations that had sprung up in the wake of Earth Day in April 1970, this notion of individual responsibility framed many assumptions about the root cause of problems such as waste disposal, linking it to such presumably individual-choice issues as population control.

This concept of individual responsibility, however, did nothing to assist public decisions about how to handle solid waste. The emphasis on technology had enormously strengthened the position of sanitary landfills over open dumps, but most local officials continued to rely on the open pit as the cheapest, simplest, and easiest method to handle wastes. As many as 14,000 communities were still using open dumps in 1972, according to EPA. A few years later, another EPA study estimated that of 17,000 to 18,000 existing landfill sites, more than 90 percent continued to fail to meet even the minimum requirements for a sanitary landfill.[55] Only new legislation in 1976, an increasing scarcity of urban landfill sites, and growing public sentiment against contamination of the land and water finally placed real pressure on the open dump system.

Yet the sanitary landfill never really had a chance to establish itself as an alternative to the open dump. By the late 1970s and early 1980s, the focus on landfill problems had begun to extend to *all* sites, including those that met the new and more stringent standards, such as clay caps and liners to prevent leaching of the wastes into the groundwater. New revelations about groundwater contamination in a wide array of communities brought increasing attention to the problem of leaking landfills. Several studies during the late 1970s pointed out that leaching was a problem facing all landfills: only the timing of contamination would be influenced by the degree of sanitary techniques utilized at a landfill.[56] The concentration and toxicity of the pollutants found at even sanitary landfills, moreover, were seen as potentially capable of causing "as great a cancer risk as those from industrial waste landfills," as one later study put it.[57] In addition, the discovery of toxic air emissions from landfills in the early 1980s increased the controversies around this increasingly problematic strategy.[58] Though still the dominant disposal method, landfilling, by the 1980s, had come under siege.

Through the decade of the 1970s, this focus on the contamination of the land, the "third pollution," as it came to be known, had become a central environmental issue that contributed to the growing solid waste dilemma. EPA's Office of Solid Waste, for example, had become one of the largest parts of the mushrooming environmental bureaucracy of the 1970s, and much of its work was devoted to reviewing and dealing with the problems of land disposal sites.[59]

But the interest in new technological solutions still prevailed at EPA, within Congress, and especially among local policy-makers most directly affected by the growing discontent around landfills. A new terminology arose to signify the new hopes of the waste managers: "resource recovery." A ubiquitous, often vague term that became increasingly prominent during the decade, it evoked the old Progressive dream of capturing part of the waste stream in some new form. This included recycling concepts, or "materials recovery," and, most significantly, to many of the public officials and waste industry executives, "energy recovery" linked to the burning of the waste. The latter, known as "waste-to-energy" or "refuse-to-energy," was little more than a variation of an old theme: burn it instead of burying it. With a potentially vast market, given the landfill controversies facing large and medium-sized cities, a new waste-to-energy industry began to take shape, portending a disposal alternative that could also be a highly lucrative business.

Yet, as the 1970s ended, landfills remained the overwhelming disposal method of nearly every community, with "waste-to-energy" a rather exotic option. But the conflicts over landfills would soon be at their peak, and a number of public officials were ready to concede that the days of land disposal were numbered. How to get rid of the garbage, this "growing cancer on the land," had become at once a national dilemma, the setting for the development of major new public works projects, and, as a consequence, a potentially rich commercial opportunity.[60]

NOTES

1. "Waste Import Bans Are Growing," Bruce J. Parker, *Waste Age*, October 1987; "Garbage Imperialism Must Stop," *Los Angeles Times*, May 18, 1987.
2. *Hazardous Waste News*, February 29, 1988, no. 66, Princeton, New Jersey, *New York Times*, March 18, 1988; "After Years of Warning, Jersey Has Sudden Trash Crisis," Ivar Peterson, *New York Times*, April 26, 1987.
3. *New York Times*, November 10, 1988, November 28, 1988; U.S. Environmental Protection Agency (EPA), "Flash Report: Philadelphia Incinerator Ash Exports for Panamanian Road Project—Potential Environmental Damage in the Making," John C. Martin, Office of the Inspector General, October 5, 1987; *Burnt Offerings 2: Greenpeace's Report on the Dumping of Philadelphia's Toxic Incinerator Ash in Haiti*, Greenpeace USA, February 15, 1988; "Move Over Garbage Barge," Jeanne Wirka, *Environmental Action*, vol. 19, no. 3, November/December 1987.
4. *New York Times*, June 27, 1988; "Norwegian Official Arrested in African Toxic Dumping Case," Associated Press, June 11, 1988, and "U.S. Toxic Waste Threatens African Island," United Press International, May 21, 1988, both reported by Greenlink, the computer news service of Greenpeace USA.
5. *Conservation by Sanitation*, Ellen Swallow Richards, John Wiley & Sons, New York, 1911; *Pollution and Reform in American Cities, 1870–1930*, Martin Melosi, ed., University of Texas Press, Austin, 1980; *Solid Waste Management*, D. Joseph Haggerty, Joseph

L. Pavoni, and John E. Heer, Jr., Van Nostrand Reinhold Company, New York, 1973. A general discussion of the history of solid waste management and resource recovery in the United States is found in *Currents in the Waste Stream: A History of Refuse Management and Resource Recovery in America,* by Daniel Thoreau Sicular, master's thesis, University of California at Berkeley, Department of Geography, 1984. Melosi's work in particular provides an important advance in developing the critical study of urban environmental history, including solid waste.

6. "Garbage Collection and Disposal," *Engineering News,* vol. 42, September 1899.
7. "Disposal of Municipal Refuse," Rudolph Hering, *Transactions of the American Society of Civil Engineers,* vol. 54-E, no. 6, pp. 265–308, American Society of Civil Engineers, New York, 1905. For comparative U.S./European data for 1959–61, see "Estimation of Solid-Waste-Production Rates," Walter Niessen, *Handbook of Solid Waste Management,* Van Nostrand Reinhold Company, New York, 1977, p. 561. Figures for the contemporary period are found in *Garbage: Practices, Problems and Remedies,* Joanna D. Underwood, Allen Hershkowitz, and Maarten de Kadt, INFORM, New York, 1986, p. 3.
8. *Garbage in the Cities, 1880–1980,* Martin Melosi, Texas A&M University Press, College Station, 1981, pp. 155, 197; "What Is the Best Way to Collect and Remove the Waste of Towns and Cities and What Does It Cost," Colonel W. F. Morse, in *Public Health: Papers and Reports,* 19th Annual Meeting of the American Public Health Association, Kansas City, Missouri, October 20–23, 1891, Concord, New Hampshire, 1892.
9. "Systems of Garbage Collection and Disposal," *Municipal Engineering,* vol. 28, no. 5, May 1905; "History of Solid Waste Management," David Gordon Wilson, in *Handbook of Solid Waste Management,* Van Nostrand Reinhold Company, New York, 1977, chap. 1; "The Utilization of City Garbage," George E. Waring, Jr., *Cosmopolitan Magazine,* vol. 24, no. 4, pp. 405–12, February 24, 1898; *Street Cleaning and Its Effects,* George E. Waring, Jr., Doubleday and McClure, New York, 1898.
10. *Garbage in the Cities,* Martin Melosi; "Modern Methods of Municipal Refuse Disposal," Robert H. Wyld, *American City,* October 1911; "The Collection of Municipal Solid Waste," William F. Morse, *American Journal of Public Health,* July 1910.
11. "Report of Committee on Disposal of Waste and Garbage," *Public Health: Papers and Reports,* 19th Annual Meeting, American Public Health Association, 1891.
12. State of New Jersey *v.* City of New York (290 U.S. 237–240), December 4, 1933; *The Politics of Pollution,* James Ridgeway, New York, 1971.
13. *Garbage in the Cities,* Martin Melosi; "The Disposal of the City's Waste," William F. Morse, *American City,* May 1910.
14. "History of Solid Waste Management," David Gordon Wilson, in *The Solid Waste Handbook,* ed. William D. Robinson, New York, 1986; "The Disposal of Municipal Wastes," William F. Morse, *Municipal Journal and Engineer,* March 1907; "Incineration Today and Tomorrow: A Survey of Incineration Practices—Present and Future," Junius W. Stephenson, *Waste Age,* May, 1970; *Refuse Disposal in American Cities: A Report,* U.S. Chamber of Commerce, Washington, D.C., 1931.
15. Between 1920 and 1972, 322 high-temperature incincerators (termed "destructors" at one time) were built in the United States. As of 1972, 193 were still operating. Source: *Environmental Assessment of Municipal Scale Incinerators,* Solid Waste Management Office, U.S. EPA, no. 530/SW-111, Washington, D.C., 1973. There were also hundreds of smaller, lower-temperature incinerators termed "crematories"; see also "Refuse Disposal in California," *Municipal Journal and Engineer,* January 25, 1917; "Incinerators for Garbage and Refuse Disposal," parts 1 and 2, H. S. Hersey, *American City,* February 1938, pp. 61–69, March 1938, pp. 85–89; "Incinerators for Garbage and Refuse Disposal," Alden E. Stilson, *American City,* April 1938, pp. 109–13.
16. By 1914, of the 29 destructors in the United States, only 8 captured the heat for use as

steam and energy; see also "Disposal of Municipal Refuse," Rudolph Hering, *Transactions of the American Society of Civil Engineers,* vol. 54-E, no. 6, pp. 265–308, especially, p. 289, American Society of Civil Engineers, New York, 1905.

17. "Refuse Collection and Disposal," *Municipal Journal and Engineer,* vol. 39, November 1915.
18. "Garbage and the European War," *Municipal Journal and Engineer,* vol. 40, no. 1, January 6, 1916; "War Profits from Garbage," *Municipal Engineering,* vol. 50, no. 3, March 1916.
19. "Refuse Disposal in Boston," Hiffert Winslow Hill, *American Public Health Association, Public Health Papers and Reports,* vol. 27, pp. 186–93, 1902; of the 45 reduction plants built by 1914, 22 were still in use. See "Recent Refuse Disposal Practice," William F. Morse, *Municipal Journal,* vol. 37, no. 24, pp. 848–51, December 10, 1914.
20. *Garbage in the Cities,* Martin Melosi; *Municipal Journal and Engineer,* vol. 32, no. 23, June 6, 1912.
21. *Refuse Disposal in American Cities,* Chamber of Commerce. For a description of hog feeding prior to, during, and immediately after World War I, see *Collection and Disposal of Municipal Refuse,* Rudolph Hering and Samuel A. Greeley, McGraw Hill, New York, 1921, p. 41; also "Garbage Piggeries," Alvah W. Brown, *American Journal of Public Health,* December 1912; "Garbage Disposal by Hog Feeding in a Small Indiana City," *American City,* November 1938, p. 54.
22. *Municipal Refuse Disposal,* prepared by the Committee on Refuse Disposal, American Public Works Association, Chicago, 1961. "Feeding Garbage to Hogs Spreads New Disease," *Public Works,* vol. 83, no. 11, November 1952. "Garbage-Fed Hogs a Health Menace," *The American City,* vol. LIX, no. 2, p. 17, February 1944.
23. *The Waste Makers,* Vance Packard, David McKay Company, New York, 1960.
24. *The Mass Consumption Society,* George Katona, New York, 1964; *All Consuming Images: The Politics of Style in Contemporary Culture,* Stuart Ewen, New York, 1988.
25. "Price Competition in 1955," Victor Lebow, *Journal of Retailing,* Spring 1955; *Channels of Desire: Mass Images and the Shaping of American Consciousness,* Stuart and Elizabeth Ewen, New York, 1982; *The Image Empire: A History of Broadcasting in the United States, Volume III: From 1953,* Erik Barnouw, New York, 1970; *The Sponsor: Notes on a Modern Potentate,* Erik Barnouw, New York, 1978.
26. *An Evaluation of the Effectiveness and Costs of Regulatory and Fiscal Policy Instruments in Product Packaging: Final Report,* Taylor H. Bingham et al., 1974, SW-74c.
27. "Profitability: Package Design in the Seventies," Walter Stern, in *32nd Annual National Packaging Forum of the Packaging Institute,* Packaging Report F-7039, Chicago, October 5–7, 1970.
28. "The Growing Package," *Time,* January 5, 1959.
29. "Packaging USA," Eric Outwater, in *Proceedings: First National Conference on Packaging Wastes,* September 22–24, 1969, SW-9g, EPA, 1971.
30. "Motivating Ourselves for Action," Norvell Gillespie, in *Proceedings: First National Conference on Packaging Wastes,* EPA, 1971.
31. *Characterization of Municipal Solid Waste in the United States: 1960–2000,* Franklin Associates, Prairie Village, Kansas, EPA PB 87–178323, US EPA Office of Solid Waste, Washington, D.C., July 25, 1986.
32. "Materials in 1950 and the Future," *Modern Plastics,* January 1950; *Petrochemicals: The Rise of an Industry,* Peter Spitz, New York, 1988; *Wrapped in Plastic: The Environmental Case for Reducing Plastics Packaging,* Jeanne Wirka, Environmental Action Foundation, Washington, D.C., August 1988.
33. "What to Do with Plastic Cars When They're Junk," A. Stuart Wood, *Modern Plastics,*

June 1988; *Plastics: The Risks and Consequences of Its Production and Use: An Industry Overview,* Patricia Lichiello and Lauren Snyder, Graduate School of Architecture and Urban Planning, UCLA, June 1988.

34. "RCRA Study of Glass and Plastic Recovery," Tom Archer and John Huls, *Municipal Solid Waste: Resource Recovery,* Proceedings of the Seventh Annual Research Symposium, Report no. 600/9-81-002c, U.S. Environmental Protection Agency, March 1981.
35. *Choices for Conservation,* Resource Conservation Committee, Final Report to the President and Congress, U.S. Environmental Protection Agency, Report no. SW-779, July 1979.
36. "RCRA Study of Glass and Plastic Recovery," Tom Archer and John Huls; "More Plastic Use Means More Waste, Which Means . . . ?" Jill W. Tallman, *Waste Age,* September 1987.
37. *Sales Management,* September 19, 1958; *Time,* January 5, 1959.
38. *One Dimensional Man: Studies in the Ideology of Advanced Industrial Society,* Herbert Marcuse, Boston, 1964; *Channels of Desire: Mass Images and the Shaping of American Consciousness,* Stuart and Elizabeth Ewen, New York, 1982; "Price Competition in 1955," Victor Lebow.
39. "The Look Is You," Naomi Jaffe and Bernardine Dohrn, in *The New Left: A Documentary History,* ed. Massimo Teodori, Indianapolis, 1969.
40. Cited in "Price Competition in 1955," Victor Lebow.
41. "Garbage Collection System Embodies High Standards of Sanitation," G. P. Manz, *American City,* October 1946, pp. 103–4; "Landfills for Refuse Disposal," Edward J. Cleary, *Engineering News-Record,* vol. 121, no. 9, September 1, 1938, pp. 270–73.
42. *Closing Open Dumps,* 530/SW-61ts, US Environmental Protection Agency, Office of Solid Waste Management, 1971.
43. Cited in *Third Pollution: The National Problem of Solid Waste Disposal,* William E. Small, New York, 1971; "Keeping a Sanitary Landfill Sanitary," Bayard F. Bjornson and Malen Bogue, *Public Works,* vol. 92, no. 9, September 1961; "Operation of Sanitary Landfills by Counties," *Public Works,* vol. 90, no. 2, February 1959.
44. "The Dual Disposal of Sewage," Morris M. Cohn, *Municipal Sanitation,* vol. 5, no. 10, October 1934; Edward T. Russell, *Engineering News-Record,* vol. 121, no. 13, September 29, 1938.
45. "Summarizing the Sanitary Fill: An American City Staff Report," W. S. Foster, *American City,* April 1947, pp. 92–93; "A Study of the Sanitary Fill at Fort Worth, Texas: Cost Analysis; Operating Methods," R. R. Lacy, *American City,* May 1946; *Third Pollution,* William Small.
46. "Landfill Ends Refuse Disposal Problems," Fred Geiser, *Public Works,* vol. 90, no. 3, March 1959; "Refuse Collection and Disposal Developments in 1945," Carl Schneider, *Public Works Engineers' Yearbook,* pp. 63–65, American Public Works Association, Chicago, 1946; "Incineration and Alternative Refuse Disposal Processes," Ralph Stone and Francis R. Bowerman, *Transactions, American Society of Civil Engineers,* vol. 121, 1956, pp. 273–310.
47. Cited in *Federal Pollution Control Programs: Water, Air, Solid Wastes,* Stanley Degler, rev. ed., Bureau of National Affairs, Washington, D.C., 1971, p. 36.; "Solid Waste Management and the Packaging Industry," Richard O. Vaughan, in *Proceedings: First National Conference on Packaging Wastes,* EPA, 1971; "The Garbage Glut: Desperate Cities Seek New Methods to Solve Growing Waste Problems," Alan Adelson, *Wall Street Journal,* February 16, 1968.
48. *Washington Post,* February 16, 1968; February 17, 1968; "From Dump to Landfill to Park," *The American City,* March 1968, vol. 83, no. 3, p. 43.
49. *An Environmental Assessment of Gas and Leachate Problems at Land Disposal Sites,*

530/SW-110-OF, EPA, Office of Solid Waste, 1973; "Overview of Resource Recovery," Carlton C. Wiles, *Municipal Solid Waste: Resource Recovery,* Proceedings of the Seventh Annual Research Symposium, EPA, March 1981; "Hard Road Ahead for City Incinerators," *Environmental Science and Technology,* vol. 6, no. 12, November 1972.
50. Solid Waste Disposal Act of 1965 (P.L. 89–272); Resource Recovery Act of 1970 (P.L. 91–512). See also *Federal Pollution Control Programs,* Stanley Degler; Resource Conservation and Recovery Act, 42 U.S.C. §6901.
51. "Consumption: Domestic Imperialism," David Gilbert, Bob Gottlieb, and Susan Sutheim, in *The New Left: A Documentary History,* ed. Massimo Teodori, supra.; "Recycling in the United States: The Vision and the Reality," Oscar W. Albrecht, Ernest H. Manuel, Jr., and Fritz W. Efaw, in *Municipal Solid Waste: Resource Recovery,* Proceedings of the Seventh Annual Research Symposium, EPA, 1981.
52. *Cans and Bottle Bills,* William K. Shireman et al., California Public Interest Research Group, Stanford Environmental Law Society, Berkeley and Stanford, 1981; *Garbage: The History and Future of Garbage in America,* Katie Kelly, Saturday Review Press, New York, 1973.
53. Cited in *Third Pollution,* William Small.
54. See introduction by Edmund Muskie, *The Politics of Pollution,* J. Clarence Davies III, Pegasus, a division of Bobbs-Merrill, Indianapolis, 1970.
55. *Closing Open Dumps,* U.S. Environmental Protection Agency, Office of Solid Waste Management, 1971.
56. *Solid Wastes: Origin, Collection, Processing and Disposal,* C. L. Mantell, John Wiley & Sons, New York, 1975; "The Politics of Resource Recovery, Energy Conservation, and Solid Waste Management," Georges Antunes and Gary Halter, *Administration and Society,* vol. 8, no. 1, May 1976; *The Prevalence of Subsurface Migration of Hazardous Chemical Substances at Selected Industrial Waste Land Disposal Sites,* Geraghty and Miller, EPA no. 530/SW-634, National Technical Information Services PB275103, 1977 (as reported in *Hazardous Waste News,* no. 71, Environmental Research Foundation, Princeton, New Jersey, April 4, 1988).
57. "An Estimation of the Risk Associated with the Organic Constituents of Hazardous and Municipal Waste Landfill Leachates," K. W. Brown and K. C. Donnelly, *Hazardous Waste and Hazardous Materials,* volume 5, no. 1, 1988 (as reported in *Hazardous Waste News,* no. 90, Environmental Research Foundation, Princeton, New Jersey, August 15, 1988).
58. "Regulation of Toxic or Hazardous Air Contaminants from Land Disposal Sites," Edward Camarena and Carol A. Coy, South Coast Air Quality Management District, for presentation at the 77th Annual Meeting of the Air Pollution Control Association, San Francisco, CA, June 24–29, 1984; *The Toxic Cloud,* Michael H. Brown, Harper and Row, New York, 1988.
59. *An Environmental Assessment of Gas and Leachate Problems at Land Disposal Sites,* 530/SW-110-OF, EPA, Office of Solid Waste, 1973; interview with Frank A. Smith, 1988.
60. Cited in *Third Pollution,* William Small.

2

THE RESURRECTION OF INCINERATION: "TURNING GARBAGE INTO ENERGY ALL OVER THE USA"

"POLITICAL GARBAGE"

Over the years, this consummate New York politician, a one-time elected official, former corporate executive, and full-time advocate, had become something of an industry pundit. As a City Council member and then mayor of Yonkers, Alfred DelBello had witnessed the decline of an earlier generation of incinerator plants buffeted by new regulations and requiring new control systems that had become just too expensive to operate. He had also put forth the position at the time that a bigger solution was possible, one that could be implemented on a regional level and involve new technologies and increased economies of scale. And when he became Westchester County executive a few years later, he was able to pursue just such a course. The big Westchester County waste-to-energy incineration project that DelBello helped package was based on the new mass burn technology. And though the plant's original vendor dropped out because it was unwilling to assume the risks, plans were back on track when another major player, the Wheelabrator-Frye Company, stepped in to push ahead with its own new mass burn patented system imported from Europe.

The Westchester plant would, as with many of the other watershed projects of the 1970s, experience a range of difficulties and often come close to being shut down permanently. But DelBello had already by then become a firm believer in the mass burn system despite its troubles. After he was elected lieutenant governor of the state in 1982, DelBello continued to promote the technology even though he no longer played as direct a role in encouraging its development.

When executives from Signal (which later merged with Wheelabrator) approached this restless and ambitious politician in 1985, DelBello, frustrated by his lack of any significant policy-making role in Albany, jumped at the chance. "It was a wonderful opportunity," DelBello recalled. Signal, a major player in waste-to-energy, had decided to set up its own subsidiary to focus on issues of refuse, water, and privatization, and they wanted DelBello to be its chief executive officer. DelBello not only had experience with one of the first mass burn systems in the country, but he also was considered adept at what Signal explained would be the primary focus of his new job: politics. "Politics is really 80% of it," DelBello remarked in a 1988 interview. "If the politics got out of the way," DelBello proclaimed to *Forbes* magazine in 1985, "we'd be turning garbage into energy all over the U.S.A."[1]

Through the early and mid-1980s, DelBello's enthusiasm for mass burn spread within the industry and among municipal officials, EPA analysts, and congressional figures. Incineration, with strong backing from the financial sector, quickly became a high-growth industry, after a shakeout period during which mass burn technology and other methods competed to achieve dominance in the potentially lucrative solid waste disposal market. Much of the basis for the growth, however, was political, not technical. Landfills were under attack, land disposal regulations were multiplying, the waste stream was getting larger. Industry executives felt secure, even smug, in their new-found confidence, despite the continuing problems in the operation of their technology. Solid waste, they declared confidently, would not go away. "We won't ever be having," another executive from Signal said in an interview, "an embargo on garbage."[2]

For DelBello (who would later leave his Signal job and become a consultant to the industry) and other trash-burning advocates, the question boiled down to political will. Given the increasing problems of landfills, the technology of waste-to-energy, particularly the mass burn incineration system, was presumed ready to fill the void. Many public officials, at both the local and federal levels, agreed. But just when the resurrection of incineration seemed inevitable, new challenges relating to the costs and hazards of the technology and its public acceptance interrupted the industry's momentum. Like the nuclear power industry of the 1970s, which it resembled in numerous ways, the newly embraced waste-to-energy option seemed ready to rise to the top of waste management agendas everywhere, only to find its ascent blocked by controversies and unresolved issues. And Alfred DelBello's vision of a country ringed with massive new incineration plants, so compelling to industry investors and supporters, would, at the moment of its broadest acceptance by policy elites, be rudely and intemperately undermined by a politics that had not been anticipated.

THE SEARCH FOR NEW TECHNOLOGIES

In its premier, May 1970 issue, a lead article in the trade journal *Waste Age* examined the prospects for the incineration industry.[3] Though incineration was a long-standing disposal technology, only recently had it begun to receive increased attention from solid waste policy-makers at both the local and federal levels, due to concern over landfill problems. While older incineration systems were becoming increasingly problematic, a handful of new incineration technologies were being developed. Hopes had been raised that a budding, reorganized incineration industry could emerge out of the shadow of the older, rather disreputable plants then in existence.

Despite the renewed interest, however, advocates worried that incineration was still a "dirty word" because of its troubled history and current operational problems. The industry's biggest difficulty, in fact, was its poor reputation. "Perhaps," the *Waste Age* author noted, "to complete the transition from a dirty, smelly, necessary evil to a modern industrial type plant, the time has come to drop 'incinerator' and the picture of the past which it frequently calls to mind, and adopt another term which will depict the type of plant we now build and foresee for the future."

The key to this change in perception centered around the revival of the technology of generating energy from waste. By 1970, a new language had made its appearance, which included the terms "resource recovery" and "waste-to-energy," referring to technologies that had indeed begun to supplement and even displace this older, "dirtier" version of burning wastes. An incinerator became a "resource recovery" plant; burning the trash was now defined as a form of energy recovery.

Despite the verbal shift, the production of energy from waste still played, in 1970, a limited role. Landfills predominated among waste disposal options, while most incinerators operated as basically trash-burning/garbage-reducing facilities. And although warning signals from places like Libya indicated a changing and volatile international oil market, the price of energy remained deceptively low and electrical utilities continued to be disinterested in the small-scale, low-volume generation of electricity represented by a potential "waste-to-energy" industry. Even the industry's own advocates, furthermore, emphasized that its facilities were best capable of offering "a method of volume reduction" of wastes, rather than generation of electricity. "The true objective of the total operation," one waste industry analyst wrote in the same *Waste Age* issue, "is the production of a concentrated waste, an end product of greater density."[4]

The interest in the new trash-burning technologies had emerged in the United States during the mid-1960s, in the midst of the growing concerns about large-volume wastes, new air quality regulations, and such polluting and highly visible disposal methods as open dumps. Though many communities had turned to landfills in the 1950s and 1960s, some older incinerators still in use were responsible for disposing of about 10 percent of the waste stream during that period. These plants had developed the reputation for being "dirty" in terms of air emissions, as well as for the vermin, putrescent smells, and blowing paper and ash.[5]

With the passage of new air emission standards in the 1967 Air Quality Act, the older incinerators were obliged to add expensive air pollution controls, such as "scrubber" systems. Since incinerators already had high capital costs, and operating costs were often at least three times as high as those for sanitary landfills, the requirements for "scrubbers" and "precipitators" priced these incinerators out of the market.[6]

In New York City, for example, it was estimated at the time that it would cost the city at least $60 million to equip its incinerators with adequate emission control devices, which represented a per unit cost of $12,000-per-ton daily disposal capacity. Within five years after the passage of the 1967 legislation, more than 100 incinerators had been shut down, and a number of others, according to EPA officials, were operating at only 50 to 60 percent of design capacity. Nearly all of these would also be closed down within the decade.[7]

With the incineration industry in serious trouble, public officials and a handful of construction companies once again looked to Europe to provide a solid waste disposal model. Their interest was focused primarily on an electricity-generating system that combined a different grate and heating process known as the water-wall combustion unit. These systems had been operational in Europe for more than 15 years and had matured in the context of scarce land and a rising per capita generation of waste. Considered "cleaner" than the older "refractory" lined units that operated in this country, the European systems were also attractive in the context of possible sources of federal funding for "resource recovery" projects.[8]

The first two major water-walled combustion unit plants introduced in this country were built in Norfolk, Virginia, and Chicago, Illinois. In 1967, the U.S. Navy, concerned about disposal problems for its Norfolk, Virginia, Navy yard, decided to model its new incineration facilities after a European system based on a water-walled boiler technology. The Navy's primary interest was its concern for waste reduction, though the two plants it designed, both 180 tons per day capacity (tpd), were also capable of producing 50,000 pounds of steam per hour.[9]

Around the same time the Navy plant became operational, the city of Chicago began its own well-publicized search for a new incineration technology. Chicago's waste disposal problems were particularly severe. Its landfill space had begun to reach capacity a good ten years before other municipalities would experience similar problems. The Chicago Department of Public Works, furthermore, was concerned that new air quality regulations derived from the federal Air Quality Act would place major constraints on any new burn system it might build without a more advanced technology and pollution control system. The new air regulations in effect would likely make construction and/or refurbishing of any of the conventional incinerator technologies, even the most advanced at the time, prohibitively expensive.[10]

The Chicago search also focused on the European waste-to-energy plants. Chicago officials eventually selected a water-walled technology that utilized electrostatic precipitators, the most advanced air pollution control technology at the time. As a consequence, the Chicago plant came to be perceived as a model incinerator that could "serve as a standard of comparison against which to measure the predicted performance of innovative alternatives," as one Columbia University report noted.[11]

Chicago's Northwest Waste-to-Energy Facility, as the plant was called, began operations in January 1971. It consisted of four separate water-wall furnaces, each designed to burn 400 tons per day of refuse. Three of the units were to be operational at any given moment, with the fourth reserved for standby capacity. The total capital cost of the plant was $23 million; an additional $2.5 million was later required for the boiler water treatment system conversion and the steam export line. Part of the costs were subsidized by a grant from the federal government, which hoped to use the facility as a demonstration project to establish guidelines for incinerators. The project was fully owned and operated by the city of Chicago, but, in a move that prefigured an eventual pattern for the industry, the city also purchased the equipment and technology from private companies, or vendors, thus helping underwrite the emergence of a newly restructured incineration industry.[12]

During the late 1960s, as waste disposal became an issue of national debate, policy-makers began to emphasize the value of economies of scale, especially for incineration. An MIT-based study group headed by Lyndon Johnson's science adviser Jerome Weisner, for example, recommended that both large investor-owned electrical utilities as well as *Fortune* 500–type construction and resource development firms participate in the design, building, and operation of a new generation of incinerators to help reshape the direction of waste disposal.[13]

By serving as models, the Chicago Northwest facility and several other refuse-burning plants under design or beginning operation at the time influenced the restructuring of incineration technology. The grate system chosen by Chicago for its Northwest plant, for example, was designed and manufactured by West Germany's Martin Company, one of the biggest and most active of the European companies. Martin had been eagerly seeking an outlet in the U.S. market for its technology, and, as part of the

Chicago selection process, had sold the licensing rights for its system to Universal Oil Products, a subsidiary of a *Fortune* 500 company, the Signal Corporation. Signal's acquisition of the Martin licensing rights and its selection by the city of Chicago catapulted the company into the front ranks of this new waste disposal industry.[14]

The activities of Signal and several other large corporate entities, such as the American Can Company, Monsanto, and the Union Carbide Corporation, appeared to herald the dawn of a new age of trash-burning technologies and industrial development. By the mid-1970s, several competing waste-to-energy refuse-burning systems had been introduced in hopes of establishing preeminence in this evolving industry. Most of the other technologies were U.S.-based, and some had been used in earlier periods, though the interest in the waste-to-energy idea had renewed and expanded interest in each of the technologies under consideration.

The leading waste-to-energy technology that emerged in the United States during this period was the water-wall combustion system known as mass burn incineration, which used unprocessed municipal solid waste as fuel, similar to the two facilities developed at Norfolk and Chicago. In this system, the waste would ordinarily first be deposited into a large pit or tipping floor. Overhead cranes would then mix and transfer the waste from the pit to the incinerator intake while the crane operators, high above the intake, would seek to remove (though not always successfully) large, unburnable items from the waste, such as refrigerators or engine blocks.[15]

Once inside the incinerator, the waste would be burned on a moving grate system. The grate is the heart of the incinerator, and differences in its design have accounted for several proprietary patents that sought to capture the major share of the market for their system. All are designed, theoretically, to insure a maximum burn. In the Martin system, for example, the grates are angled toward the feed chute so that the burning refuse ignites the incoming garbage and the combustion process lasts longer. In this system and most others, oxygen (air) is forced through the grates to control and enhance the combustion process. The furnace temperature is designed to be maintained at 1,800 degrees Fahrenheit. The Martin plants and many other mass burn facilities have utilized the water-wall technology, in which the furnaces are enclosed by walls constructed with closely spaced water-filled tubes. The water circulating through the tubes recovers the heat generated from the burning waste. The steam that is generated is then utilized for heating purposes or for generating electricity (see figure 2.1).

By the mid-1970s, several mass burn systems had been built, or were in construction or planning stages. Aside from the Chicago Northwest Facility, the Martin grate system was applied in a number of other plants,

Figure 2–1

MASS BURNING WASTE POWER PLANT OF WIDMER AND ERNST, BIELEFELD-HERTFORD, FEDERAL REPUBLIC OF GERMANY

1. Waste bunker
2. Crane
3. Charging hopper
4. Grate
5. Combustion chamber
6. Steam boiler
7. Electrostatic precipitator
8. Flue gas fan
9. Wet scrubber
10. Stack
11. Turbine generator
12. Fly ash conveying system
13. Residue discharging system
14. Residue bunker
15. Primary air system with preheater
16. Secondary air system

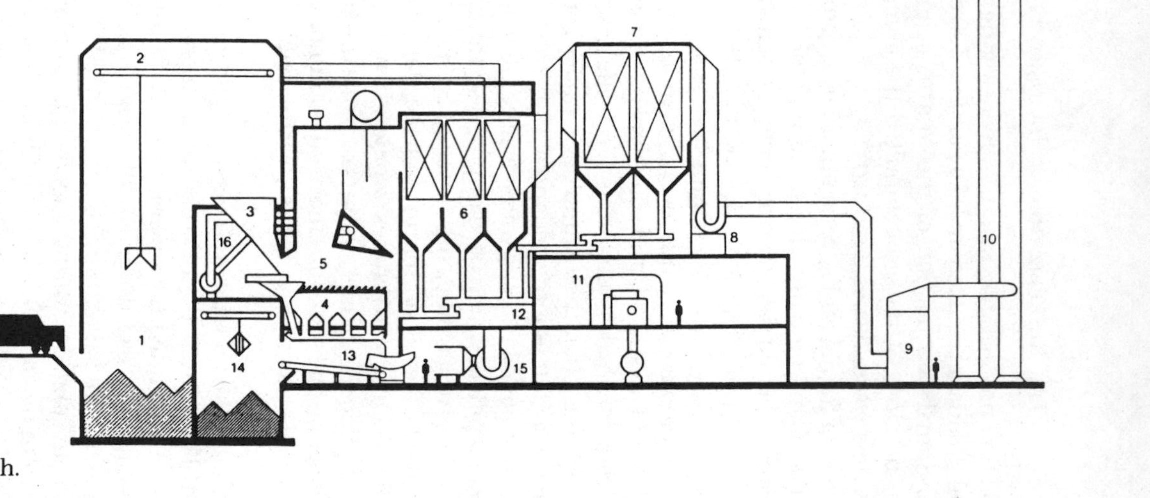

Longitudinal section of the waste power plant.
Combustion capacity: 3 × 385 t/24h.
Steam production: 3 × 52.4 t/h.

SOURCE: EPA, "Combustion Control of MSW Incinerators to Minimize Emission of Trace Organics," *Municipal Waste Combustion Study*, January 1987, section 5, page 2.

including two important early facilities located in Harrisburg, Pennsylvania, and Braintree, Massachusetts. Another key European-based waterwall technology, the Von Roll incinerator/grate system, which had been acquired by Wheelabrator-Frye for use in a 1,200-ton-per-day facility in Saugus, Massachusetts, was also introduced into the United States around the same time as the Martin system. Though the European-imported mass burn system was already establishing itself as the leading technology in this country, each of these plants experienced significant start-up and operational difficulties. These included large volumes of ash residue, continuing stack emissions such as nitrogen oxides, high capital investment resulting in a high net amortized operating cost, and difficulties in marketing the steam, or energy, generated.[16]

Aside from the mass burn incineration system, three other technologies emerged as the main, competitive waste-to-energy systems during the 1970s.[17] One of the oldest and most widely used was the small modular combustion unit designed primarily for industrial sites and large retail or commercial complexes and apartment houses. Over the years, these burn systems had been designed exclusively for volume reduction of wastes, but in the late 1960s and early 1970s, a heat-recovery unit was added to allow for the recovery of energy either as hot water, steam, or hot air. Though primary application of this new, "waste-to-energy" system was intended for industrial sites, the producers of these units hoped that smaller-sized cities and communities might find the technology appropriate to their situation.

Another older, U.S.-based waste-to-energy technology that was redesigned during the 1970s was the pyrolysis method. This system differed from the combustion technologies in that it was heat absorbing (endothermic) rather than heat generating (exothermic). The absorption occurred by exposing the organic material in the waste to heat in the absence or near absence of oxygen. The heat was then recovered as energy by the transformation of the solid waste into steam or a gaseous or liquid fuel. Most of the pyrolysis projects during the 1970s remained at an experimental or demonstration stage, while the largest (1,000 tpd) of these facilities, located in Baltimore and originally developed by Monsanto, was continually fraught with operational difficulties after the plant was completed in 1975.[18]

The strongest competitors to the mass burn incinerators, however, were the refuse-derived fuel (RDF) systems, a new incineration technology introduced shortly after the mass burn system made its appearance in this country in the late 1960s. Unlike the mass burn technology, which used unprocessed solid waste as its fuel source, RDF systems involved a shredding and/or separation-oriented method to reduce the particle size of the waste as well as to remove certain components of the waste stream, such as ferrous metals. This smaller, more combustible fraction of the waste would be ground down into dense, pelletlike units suitable as fuel in a stoker-fired fuel-burning system such as those using coal. The RDF technology would thus be effective both as a waste disposal option and as a fuel supplement for other electricity-generating systems. Some of the first RDF plants, such as those in Ames, Iowa, which began operation in 1975, and Milwaukee, Wisconsin, which was developed by the American Can Company, also experienced initial operational difficulties. The RDF option was also found to be economically unattractive by TVA for its coal-fired plants after a 1976 evaluation.[19]

During the early and mid-1970s, all of the new waste-to-energy operations, including the mass burn plants, were plagued by cost overruns and major operational problems. The Saugus, Massachusetts, plant, for example, the first to use the Von Roll grate, ultimately needed to turn to the federal government to pay for $11 million worth of repairs. The Chicago Northwest plant similarly experienced immediate cost overruns and time delays during plant construction as a result of the discovery of an underground stream during excavation of the site, which in turn required a range of remedial measures. Even the potential for revenue from energy sales had faded. Some of the plants, including Chicago Northwest, started up without any electrical generating capacity because of low energy prices at the time, and no plan or mechanism to market their energy if such capacity were installed. Other facilities derived only limited income from energy sales prior until 1978. Then, new legislation governing the sale of energy to utilities from "alternative" sources provided a new framework for the marketing of the self-generated electricity from waste-to-energy plants.[20]

Many of the plants were also repeatedly shut down or were operating well below capacity, as unanticipated problems developed after the facilities became operational. Such problems included the incomplete combustion of the wastes, which damaged and clogged the grates, and the corrosion of the water-wall tubes from corrosive compounds like hydrogen chloride, formed, in part, by the combustion of plastics. Through much of the 1970s, despite increased interest, the various waste-to-energy technologies were still considered experimental or, at the least, in need of much greater on-line experience. Mass burn incinerators and RDF plants, the two favorite waste-to-energy options, also remained considerably more expensive than even the most advanced sanitary landfill at the time.[21]

Incineration as a whole, furthermore, was in sharp decline through the 1970s. Of 364 municipal waste incinerators operating in 1969, nearly 90 percent of which were refractory-lined units, only 67 remained in operation in 1979.[22] By the end of the decade, the combination of the few remaining, more conventional refractory units, plus the handful of new waste-to-energy plants, handled only a minuscule proportion, less than 1 percent, of all municipal waste. In July 1979, an industry consultant, writing in a trade publication, lamented this state of waste-to-energy. "Our dreams have taken us into what now seems to be the clouded skies of resource recovery—of refuse to energy—of burning to earning," the consultant warned. "We are not facing up to the truth about the expense, inefficiency, and impracticability of the waste recovery projects that make Monday morning headlines." Other industry supporters and consultants, who also acknowledged the range of problems, feared that these early indications of trouble would force the waste-to-energy industry to stay more experimental than commercial, and ultimately erode the initial interest expressed by both the financial community and public agency officials. The new era of waste-to-energy, despite its high-tech systems and large corporate presence, appeared to remain in limbo.[23]

THE CRISIS OPTION

The 1970s, a decade of turmoil but only limited change in the solid waste area, came to represent a transitional period among various solid waste disposal options. By the late 1970s, a sharp contrast had developed between the promise of waste-to-energy and its unmet expectations. At the same time, the increasing difficulties facing landfills assumed an intensity that took public officials, especially at the municipal level, by surprise. The broad concerns but just limited policy initiatives of the 1970s gave way to a growing crisis atmosphere that became sharper and more visible during the 1980s. Moreover, the focus on solid waste was being displaced, particularly at the federal level, with new concerns about hazardous wastes, stimulated by the incidents of toxic contamination in such communities as Love Canal, New York, Times Beach, Missouri, and Glen Avon, California, the site of the Stringfellow Acid Pits. As part of this shift in emphasis, EPA's Office of Solid Waste, after 1979, shifted nearly its entire budget and personnel in less than a couple of years from the solid waste to the hazardous waste area. By the early 1980s, the federal solid waste research efforts, demonstration projects, and grant programs in the area of resource recovery, as well as funding to states for the enforcement of performance standards, had been fully eclipsed by the crisis management programs devoted to hazardous wastes and their cleanup.[24]

At the local level, however, the solid waste disposal problem preoccupied public officials more than ever during the 1980s. The new timetable established by federal legislation to close down all open dumps and landfill sites not meeting regulations transformed initial concerns around an eventual squeeze on landfill space into an actual crisis of capacity shortage in several areas of the country, particularly in the Northeast. Municipal officials were forced to either undertake a protracted (and often losing) political battle to keep existing landfills open—let alone obtain additional landfill space—or to otherwise seek out alternative disposal methods. The enormous public concerns over landfill problems further narrowed the political choices available. By the early 1980s, a mood of urgency had begun to settle over the policy discussions and debates on the waste issue.

In this setting, the incineration option received renewed attention. As opposed to its tentative reappearance in the 1970s, characterized by multiple complications and uncertainties, incineration came to be viewed as more viable politically and economically in the changing circumstances of the early 1980s. A number of large private vendors, moreover, had begun to view their participation not as merely a speculative interest in a new and experimental technology but as a profitable investment in a dynamic and rapidly expanding business. While the primary reason for this change of perspective remained the dilemma caused by the escalating troubles of landfills and the steady growth of the waste stream, the questions of energy, privatization, and environmental politics also contributed to the heightened interest in incineration.

The energy issue was important in that it helped secure what appeared to be a significant source of income for waste-to-energy plants. By the late 1970s, the cumulative impact since the 1973–74 "energy crisis" of increased oil prices and new government policies encouraging alternative, self-generated sources of energy came to have direct bearing on the economics of incineration. With the assumption that oil prices would continue to climb, alternative energy producers, including the waste-to-energy industry, explored the possibility of pegging the sale of their energy to utilities at a price basically equivalent to the price of oil-generated electricity. This provision for an "avoided cost" sale became a key component of the push for new legislation to effectively force the utilities to purchase energy from alternative sources.

The purchase of incinerator-generated energy as well as other "self-generated" or "co-generated" alternative energy sources had long been discouraged by the electrical utility industry. As late as the turn of the twentieth century, when "waste-to-energy" incineration technology was first explored, 60 percent of the electricity used in the United States was self-generated (through such sources as wind and wood) rather than purchased from utilities. By the 1970s, however, that figure had shrunk to only 4.2 percent of U.S. capacity.[25]

That extraordinary decline in self-generated or co-generated energy had both technological and political roots. The invention of the turbogenerator in 1903 permitted vast economies of scale in electric generation, which in turn enabled utilities to shift from direct to alternating current and allowed electricity to be transported over greater distances. The utilities that controlled this source of electric power came to be seen as "natural" monopolies and regulations were structured to discourage competition from independent generators. Regulated utilities, furthermore, were allowed a rate of return based on their capital investment, so that the development of alternative (nonutility) generated sources reduced the need for capital expenditures and thus limited the utilities' rate of return. As a result, utilities did everything in their power to discourage self-generated energy, including waste-to-energy sources.[26]

During the 1970s, interest in alternative energy produced outside the regulated utility system increased substantially. The Carter administration, as part of its overall energy policy designed to reduce dependence on foreign oil, proposed the Public Utilities Regulatory Policies Act (PURPA), passed in 1978. PURPA was structured in part to allow for greater competition among independent generators (and co-generators) producing alternative or non-oil-based energy. The legislation provided that utilities had to purchase power from these generators at a price to be set as high as the "cost to the electric utility of the electric energy which, but for the purchase from such co-generator or small power producer, such utility would generate or purchase from another source." On that basis, the sale price would be the equivalent of the cost to produce the most expensive source of electricity, which was likely to be from an oil-based generator.[27]

This concept of "full avoided cost" appeared to provide a crucial economic advantage to the waste-to-energy plant. Between 1970 and 1978, every incinerator facility that became operational had received only limited income from the sale of electricity (or, in some instances, no income at all), partly due to inability to sell the energy directly to utilities. Most plant operators, however, believed that the revenues derived from such a sale would ultimately be crucial to the economic success of these plants. PURPA created new opportunities in both economic and political terms for waste-to-energy facilities by expanding one key income source as well as providing an additional "environmental"-oriented reason—substituting for oil—to support a waste disposal technology that could be touted as a more benign and environmentally acceptable way out of the landfill impasse.

By the early 1980s, support for incineration reached a peak. Landfill regulations were increasingly tied to environmental considerations such as potential groundwater contamination and/or toxic air emissions, and several environmental and community activists who were engaged in their own bitter and protracted conflicts with public officials over landfill sites were more willing to see waste-to-energy as a possible alternative, a way to capture the "energy resource" in municipal solid waste. Municipalities themselves were attracted to the notion that the new waste-to-energy facilities represented a "state-of-the-art" technology capable of minimizing environmental risks, while providing an alternate strategy for waste disposal. Politically, waste-to-energy appeared capable of attracting more broad-based support than landfills, an additional boon to harried public officials who were becoming more willing to trade off the still cheaper but politically risky land disposal option with what appeared to be an acceptable, albeit relatively untested, alternative.[28]

The incineration option during the early 1980s was further enhanced by the Reagan administration's evolving strategy of *privatization*. Reagan entered the White House vowing to reduce the role of government, not only in terms of the federal bureaucracies but by paring down federal support of state and local government activity as well. The Reagan administration quickly proposed, through a series of commissions and legislative initiatives, a range of new economic incentives and programs. These were designed to encourage the private sector to engage in activities such as solid waste collection and disposal that had traditionally been the province of local government. These privatization initiatives included changes in the tax laws, the increasing use of such devices as industrial development bonds, and the growing EPA emphasis in favor of private sector involvement in solid waste management. The thrust toward privatization, furthermore, helped create a regulatory and economic environment for incineration in which private vendors and the large supporting array of private sector companies could flourish. Publications with such names as the *Privatization Review* were filled with stories about the possibilities of an expanding private sector role within a public-private partnership, complete with joint financing arrangements, turnkey contracts (building a complete plant at a guaranteed price), reduced labor costs, and a wide variety of economic and operational benefits for both private company and public agency alike.[29]

The impact of this push for privatization in the solid waste area was immediate and substantial. It provided additional economic incentives for a variety of companies to aggressively pursue incineration to secure a share of the market. At the same time, many municipalities became convinced that by contracting out for the construction and/or operation of an incineration plant, they would be able to reduce their risks and obtain that alternative to landfills so urgently sought. Privatization, along with PURPA and a broader support base, provided additional momentum to an industry already poised for rapid growth because of the changing circumstances for landfilling. For municipalities, incineration was ready to become their crisis option.

INDUSTRY TAKE-OFF

The increase in the number of proposed incineration projects in the early and mid-1980s was phenomenal. Just 60 plants were either on line, proposed, or under construction in 1980 (several of them small in scale and considered experimental or demonstration projects), when municipalities began to make the move toward incineration. A kind of bandwagon effect was created, with trade publications, investment banks engaged in municipal financing, and various municipal organizations such as the U.S. Conference of Mayors and the National League of Cities becoming highly committed partisans of incineration. "Waste-to-energy is no longer a newfangled technology or a tinkerer's dream," proclaimed the executive director of the National League of Cities in 1987. "It is a compelling force, driven by the difficulties many communities are experiencing in obtaining new landfills and spurred by genuine desires to manage wastes in the best possible ways."[30]

Within five years, the number of incineration plants already built or under consideration had, according to a 1985 League of Cities estimate, more than tripled to nearly 200. Just a couple of years later, the trade publication *Waste Age*, in a more detailed, state-by-state analysis of actual and projected refuse incineration capacity, pointed to 157 plants operating or under construction with nearly 80,000-tons-per-day capacity, and another 186 currently planned and sited plants with an additional 145,000-tons-per-day capacity. Another 1987 study by the EPA, which cited 110 plants in operation but another 220 in the planning or construction stage, declared that the incineration of solid waste was "expected to grow at an astonishing rate . . . significantly faster than the growth rate for municipal refuse generation."[31]

The interest in waste-to-energy spread throughout the country (see table 2.2). Forty-four of the 50 states in the *Waste Age* survey had at least one facility planned or under construction and 14 states had at least 10 such facilities in the planning or construction stage. A few states, such as New York, New Jersey, Pennsylvania, Florida, Massachusetts, and California, had made a significant move toward waste-to-energy with the expectation that in some communities, these facilities would displace landfills as the *primary* method of disposal. "Is there a waste-to-energy facility in your area? If there isn't, will there be?"—the executive director of the National League of Cities asked a theoretical question of his constituents. "Ask those two questions of any mayor, city council member, city manager, or solid waste staff person in almost any community today, and you'll get an answer with a very positive ring." The director answered his

Table 2–2

SUMMARY OF PLANNED MUNICIPAL WASTE COMBUSTION FACILITIES

DESIGN TYPE	DESIGN CAPACITY RANGE (TPD)	TOTAL DESIGN CAPACITY (TPD)	NUMBER OF PLANNED FACILITIES	LOCATION: NEW ENGLAND AND MID-ATLANTIC	NORTH CENTRAL	SOUTH ATLANTIC	SOUTH CENTRAL	MOUNTAIN AND PACIFIC
Mass burn[a]	<250	3,055	18	9	2	2	1	4
Modular combustor[b]		1,377	14	7	3	0	2	2
RDF		450	3	1	1	0	0	1
UDT[c]		1,225	7	3	2	0	1	1
Mass burn	250 to <500	6,155	17	7	2	3	0	5
Modular combustor		3,730	10	6	0	1	0	3
RDF		730	2	2	0	0	0	0
UDT		3,220	9	3	2	2	0	2
Mass burn	500 to <1,000	21,653	33	17	2	4	3	7
Modular combustor		0	0	0	0	0	0	0
RDF		8,544	11	3	2	2	0	4
UDT		3,700	6	3	0	1	2	0
Mass burn	≥1,000	82,532	50	25	2	6	1	16
Modular combustor		0	0	0	0	0	0	0
RDF		29,150	15	2	2	3	0	8
UDT		27,850	15	8	0	3	1	3
Total		193,371	210	96	20	27	11	56

SOURCE: *Municipal Waste Combustion Study,* EPA. 1987.

[a] Includes both overfeed stoker and rotary combustor designs.

[b] Includes both starved air and excess air designs.

[c] Design type has either not been specified or data on design type were not provided in the references.

own question. "I would venture that eight out of ten would answer 'not yet' to the first question and 'very likely' to the second."[32]

The interest in waste-to-energy focused primarily on mass burn incinerators, many designed as large economies of scale facilities, ranging in size from 1,000 to 2,000 tons per day (tpd) to as large as 3,000 or 4,000 tpd. Several of these plants have been defined as second- and even third-generation incinerators, incorporating some of the newer, more technically advanced and sophisticated equipment. They have attracted some of the larger vendors and major investment banks, who covet the estimated $17 billion to $20 billion of municipal financing they will require. In one direct appeal to municipal officials, for example, the investment firm of Smith Barney proclaimed that in 1984 and 1985 alone, Smith Barney had become the "Number One in Resource Recovery Financing," with more than $2.7 million worth of activity. In a 1986 study, Kidder, Peabody cited the large incinerator plants as one of the few growth areas for engineering and construction firms. According to another major investment house, Shearson Lehman Brothers, incineration had become a significant growth industry.[33]

Several municipalities and other regional entities also developed an interest in refuse-derived fuel (RDF) technology. Although fewer than 25 communities had developed active plans in 1987 to build RDF facilities, according to a *Waste Age* survey, these plants were designed to cumulatively handle as much as 40,000 tons per day of solid wastes, nearly a quarter of the potential capacity of all waste-to-energy facilities then either on line, proposed, or under construction. RDF's market share varied significantly with the size of the plant: while only holding 7 percent of the market for plants with a 100-to-300-tons-per-day (tpd) capacity in the 1987 estimate, RDF accounted for 19 percent of the market for plants sized between 201 to 800 tpd, and 41 percent for plants above 801 tpd. This tendency toward economies of scale attracted such companies as General Electric, Bechtel, and Westinghouse, several of whom became joint-venture partners with smaller RDF operators and designers who were seeking financial backing. Including both mass burn and RDF, then, incineration had emerged by the mid- to late 1980s as the favored and, in several instances, the *only*, alternative to landfills for public officials eagerly seeking to diversify their waste disposal options.[34] (See table 2.3 for waste-to-energy facilities ordered as of October 1986.)

The new waste-to-energy field was quickly recognized by the investment community as a lucrative and rapidly expanding market. In 1987, *Forbes* estimated that the waste-to-energy market was worth $35 billion, with most of that ($25 billion) concentrated among the larger plants with capacities of 1,000 tons per day or greater. That same year, *Newsday* reported that investment bankers had already earned nearly $200 million

Table 2–3

EXISTING FACILITIES ORDERED BY STATE AND DESIGN TYPE AS OF OCTOBER 1986

LOCATION		COM-BUSTOR	HEAT	NO. OF COM-	TOTAL PLANT CAPACITY	TYPE OF	START-UP	
CITY	STATE	TYPE	RECOVERY	BUSTORS	(TDP)	CONTROL(S)	DATE	REFERENCES
Sitka	AK	MI/SA	Yes	2	25	ESP	1985	City Currents, October 1986
Tuscaloosa	AL	MI/SA	Yes	4	300	ESP	1984	City Currents, October 1986
Hope	AR	MI/SA	No	3	38	None	NA	
Batesville	AR	MI/SA	Yes	1	50	None	1981	
Blytheville	AR	MI/SA	No	2	70	None	1983	Direct contact to facility, February 1987
Osceola	AR	MI/SA	Yes	2	50	None	1980	
North Little Rock	AR	MI/SA	Yes	4	100	None	1977	City Currents, October 1986
Stuttgart	AR	MI/SA	No	3	60	None	NA	MRI
Hot Springs	AR	MI/SA	No	8	100	None	NA	
New Canaan	CT	MB/OF	No	1	108	VWS	NA	
Stamford 1	CT	MB/OF	Yes	1	200	ESP	1974	Direct contact to facility, March 1987
Stamford 2	CT	MB/OF	Yes	1	360	ESP	1974	Direct contact to facility, March 1987
Windham	CT	MI/SA	Yes	3	108	BAG	1981	
Washington (Solid Waste Reduction Center, 1)	DC	MB/OF	No	4	1,000	ESP	1972	Direct contact to facility, February 1987
Dade County	FL	RDF	Yes	4	3,000	ESP	1982	City Currents, October 1986
Pinellas County	FL	MB/OF	Yes	2	2,000	ESP	1983	City Currents, October 1986
Pinellas County (expansion)	FL	MB/OF	Yes	1	1,150	ESP	1986	City Currents, October 1986

Tampa	FL	MB/OF	Yes	4	1,000	ESP	1985	
Mayport Naval Station	FL	MI/EA	Yes	1	48	C	1978	City Currents, October 1986
Lakeland	FL	RDF/C	Yes	3	300	ESP	1981	City Currents, October 1986
Honolulu	HA	MB/OF	No	1	600	ESP	1970	Direct contact to facility, March 1987
Ames	IA	RDF/C	Yes	2	200	ESP	1975	Direct call to facility, March 1987
Cassia County	ID	MI/SA	Yes	2	50	None	1982	
Chicago (N.W. Waste to Energy Fac)	IL	MB/OF	Yes	4	1,600	ESP	1970	
East Chicago	IN	MB/OF	No	2	450	VWS	1971	Direct contact to facility, March 1987
Louisville	KY	MB/OF	No	4	1,000	WS	NA	MRI
Simpson County (Franklin)	KY	MI/SA	Yes	2	77	None	NA	State of Kentucky
Shreveport	LA	MB/OF	No	1	200	VWS	NA	
Haverhill/Lawrence	MA	RDF	Yes	3	1,300	ESP	1984	City Currents, October 1986
Fall River	MA	MB/OF	No	2	600	WS	1972	Direct contact to facility, March 1987
Framingham	MA	MB/OF	No	2	500	DS/BAG	1970	Direct contact to facility, March 1987
North Andover	MA	MB/OF	Yes	2	1,500	ESP	1985	City Currents, October 1986
Saugus	MA	MB/OF	Yes	2	1,500	ESP	1985	
Pittsfield	MA	MI/EA	Yes	3	240	EGB	1981	
Baltimore (Pulaski)	MD	MB/OF	No	4	1,200	ESP	NA	
Baltimore (RESCO)	MD	MB/OF	Yes	3	2,250	ESP	1985	
Harpswell	ME	MI/SA	No	1	14	None	NA	
Auburn	ME	MI/SA	Yes	4	200	BAG	1981	
Clinton (Grosse Pointe)	MI	MB/OF	No	2	600	ESP	NA	Michigan APC

Table 2–3 (*Continued*)

LOCATION		COM-BUSTOR	HEAT	NO. OF COM-	TOTAL PLANT CAPACITY	TYPE OF	START-UP	
CITY	STATE	TYPE	RECOVERY	BUSTORS	(TDP)	CONTROL(S)	DATE	REFERENCES
S.E. Oakland County	MI	MB/OF	No	2	600	WS	NA	Michigan APC
Duluth	MN	RDF	Yes	2	400	VWS	NA	
Savage	MN	MI/SA	Yes	1	60	ESP	NA	Department of Air Quality (MN)
Purham	MN	MI/SA	Yes	2	80	ESP	1986	Department of Air Quality (MN)
Red Wing	MN	MI/SA	Yes	1	72	ESP	1982	
Collegeville (St. Johns)	MN	MI/SA	Yes	1	50	WS	1981	City Currents, October 1986
St. Louis (1 and 2)	MO	MB/OF	No	4	800	WS	NA	
Fort Leonard Wood	MO	MI/SA	Yes	3	75	None	NA	
Pascagoula	MS	MI/SA	Yes	2	150	ESP	1985	
Livingston	MT	MI/SA	Yes	2	75	None	1982	
Wilmington	NC	MB/OF	Yes	2	200	ESP	1984	
Wrightsville	NC	MI/SA	No	2	50	None	NA	Trip report
Litchfield	NH	MI/SA	No	1	22	None	NA	
Durham	NH	MI/SA	Yes	3	108	C	1980	Direct contact to facility, February 1987
Wilton	NH	MI/SA	No	1	30	None	NA	
Auburn	NH	MI/SA	No	1	5	None	NA	
Pittsfield	NH	MI/SA	No	1	48	None	NA	
Meredith	NH	MI/SA	No	2	31	None	NA	
Groveton	NH	MI/SA	Yes	1	24	None	NA	
Portsmouth	NH	MI/SA	Yes	4	200	BAG	1982	
Nottingham	NH	MI/SA	No	1	8	None	1972	Direct contact to facility, March 1987
Candia	NH	MI/SA	No	1	15	None	NA	

Wolfeboro	NH	MI/SA	No	2	16	None	1975	Direct contact to facility, March 1987
Canterbury	NH	MI/SA	No	1	10	None	NA	
Albany	NY	RDF	Yes	2	600	ESP	1981	
Niagara Falls	NY	RDF	Yes	2	2,200	ESP	1981	
Brooklyn (S.W.)	NY	MB/OF	No	3	750	ESP	NA	
Glen Cove	NY	MB/OF	Yes	2	250	ESP	1983	State of New York
Westchester County	NY	MB/OF	Yes	3	2,250	ESP	1984	
Brooklyn (N. Henry St.)	NY	MB/OF	No	1	1,000	ESP	NA	
Huntington	NY	MB/OF	No	3	450	WS	NA	
New York (Betts Avenue)	NY	MB/OF	Yes	4	1,000	ESP	NA	
Skaneateless	NY	MI/SA	No	1	13	None	1975	Direct contact to facility, March 1987
Oneida County (Rome)	NY	MI/SA	Yes	4	200	None	1985	
Cattaraugus County (Cuba)	NY	MI/SA	Yes	3	112	None	1983	City Currents, October 1986
Oswego County (Volney)	NY	MI/SA	Yes	4	200	ESP	1985	
Akron	OH	RDF	Yes	3	1,000	ESP	1979	
Columbus	OH	RDF	Yes	6	2,000	ESP	1983	
N. Dayton	OH	MB/OF	No	2	600	ESP	1970	
S. Dayton	OH	MB/OF	No	2	600	ESP	1970	
Euclid	OH	MB/OF	No	NA	200	ESP	NA	State of Ohio
Tulsa	OK	MB/OF	Yes	2	750	ESP	1986	
Miami	OK	MI/SA	Yes	3	108	None	1982	Direct contact to facility, February 1987
Marion County	OR	MB/OF	Yes	2	550	DS/BAG	1986	
Philadelphia (N.W. Unit)	PA	MB/OF	No	2	750	ESP	1957	Direct contact to facility, March 1987

Table 2–3 (*Continued*)

LOCATION		COMBUSTOR	HEAT	NO. OF COM-	TOTAL PLANT CAPACITY	TYPE OF	START-UP	
CITY	STATE	TYPE	RECOVERY	BUSTORS	(TDP)	CONTROL(S)	DATE	REFERENCES
Philadelphia (E. Central Unit)	PA	MB/OF	No	2	750	ESP	1965	Direct contact to facility, March 1987
Harrisburg	PA	MB/OF	Yes	2	720	ESP	1973	
Johnsonville	SC	MI/SA	Yes	1	50	None	NA	
Hampton	SC	MI/SA	Yes	3	270	ESP	1985	Consumat
Nashville	TN	MB/OF	Yes	2	720	ESP	1974	City Currents, October 1986
Nashville (Expansion)	TN	MB/OF	Yes	1	400	ESP	1986	City Currents, October 1986
Gallatin	TN	MB/RC	Yes	2	200	ESP	1981	Direct contact to facility, February 1987
Dyersburg	TN	MI/SA	Yes	1	100	None	1980	City Currents, October 1986
Lewisburg	TN	MI/SA	Yes	1	60	WS	1980	City Currents, October 1986
Cleburne	TX	MI/SA	Yes	3	115	ESP	1986	State of Texas, City Currents, October 1986
Carthage City	TX	MI/SA	Yes	1	36	None	1985	State of Texas
Gatesville	TX	MI/SA	Yes	1	20	None	NA	
Center	TX	MI/SA	Yes	1	36	None	1985	State of Texas
Palestine	TX	MI/SA	Yes	1	28	WS	NA	
Waxahachie	TX	MI/SA	Yes	2	50	WS	1982	City Currents, October 1986
Ogden	UT	MB/OF	Yes	3	450	ESP	NA	
Portsmouth	VA	MB/OF	Yes	2	160	ESP	1971	
Norfolk (Navy Station)	VA	MB/OF	Yes	2	360	ESP	1967	
Hampton	VA	MB/OF	Yes	2	200	ESP	1980	
Harrisonburg	VA	MB/OF	Yes	2	100	ESP	1982	

Galax	VA	MB/RC	Yes	1	56	BAG	NA	City Currents, October 1986
Salem	VA	MI/SA	Yes	4	100	None	1970	
Newport News (Ft. Eustis)	VA	MI/SA	Yes	1	35	None	1980	
Bellingham	WA	MI/EA	Yes	1	100	None	1986	
Bellingham	WA	MI/SA	Yes	2	100	None	1986	
Sheboygan	WI	MB/OF	No	2	240	WS	NA	State of Wisconsin
Waukesha	WI	MB/OF	Yes	2	175	ESP	1971	
Barron County	WI	MI/SA	No	2	80	ESP	1986	State of Wisconsin
Madison	WI	RDF/C	Yes	2	400	ESP/C	1979	City Currents, October 1986

Key

Combustor Types:
MI/SA = Modular combustor with starved air
MI/EA = Modular combustor with excess air (Vicon)
RDF = Refuse-derived fuel fired in dedicated boiler
RDF/C = Refuse-derived fuel/coal co-firing
MB/OF = Mass burn with overfeed stoker
MB/RC = Mass burn in rotary combustor

Types of Controls:
C = Cyclone
ESP = Electrostatic precipitator
WS = Wet scrubber
DS = Dry scrubber
VWS = Venturi wet scrubber
BAG = Baghouse
EGB = Electrostatic Gravel bed

NA = Data not available or technology undecided

in fees and commissions for plants whose contracts had either been signed or were under consideration. The tendency toward concentration had also become more pronounced among the private vendors, or developers, contracting with public agencies to build and operate these facilities.[35]

By the late 1980s, a process of consolidation and expansion of private sector activity within the incineration industry had increased significantly. The industry itself had evolved into two distinct sets of players. On the one hand were the myriad of firms including the boiler manufacturers, construction and engineering service companies, the waste haulers, investment banks and financial underwriters, consulting firms, and a wide range of companies that provide equipment and components used in the incineration process, from cranes and forklifts, to boiler and turbine parts, to air pollution control equipment and computer monitoring. This new system of "resource recovery," *Newsday* commented, "is a kind of municipal-industrial complex."[36]

But it has been a select group of players—waste-to-energy developers, including the companies that hold the licenses for the incinerator technology—who have become the driving force in the industry. Four of these companies have for the past several years controlled slightly more than 50 percent of the entire market. As of 1988, 19 companies in total served as project managers of waste-to-energy plants and 36 companies were considered facility developers. Although the industry was considered competitive, analysts anticipated further market concentration, given especially the trend toward construction of the larger and more complex systems increasingly favored by municipalities and regional public entities. Nearly all the project managers were new to the business and only a few had any operational experience. Even among those in the front ranks of the industry, several had no experience in either running or even completing construction of a facility. Market share, in fact, was determined by the number of operating plants as well as those in the planning or construction stage.[37]

The four industry leaders—Ogden Martin, Wheelabrator Environmental Systems (formerly Signal), American Ref-Fuel, and Combustion Engineering—are large, diversified corporations. With the exception of American Ref-Fuel (a joint venture operation of Browning-Ferris and Air Products), the waste-to-energy activities of these companies, and several of the other developers and project managers in the field, are only one part of their corporate parent's overall operations.

Many of these corporations are ranked within *Fortune*'s list of the 500 largest industrial companies in the country. They include such giants as Westinghouse, Bechtel, Fluor, Foster Wheeler, Raytheon, and Katy-Seghers. These companies are identified more as construction, engineering, resource development, or consumer service firms than as waste-to-energy or waste management corporations. Many of the larger companies, furthermore, are involved in a broad range of incineration activities beyond the burning of municipal solid wastes, such as tire burning, hazardous waste and hospital waste incineration, and various burn technologies that deal with agricultural wastes, wood chip wastes, and sewage sludge. The key exceptions to this industry tendency are Waste Management, Browning-Ferris, and Laidlow, whose primary businesses have been defined as waste hauling and landfill operations.[38]

The two companies with the largest share of the market among project managers (both with about 20 percent) are Ogden Martin and Wheelabrator, formerly Signal Environmental Systems. Wheelabrator/Signal is, in one form or another, the oldest mass burn incineration company in the country. The Massachusetts-based company had been the industry leader in both tonnage handled and facilities built and operated. Signal's acquisition of the West German Martin technology played a significant role in its early ascendancy, and it became the vendor for several of the new generation of incinerator plants in the early 1970s, such as Chicago Northwest and the Saugus and Harrisburg plants. In 1983, Signal acquired one of its main competitors, Wheelabrator Technologies, license holder of the Swiss Von Roll technology, and created Signal Environmental Systems, the subsidiary that Alfred DelBello was recruited to head. The parent company, which had just previously merged with the multinational Allied Corporation to become Allied-Signal, Inc., was reorganized within the Henley Group, which established Wheelabrator (formerly Signal Environmental Systems) as a separate operating unit in Henley in June 1986. All of Wheelabrator's businesses were based on the Signal operations and the Wheelabrator name was itself assumed after the Federal Trade Commission forced Wheelabrator/Signal to sell one of its incinerator licenses and the new company decided to keep the Von Roll technology and sell the Martin license instead.[39]

Von Roll is one of the oldest of the European incinerator companies. Its technology has been used in more than 200 plants worldwide, making it the leading money-maker for its corporate parent. When Wheelabrator/Signal, now dependent on the Von Roll technology, went through its complex corporate restructuring process in the mid-1980s, it slowed its acquisition of new projects, which prevented it from expanding its market share, causing it to fall to second place after Ogden Martin in 1987.[40]

Wheelabrator, however, still plays a leading role in the industry, with 14 plants either in operation, under construction, or in the planning stage in 1989. "The driving force behind this company is the garbage crisis and the need to dispose of municipal waste," Wheelabrator president Rodney Gilbert told *Forbes* magazine.[41] Moreover, the company has continued to explore new arrangements and developments, the most striking of which involved a 1988 transaction between Wheelabrator and Waste Management, the largest solid, hazardous, and radioactive waste management company in the nation. The deal centered around Waste Management's purchase of 23 percent of the stock that Wheelabrator Technologies had decided to offer to the public, thus making it the second largest stockholder in the company after Henley itself. In return, Waste Management agreed to guarantee landfill space at the market price for disposal of the ash from Wheelabrator's incinerators. At the same time, Wheelabrator has the exclusive right to build waste-to-energy plants next to Waste Management landfills.[42]

The Wheelabrator/Waste Management linkage of landfill and privately developed incineration facilities through what have been called "merchant plants," or "landfill extenders," could potentially alter the structure of the industry. "It's a natural marriage," Alfred DelBello proclaimed, speculating that, besides Waste Management, other waste haulers and landfill operators such as Browning-Ferris and Laidlaw could, by adopting such an approach, significantly expand their own operations directly into the waste-to-energy market. For Wheelabrator, the merchant plant concept had the advantage that the company might more easily overcome any future resistance to specific incineration projects by basically eliminating the siting issue, a key potential obstacle in securing support for the project.[43]

Ogden Martin, the company that now leads the market, is a relative newcomer to the waste-to-energy market. Ogden, a wholly owned subsidiary of the Ogden Corporation, has grown quickly since it entered the field in 1983 with its purchase from Signal of the West German Martin process. The Martin process became the leading system on a worldwide basis, used in approximately one-third of all incinerator systems internationally. By 1988, in fact, Ogden had more large incinerators in the planning stage than its competitors, with several more nearing completion or under construction. All of these were mass burn plants except for one RDF facility in Lawrence, Massachusetts. Ogden operated plants located in Oklahoma, Oregon, and Florida, with others close to completion in Virginia and Connecticut, and more than a dozen other facilities in various stages of development.[44]

Unlike several of the other large vendors, Ogden has not had its own construction firm. A corporation that defines itself as a "full-service" company, Ogden emphasized its ability to undertake the long-term operation of the plants. More than other developers, Ogden pursued the option of private ownership of waste-to-energy facilities in attempting to link its service role with the economic advantages of privatization. After 1986, however, when privatization arrangements were circumscribed by a new tax law, Ogden's parent company ultimately established a new subsidiary, Ogden Financial Services, to help Ogden Martin and other company subsidiaries obtain low-cost equity funds for new plants. Despite these obstacles, Ogden aggressively pursued new project opportunities, allowing it to keep pace with the rapid expansion of the industry.[45]

American Ref-Fuel and Combustion Engineering are Ogden and Wheelabrator's closest competitors, with approximately 8 and 7 percent of the market respectively, through 1987. Of the four major waste-to-energy firms, American Ref-Fuel was organized to directly concentrate on the waste-to-energy market. The company is structured as a partnership between its co-owners, Air Products and Browning-Ferris. Browning-Ferris is one of the nation's largest waste collection and disposal companies, second only to Waste Management in its range and volume of activities. The other half of the partnership, Air Products, is experienced in heavy industrial processes, and owns Stearns Catalytic, a major engineering and construction firm. The partnership structure was conceived in part to develop breadth and flexibility in its negotiations with local governments to construct and operate waste-to-energy facilities. In its most notable project in Houston, for example, the joint venture sought to incorporate hauling arrangements as part of its scope of work. And while American Ref-Fuel did not have the waste supply contract with the city, it was given the task of negotiating private contracts with local haulers, the biggest of which is Browning-Ferris. As some analysts have commented, a logical next step for the partnership would be to pursue the merchant plant idea to link existing landfill sites with future waste-to-energy facilities.[46]

The fourth of the waste-to-energy firms, Combustion Engineering, has had a long history as a designer of conventional electric power plants and as a builder of boiler systems used in incinerator plants. Unlike the other industry leaders who are primarily involved with mass burn technologies, Combustion Engineering has focused on RDF. Combustion Engineering will be project manager for several of the largest of these facilities, including a 4,000-ton-per-day facility in Honolulu and a 2,000-ton-per-day facility in Detroit, and is also involved with a 750-tpd mass burn plant in Huntington, New York, scheduled to become operational in 1990.[47]

Most strikingly, Combustion Engineering is best known for its role as one of the Big Four firms involved in the construction and operation of nuclear power plants. All the other members of the Big Four—Westinghouse (which began exploring acquisition of American Ref-Fuel in 1989), Babcock & Wilcox, and General Electric—have also undertaken in recent years a development role in waste-to-energy and are likely to expand their level of participation in the future.[48]

There are, as both industry and community opposition groups have noted, obvious parallels between the nuclear and waste-to-energy industries. Both were initially heralded as "alternative," technologically advanced systems that would supplant their major rivals (oil-generated electricity and landfills), which were experiencing long-term decline. Both nuclear power and waste-to-energy plants are expensive, capital-intensive, economies of scale facilities regarded initially with some hesitancy by their buyers (utilities and municipalities). As a consequence, the large vendors, who quickly came to dominate their respective industries, sought to allay fears by initially offering risk-free guarantees, such as turnkey contracts. The turnkey arrangements provided that the construction of a plant would be set at a given price, with the buyer simply paying that sum, no matter what the actual cost of construction, and then "turning" the key once the facility had been completed.[49]

Both nuclear power and waste-to-energy, furthermore, have received crucial subsidies and backing from the federal government as a means to stimulate and further the transition to these new technologies. As a result, early doubts about cost and performance were transformed into support and enthusiasm from each of the parties involved—the vendors, the buyers, and the federal government—helping to create in the process a huge and rapidly growing market. Like nuclear power two decades earlier, mass burn incineration had come, in a few short years after its take-off in the 1980s, to be touted as the technology of the future. And, like nuclear power, this future-oriented technology would almost as quickly see the bottom begin to drop out of its market, with issues like cost, safety, and environmental consequence shaping the debates over its future role. Less than a decade after its take-off, the industry found itself under siege.

WARNING SIGNS

Despite all the heady talk of a waste-to-energy future in this country, the position of the industry by the late 1980s is not as secure as had been anticipated just a few years before. Costs, for one, have become a more volatile issue, a function partly of the changing picture around the energy aspect and privatization arrangements that had previously stimulated the industry. Environmental and health considerations, which played such a crucial role in the decline of landfills, have again had a bearing on the waste debate. Air emissions and ash residue, for example, raised questions about the future regulatory framework for these projects, a factor with an economic dimension as well. As a consequence, even though the tipping fees for landfills continued to dramatically escalate, waste-to-energy tipping fees have been going up as well.[50]

The limited operational history of the industry has also become a factor. By 1988, mechanical failures had caused unscheduled shutdowns at more than half of the operational plants, while forcing the closure of $720 million worth of operating incinerators. The enthusiasm of the financial community came to be tempered by their recognition that the technology, in the words of Moody's Investors Service, was "not yet proven" and that an incineration project "clearly entails major risks."[51]

Economic concerns have been further exacerbated by new and potentially sharp political conflicts. By the late 1980s, both mass burn incinerators and RDF facilities had begun to generate significant community opposition. Ad hoc, grass-roots groups opposing local incinerator projects literally appeared overnight in dozens of communities where facilities were planned. They raised questions around the costs, subsidies, and effectiveness of such programs, and challenged the assumptions about public health and environmental impacts. This opposition undercut the claim of broad public support, which these projects had relied upon to help promote waste-to-energy as an alternative to landfills.[52]

Local officials and private vendors at first dismissed the opponents for their lack of access to and influence over various decision-making bodies. Many of these groups, however, became adept at developing a high level of expertise and mobilizing large numbers of people, all of which made it difficult for public officials to proceed. As several major plants were either postponed or, in some instances, canceled due to the effective opposition of the community groups, incineration advocates finally realized they were in for a significant fight. It was not only a matter of political will, as Alfred DelBello suggested in 1985; it was a matter of political *survival* for the future of the incineration industry, as the *Wall Street Journal* claimed just three years later.[53]

With conflicts over incineration spreading rapidly from community to community, the focus of debate around solid waste management shifted from the question of the best technology and the role of the industry to the policy-making process within the public sector. Solid waste management became a battleground, in the planning arena where options were selected, and in the legislative and executive branches, which became increasingly preoccupied with solid waste. The front lines of this battleground, furthermore, were no longer limited to local jurisdictions where waste problems had been tackled for nearly a century, but they stretched to state and federal levels. The issue became regional as well as local, one hotly debated in both national forums and town councils. The question of what regulations and policies were needed had become directly tied to the question of who would decide and on what basis such decisions would be made. Garbage, an issue that had generated a multi-billion-dollar industry, had also become a matter of politics.

NOTES

1. Interview with Alfred DelBello, 1988; "Political Garbage," James Cook, *Forbes*, July 1, 1985, vol. 136, no. 1; "DelBello: Resource Recovery May Be the Best Alternative," William M. Wolpin, *World Wastes*, February 1986.
2. Interview with Kevin Stickney, Vice President, Marketing, Signal Environmental Systems, April 29, 1987.
3. "Incineration Today and Tomorrow: A Survey of Incineration Practices—Present and Future," Junius W. Stephenson, *Waste Age*, May 1970.
4. "Another Point of View," Richard W. Eldredge, *Waste Age*, May 1970.
5. *Environmental Assessment of Municipal Scale Incinerators*, 530/SW-111, EPA, Office of Solid Waste, 1973; "Hard Road Ahead for City Incinerators," *Environmental Science and Technology*, vol. 6, no. 12, November 1972.
6. "Recovering Energy from Municipal Solid Waste," David B. Sussman and Stephen J. Levy, in *Fourth United States–Japan Governmental Conference on Solid Waste Management*, March 12–13, 1979; Environmental Protection Agency, SW-789, 1979; "Incineration Today and Tomorrow," Junius W. Stephenson.
7. *A Pollution-Free System for the Economic Utilization of Municipal Solid Waste for the City of New York; Phase 1: A Critical Assessment of Advanced Technology*, Helmut W. Schulz, School of Engineering and Applied Science, Columbia University, New York, April 15, 1973; *Environmental Assessment of Municipal Scale Incineration*, EPA, 1973; "Hard Road Ahead for City Incinerators," *Environmental Science and Technology*, November 1972.
8. "Refuse Is Reusable: US Experts Study European Disposal Plants Which Turn Trash into Heat and Electric Power," Harland Manchester, *National Civic Review*, February 1968; "Refuse-to-Energy Plant Uses Von Roll Incinerators in U.S.," *Environmental Science and Technology*, vol. 8, no. 8, August 1974.
9. "Incineration Today and Tomorrow," Junius W. Stephenson, *Waste Age*, May 1970.
10. Letter from Dr. Emil Nigro, Coordinating Engineer, City of Chicago, Department of Streets and Sanitation, to Terry Bills, March 4, 1987; "Plants Burn Garbage, Produce Steam," *Environmental Science and Technology*, vol. 5, no. 3, March 1971.
11. *A Pollution-Free System for the Economic Utilization of Municipal Solid Waste for the City of New York*, Helmut W. Schulz.
12. Letter from Dr. Emil Nigro to Terry Bills, March 4, 1987; "Northwest Waste to Energy Facility," Department of Public Works and the Department of Streets and Sanitation, City of Chicago, n.d.
13. Cited in *Third Pollution*, William Small.
14. Interview with Kevin Stickney; *Environmental Science and Technology*, August 1974.
15. "Recovering Energy from Municipal Solid Waste," David B. Sussman and Stephen J. Levy; *Waste-to-Energy*, California Waste Management Board, Sacramento, California, June 1983; "Air Pollution Control at Resource Recovery Facilities," California Air Resources Board, May 24, 1984.
16. *Environmental Science and Technology*, vol. 8, no. 8, August 1974; *A Pollution-Free System for the Economic Utilization of Municipal Solid Waste for the City of New York*,

Helmut W. Schulz; *Corrosion Studies in Municipal Incineration,* P. O. Miller, 530/SW-72-3-3, EPA, Office of Research and Monitoring, National Environmental Research Center, Cincinnati, Ohio, 1974.

17. For a description of the different technologies, see "Municipal Waste Combustion Study," 530/SW 87-021, EPA, 1987; "Recovering Energy from Municipal Solid Waste," David B. Sussman and Stephen J. Levy; "The Place of Incineration in Resource Recovery of Solid Waste," J. H. Fernandes and R. C. Shenk, *Combustion,* October 1974.
18. "Markets for Refuse-Derived Energy in New York City," Dennis Ischia et al., *Resource Recovery and Energy Review,* July–August 1976; "Recovering Energy from Municipal Solid Waste," David B. Sussman and Stephen J. Levy; *An Environmental Assessment of Gas and Leachate Problems at Land Disposal Sites,* 530/SW-110.of, EPA, Office of Solid Waste, 1973.
19. "Energy from Waste: Recovering a Throwaway Resource," Anne Knight, *Solid Waste and Power,* February 1989, vol. 3, no. 1; "Resource Recovery Program Development and Experience Profile," D. E. (Ted) Hill, Vice President, National Ecology, Pimonium, Maryland, March 3, 1987; "Resource Recovery State-of-the-Art: A Data Pool for Decision-Makers in New Jersey," Institute for Local Self-Reliance, November 1985; presentation by Eugene Tseng, American Ecology, UCLA, April 18, 1987.
20. "Waste Reduction and Resource Recovery: Second Report to Congress," EPA, 1975; letter from Dr. Emil Nigro to Terry Bills, March 4, 1987.
21. "*Waste Age* Tipping Fee Survey," *Waste Age,* March 1980; March 1986.
22. *COPPE Quarterly,* vol. 1, no. 2, Fall 1987, Council on Plastics and Packaging in the Environment; *Environmental Assessment of Municipal Scale Incinerators,* EPA, 1973.
23. "Resource Recovery: Fact or Fiction?" Eugene L. Pollock, National Solid Waste Management Association *Reports,* July 1979; *Municipal Waste Combustion Study,* 530/SW-86-054, EPA, Office of Solid Waste, June 1987.
24. Interviews with Frank Smith, EPA, 1988; Susan Sakaki 1988; also "Status of the Solid Waste Effort in the United States," John W. Thompson, in *Fourth United States–Japan Governmental Conference on Solid Waste Management,* March 12–13, 1979, EPA.
25. "The Growing Role of Independent Electricity Producers," Richard Munson, *Public Utilities Fortnightly,* May 30, 1985; "Encouraging Decentralized Generation of Electricity: Implementation of the New Statutory Scheme," Reinier H. J. H. Lock, *Solar Law reporter,* vol. 2, no. 4, November/December 1980.
26. *The Power Makers,* Richard Munson, Rodale Press, Emmaus, PA, 1985; "Encouraging Public Utility Participation in Decentralized Power Production," Bradford S. Gentry, *Harvard Environmental Law Review,* vol. 5, no. 2, 1981, pp. 297–344.
27. Public Utility Regulatory Policies Act of 1978, P.L. 95–617, November 9, 1978; "Alternative Energy Power Production: The Impact of the Public Utility Regulatory Policy Act," R. Alta Charo et al., *Columbia Journal of Environmental Law,* vol. 2, no. 2, 1986, pp. 448–93; "Cogeneration: Revival through Legislation?" Claire A. Wooster, *Dickinson Law Review,* Summer 1983, pp. 705–77.
28. "Responsible Waste Management in a Shrinking World," Ariel Parkinson, *Environment,* vol. 25, no. 10; testimony presented by Betsy Laties to the California Senate Select Committee on Investment Priorities and Objectives, November 29, 1978.
29. "Privatizing the Environment," Robert J. Smith, Policy Review, Spring 1982; *The Privatization Report,* February 1988; New Jersey Senate, Energy and Environment Committee, Public Hearing on Senate Bill 1762 and Assembly Bill 1778 (Private Contracting for Resource Recovery Facilities), July 16, 1984, Trenton, New Jersey; "Privatization: Trend of the Future," *Journal of the American Water Works Association,* February 1986.
30. "Planning for the Waste-to-Energy Facility in Your Future," Alan Beals, in *Waste-to-Energy Facilities: A Decision-Maker's Guide, League of Cities,* Alexandria, Virginia,

June 1986; "Resource Recovery Activities," *City Currents*, U.S. Conference of Mayors, October 1985, October 1986, October 1987, Washington, D.C.

31. "Waste Age 1987 Refuse Incineration and Waste-to-Energy Listings," *Waste Age*, November 1987; *Municipal Waste Combustion Study*, EPA, 1987; *Report to Congress*, EPA/530/SW-87-021a-i, Office of Solid Waste and Emergency Planning, EPA, 1987.
32. "Planning for the Waste-to-Energy Facility in Your Future," Alan Beals.
33. "Resource Recovery Outlook," Kidder Peabody, January 28, 1986; "Waste-to-Energy: Cash from Trash," Shearson Lehman Brothers, June 1986; "Burning Trash Is Becoming Big Business," *Wall Street Journal*, October 13, 1986; "Solid Waste Bond Issue Data Base," Table 2 and Table 6, Securities Data, *Waste Age*, November 1988, p. 196.
34. "WTE Developers: What Are They? Who Are They?," Mike Kilgore, *Solid Waste and Power*, vol. 2, no. 4, August 1988; "Waste Age 1987 Refuse Incineration and Waste-to-Energy Listings"; presentation by Eugene Tseng, Director of International Business Development, American Ecology, UCLA, April 18, 1987.
35. *The Rush to Burn: America's Garbage Gamble*, Special Report from *Newsday*, Long Island, New York, 1988; *Forbes*, July 1, 1985.
36. "Waste-to-Energy: Analyzing the Market," *World Wastes*, June 6, 1986; *The Rush to Burn, Newsday*.
37. *A Status Report on Resource Recovery as of December 31, 1987*, Kidder, Peabody Industry Analysis, April 29, 1988; *Waste Age*, April 1987, pp. 85–86.
38. "WTE Developers," Mike Kilgore, *Solid Waste and Power*, August 1988; "The *Fortune* 500 Special Issue," *Fortune*, July 1988; "Tough Alchemy," James Walsh, *California Business*, December 1988.
39. Wheelabrator Technologies *Annual Report*, 1987; Wheelabrator Technologies Form 10-K, 1987.
40. *A Status Report on Resource Recovery*, Kidder, Peabody, April 29, 1988; Wheelabrator Technologies Form 10-K, 1987.
41. "Burn This," Claire Poole, *Forbes*, April 25, 1988.
42. "Waste Management," Leone Young, Drexel Burnham Lambert Research, New York, May 1988; Wheelabrator Technologies Form 10-K, 1987.
43. Interview with Alfred DelBello; "Waste Management," Drexel Burnham Lambert, May 1988.
44. Ogden Corporation 1987 Form 10K, Commission File No. 1-3122; Ogden *Annual Report*, 1986 and 1987; presentation by John Phillips, Vice President, Marketing, Ogden Corporation, UCLA, March 18, 1987.
45. Ogden Corporation 1987 Form 10K; Ogden *Annual Report*, 1987.
46. "Q&A with Cliff Jessberger, President, American Ref-Fuel," *Waste Age*, April 1987, pp. 81–88; "Par for the Course," James Cook, *Forbes*, vol. 136, no. 6, September 9, 1986. See also Browning-Ferris 1987 Form 10K, Commission File No. 1-6805, especially pp. 8–10.
47. Combustion Engineering 1987 Form 10K, Commission File No. 1-117-2; Combustion Engineering *Annual Report*, 1987; "Combustion Engineering Awarded Contract to Build Minnesota Plant," *Solid Waste and Power*, vol. 2, no. 4, August 1988.
48. "WTE Developers," Mike Kilgore, *Solid Waste and Power*, August 1988; "Top of the Heap," *Solid Waste and Power*, volume 3, no. 1, February 1989, p. 4.
49. *The Nuclear Barons*, Peter Pringle and James Spigelman, New York, 1981; *Nuclear Inc.*, Mark Hertsgaard, New York, 1983.
50. *Waste Age*, November 1987; November 1986; April 1987, p. 85.
51. *The Rush to Burn, Newsday*, 1988; *Waste Age*, November 1988, p. 196, which reports total dollar amount of resource recovery issues at $1.566 billion for 1987, down from its peak of $4.707 billion in 1984.
52. *Waste Not*, Ellen and Paul Connett, eds., issues no. 1, 5, 6, 12, Canton, New York;

Everyone's Back Yard, Citizen's Clearinghouse for Hazardous Waste, issues 6:1, Spring 1988; 6:2, Summer 1988, Arlington, Virginia; "WTE Development: A Zigzag Course," Harvey W. Gersham and Nancy M. Petersen, *Solid Waste and Power,* August 1988.

53. "Burning Issue: Energy from Garbage Loses Some of Promise as Wave of the Future," Bill Richards, *Wall Street Journal,* June 16, 1988.

3

THE LIMITS OF THE PLANNING AND LEGISLATIVE PROCESS: POLICY AS POLITICS

A POINT OF LEAST RESISTANCE

For the Waste Management Board, the choice of consultant seemed appropriate and compelling. Cerrell Associates was a well-known political consulting firm, especially in Southern California. It was here that the art of political and issue-related public manipulation initially evolved more than 50 years ago, when Socialist writer Upton Sinclair and his End Poverty in California movement for governor of California were derailed by the first "media campaign" of the times.[1] Joseph Cerrell himself, who had been both campaign manager and public relations consultant for a variety of candidates seeking elected office, had also developed a reputation as an effective strategist and lobbyist for industry interests. He had advised oil companies fighting air emission regulations, beverage container interests opposing bottle bills, and the real estate developers battling slow-growth groups. He and his firm were particularly well known for developing strategies to try to neutralize or redirect popular sentiment around such issues as toxic wastes. Now, with the beginning of erosion for support of incineration as a primary waste disposal option, the California Waste Management Board had decided they needed someone like Cerrell to provide guidance and insight about how to handle this growing opposition. Their very waste management strategy, the CWMB had decided, was itself at stake.[2]

The board's concerns were substantial. Community groups—from Portland, Oregon, to Onondanga County in New York and Springfield, Massachusetts—had become increasingly successful in opposing new incinerators. While public agencies like the California board had shifted from moderate but skeptical interest to strong enthusiasm for the new technology, particularly in light of the troubles facing landfills, agency officials were finding they couldn't count on public acceptance of the new projects. Landfill projects had been and continued to be opposed by similar kinds of groups, but the public agencies felt they could build a stronger political case for waste-to-energy. During much of the 1970s, the "resource recovery" concept, including "energy recovery," had achieved a fair degree of popular support. This had been "boosted by a sense of American ingenuity in overcoming foreign oil cartels," as two analysts put it.[3] But now that initial acceptance of resource recovery had begun to fade, more and more projects were being greeted with greater public skepticism and the tactics of political confrontation.

Cerrell Associates was commissioned by the California board to address this problem directly. The consulting firm prepared a report, "Political Difficulties Facing Waste-to-Energy Conversion Plant Siting," designed to provide an understanding of the opposition by profiling their leaders and constituents and analyzing their issues, as well as providing a plan of action to minimize or avoid political confrontation and polarization. Public opposition, the Cerrell report argued, had indeed become today's "most formidable obstacle to Waste-to-Energy facilities," and it was clear that the siting process, for one, had to be structured so as to "offer the least amount of political resistance to the project."[4]

It was crucial, the Cerrell report noted, to understand how different communities and different social groups responded when environmental issues or related concerns were at stake. Even though such problems as contamination associated with waste facilities had occurred at a wide variety of locations affecting a range of social or class-based groups, not all were able to mount sustained opposition. The consultants thus emphasized the importance of locating which communities were least capable of protest.

The answer, the consultants proclaimed, was readily available, based on incineration's limited history and analogous development projects. It was not just a matter of what needed to be sited, but where it would be located and how it was to be presented. The report outlined two sharply divergent poles of public response: active community opposition at one end and passive community acceptance at the other. At the point of least resistance were rural, low-income neighborhoods, whose residents had limited education, and tended to be older in age; conversely, greater resistance could be located among urban, middle- or upper-income residents, who were younger and better educated. The framing and presentation of a project facilitated by the establishment of a public affairs program with a trained public affairs coordinator and a credible advisory committee, the report argued, would be critically important in "managing community acceptance."

This kind of "concerted public participation program," as the Cerrell report called it, became increasingly characteristic during the 1980s of the methods agencies employed to incorporate and contain the public within the policy process. The take-off of the incineration industry at the beginning of the decade had been made possible in part by an evolving planning and legislative framework that came to favor the waste-to-energy option and increasingly relied on "state-of-the-art" technology promised by the industry. These positions had been formulated largely removed from any public debate and on the presumption that the politics guiding the policy-making would reflect this emerging industry–public agency alliance. The opposition movements that formed during the decade to challenge that alliance had caused agencies like the California board to seek help in making sure the decision-making process was kept on track and that a waste-to-energy future could be secured. The PR-linked participation programs and the class-motivated insights about siting were widely adopted, with groups like Cerrell Associates much sought-after as project advisers. These efforts, however, had only limited success in eliminating controversy and confrontation over siting decisions and overall waste management plans. By the end of the decade, policy-making increasingly resembled a politics out of control, while the thrust of waste management seemed up for grabs.

ENTER THE FEDS

By the 1980s, the problem of waste disposal had emerged as a national issue, permeating all levels of government in the United States. In less than 20 years after Congress in 1965 passed its first major piece of federal solid waste legislation, the Solid Waste Disposal Act,[5] the uniquely local nature of municipal garbage collection and disposal had been transformed. Waste policies now had a regional, statewide, and national focus, with special attention directed at the competing technologies of disposal: landfill and incineration.

While much of the legislation of the 1970s attempted to address the environmental hazards of landfills, the 1980s saw a mushrooming of interest in waste-to-energy, and incineration in particular. Much of this legislation was designed to facilitate the development of burn technologies, reflecting the growing perception among public agencies and elected officials that waste-to-energy was the preferred and long-term solution to the growth of the waste stream. As a result, a broad range of regulations and legislative programs were established to deal with the planning, construction, implementation, and operation of waste-to-energy plants.

These measures covered a wide number of policy concerns. Solid waste and land use laws, for example, were introduced to control siting and contractual obligations; energy laws to regulate the production and sale of the energy such plants generate; tax laws to deal with the financing they require; and environmental laws to address the pollutants they emit. The large number of laws in turn created a complex regulatory environment in which distinct federal, state, regional, and local agencies were called upon to review and permit these facilities. Many of these provisions ultimately were designed to encourage and subsidize waste-to-energy, reinforcing its position as the primary, and occasionally the exclusive, option for waste managers and local officials.

These phenomena of multiple jurisdictions and shifting priorities had evolved out of the crisis management environment that framed the waste issue in the 1970s and 1980s, with its roots in the garbage debates of the 1960s. Progressive Era reforms that emphasized engineering solutions and technologically based strategies, and governmental action based on "rational management" and scientific principles—as well as the localized nature of the waste problem—were eventually recognized as inadequate to deal with the rapid increases in per capita waste generation and concerns about environmental hazards. During the 1950s and 1960s, municipal and engineering trade publications promoted new scientific and engineering approaches such as sanitary landfills to replace the hazardous open dumps and the outmoded, overtaxed incinerators still widely used by many local agencies. But these agencies, at the same time, were confronted by problems that were beginning to overwhelm their limited resources. Consequently, there had emerged by the mid-1960s a tentative consensus within Congress and among waste managers that highlighted the need for additional financial and technical support, particularly at the federal level, to alleviate the budding waste crisis.[6]

Prior to the 1960s, the role of the federal government in the waste area had been primarily limited to public health concerns—in particular, the issue of waste-related discharges and their impacts on water systems. The Rivers and Harbors Act of 1899, for example, gave the Army Corps of Engineers jurisdiction over waste discharges into navigable waters. The concern over water pollution similarly played an important role in restricting ocean dumping of municipal wastes and ultimately led to the prohibition of any municipal discharges into streams, lakes, and rivers. These restrictions drove the shift to land disposal as the dominant waste disposal option.[7]

During the 1950s, the surgeon general was granted authority through the Public Health Service Act to develop regulations and undertake research relating to potential communicable diseases. These efforts addressed such problems as rats and flies at open dumps. But, constrained by the program's small budget of less than $500,000, solid waste research and management programs made little headway. Aside from this limited intervention, Congress generally remained reluctant to extend federal jurisdiction into what continued to be seen as a local concern.[8]

The passage of the Solid Waste Disposal Act signified a modest policy shift by providing a limited federal role in waste management. The 1965 legislation authorized federal research, training programs, and grant support for local demonstration projects to wean municipalities away from "the open burning of solid wastes." The grants, similar to those encouraging construction of sewage treatment plants, were structured so that the federal government would pay two-thirds of the cost of local demonstration programs and three-quarters of the cost of regional projects. The act also provided 50 percent funding for the development of state waste management plans in recognition that waste facilities, particularly in some of the major metropolitan areas, were needed on a regional rather than purely local or community basis.[9]

Until 1965, only two states had developed statewide solid waste management programs, and just 12 states reported any specific solid waste activities to the U.S. Public Health Service. Thirty-one states, in fact, reported no existing statewide program. Waste management programs were not much more extensive on the local level either. Most waste management regulations in local communities consisted of general health and safety ordinances and less than half of the cities and towns in the United States with populations over 2,500 reported any kind of program for the sanitary disposal of solid waste. Finding the cheapest land and just dumping the garbage became the most prevalent plan for the management of municipal solid waste. This failure to develop regional and statewide planning was seen by federal officials as a willingness on the part of municipalities to continue to rely on older, cheaper, simpler, and environmentally hazardous approaches. With the cost factor limiting the shift to the more advanced technologies of sanitary landfilling and incineration, the federal legislation was developed in part to provide research and development subsidies to help stimulate this desired transition toward new solutions and expanded management capabilities.[10]

Federal intervention, however, remained small and cumbersome. Primary jurisdiction resided with the Bureau of Solid Waste Management, housed in the Department of Health, Education, and Welfare (HEW), and with the Bureau of Mines in the Interior Department, which maintained responsibility for mining wastes and the recycling of metals and materials. The Bureau of Solid Waste Management proved to be an inadequate and underfunded bureaucracy, limiting the development of any consistent and influential federal role. Between 1965 and the passage of new solid waste legislation in 1970, the bureau experienced a heavy turnover in personnel and shifting locations, moving from HEW headquarters in Washington to offices in Maryland, then to Cincinnati, and finally back to Maryland. When the Environmental Protection Agency was established in 1970, strong sentiment developed within Congress and the federal bureaucracy to shift solid waste programs from the HEW bureau to an Office of Solid Waste within EPA. This was accomplished specifically to implement programs established by new solid waste legislation, the Resource Recovery Act, which had been passed in October of that same year.[11]

The 1970 act was an amalgam of the earlier approach of federal financial support for local projects and waste management programs with the newer, multifaceted concept of resource recovery. The act relied heavily on the language of "conservation," linking the recovery of energy and materials to more effective waste management. There were no specific regulations or provisions to mandate either waste-to-energy or recycling and reuse programs, except for the expanded support for demonstration projects. Funding for up to eight major resource recovery projects was provided, including several waste-to-energy facilities. There was also an emphasis on providing the private sector the opportunity to develop the new technologies and thereby restructure the waste business.[12]

The 1970 legislation, however, also provided for the promulgation of guidelines for conservation, recovery, and disposal systems to help further develop state waste management programs initiated under the prior legislation. By 1975, in fact, 48 states had devised some form of waste management program, including disposal regulations. Thirty-eight of those states had developed various kinds of resource recovery programs, including plans for waste-to-energy, based on federal assistance. The nature of these programs varied widely, with budgets ranging from zero to $1.2 million. Most tended to be minimal, structuring themselves around the federal support programs rather than using the federal assistance as a foundation for a more comprehensive effort.[13]

Despite the significant concern about the issue in Congress, expressed in part through major increases in funding for the federal programs, the federal role in solid waste planning and management remained inconclusive during the early and mid-1970s. Most federal officials, including EPA administrators, continued to define solid waste as a local or regional issue and saw the federal role as informing or at best helping influence municipalities about their technical options. EPA's position on waste-to-energy, moreover, remained ambiguous. On the one hand, by controlling R&D funds and directing the flow of information, the federal government made an effort to influence the states in their choice of management options, which included various burn technologies. At the same time, however, most federal officials rejected any direct federal funding of waste-to-energy facilities, citing both overt and possible hidden costs as well as a lack of cost effectiveness. This ambivalence was reinforced by the wariness of local officials to invest in a more expensive, capital intensive, and technologically unproven system, especially while the tipping fees at landfills remained substantially lower.[14]

The federal interest in "resource recovery" systems other than waste-to-energy, such as composting, recycling, or reuse programs, were also constrained by substantial opposition from industry groups to such proposals as a federal deposit on beverage containers. Much of the federal effort regarding these programs remained once again in the research and education areas, with some modest, mostly preliminary studies laying out the possibilities for future action. While federal officials recognized the value of recycling, reuse, and waste reduction efforts such as product design regulations, they ultimately shied away from recommendations for government intervention, suggesting instead that the market could bring about such changes. Most actual policy initiatives, as a result, took place at the local and state level, often pursued by community and environmental activists through volunteer programs, the use of the initiative and referendum process, and occasionally by successfully lobbying a reluctant legislature. The EPA, meanwhile, tended to steer clear of the formal policy process; its recommendations, made primarily through annual reports to Congress mandated by the Resource Recovery Act, were more descriptions of options and reports of ongoing activities than actual plans or comprehensive management strategies for states and municipalities. Even the financial and technical assistance programs, though funded at much higher levels than HEW's efforts during the late 1960s, had no explicit direction or approach associated with them. During the Nixon and Ford administrations, EPA consistently spent less on solid waste programs than the level of appropriations made available through Congress.[15]

In the days preceding the 1976 election, with environmental issues including the land disposal of waste looming large on local political agendas, Congress passed its most far-reaching legislation on the waste issue to date. The Resource Conservation and Recovery Act (RCRA) focused on the deepening problems associated with municipal solid waste, primarily landfill disposal, as well as with new concerns related to hazardous waste. Once again, federal programs were designed to provide technical and financial assistance to states and related local subdivisions as an incentive to develop a solid waste management plan emphasizing sanitary landfills, resource recovery (including waste-to-energy), and various conservation programs. This time, moreover, EPA assistance was predicated on the development and implementation of such plans, although some at the EPA remained skeptical of the future of such programs.[16]

The key to RCRA's solid waste provisions, as spelled out in its crucial Subtitle D, was the involvement for the first time of the federal government in the permitting and regulation of waste facilities, specifically open dumps. EPA was required to set standards for landfills, while the states were obliged to inventory their open dumps and either bring them up to those standards or shut them down. EPA was also directed to establish criteria to distinguish between an acceptable "sanitary" landfill and an unacceptable dump site by reference to any "adverse effects on health or the environment from the disposal [of solid waste] at such a facility." Once the guidelines and the criteria were established and the state and local solid waste management plans were developed, a time limit of five years would be set to phase out all unacceptable dumps within a given area.[17]

The broad interpretation of what constituted an open dump (estimated as high as 94 percent of all existing land disposal sites) and the widespread use of land disposal methods (handling up to 90 percent of the waste stream) set in motion a crucial chain of events that changed the character of the waste issue. Prior to RCRA, solid waste problems were defined largely by the growing recognition of environmental and public health concerns; after RCRA, these became, particularly for municipalities, *management* problems, with a significant economic dimension. The closing of open dumps and the new requirements aimed at eliminating adverse health and environmental effects of landfills played a much greater role in redirecting municipalities than all previous federal programs, even RCRA's substantial federal assistance programs, with upwards of $45 million and $55 million authorized for the two years following passage of the legislation. With Subtitle D in place, nearly two-thirds of all existing landfills would close in less than a decade.[18]

Aside from the provisions regarding open dumps, Congress failed to include several other potentially effective programs and regulations regarding the solid waste problem, such as federal container-based collection fees and the much-debated national 5-cent beverage container tax. In the course of the debate over RCRA, the role of "waste-to-energy" had also been raised, with some interest in designating it as the preferred disposal alternative for solid wastes. One proposal had called for the creation of a U.S. Resource Recovery Corporation to guarantee investments in waste recovery plants as well as to sell insurance to cover the risks involved. This idea was successfully opposed both by the Ford administration, which was leery of another federal loan guarantee program (such as the one the Ford administration itself proposed for synthetic fuels), and a coalition of environmental and public interest organizations that worried that such a program would ultimately detract from conservation-based initiatives such as recycling.[19]

The RCRA debates over federal intervention in the solid waste area reflected deep divisions and failed to resolve some key questions of jurisdiction and regulation. These conflicts contrasted with the low-key discussions and relative lack of controversy over the hazardous waste sections of RCRA. This was partly the result of the decisions by both the Nixon and then the Ford administrations as well as industry groups to focus their efforts on modifying and influencing (or possibly blocking) other specific legislation dealing with hazardous *substances*, particularly the Toxic Substances Control Act then also being debated in Congress. RCRA provided a different kind of emphasis with its provisions concerning hazardous *wastes;* an emphasis that ultimately helped stimulate a hazardous waste management industry and a preoccupation with treatment technologies rather than the elimination or reduction of hazardous substances at the point of production.[20]

Nevertheless, RCRA, by providing an expanded definition of hazardous waste, and with its "cradle-to-grave" manifest system to track the generation, transport, storage, and disposal of such wastes, among other features, came to be defined as the country's first significant piece of hazardous waste legislation. While the RCRA provisions for solid waste—with the important exception of the procedures and timetable for closing open dumps—largely avoided federal regulation and instead relied on incentive systems and assistance programs, its hazardous waste provisions contained a number of specific regulations and a complex and detailed set of procedures and mandates.[21]

Within three years after RCRA had been signed into law, EPA's Office of Solid Waste, which had been given the authority to implement the legislation, completely shifted its focus from the solid waste to the hazardous waste arena. Hazardous waste had from the outset become preeminently an area of federal jurisdiction, particularly as the problems of liability, abandoned dump sites, and groundwater contamination came to the fore. The adverse health and environmental effects from hazardous waste sites were potentially more serious and quite likely to manifest themselves far more quickly than those from other sites and thus increased public pressure for immediate action.[22]

These were costly problems, furthermore, far more costly than the issues of solid waste disposal. Although hazardous wastes were not considered at the time to represent a large percentage of the overall waste stream, they still came to fully preoccupy the federal effort.[23] By 1979, the range of research programs, studies, task forces, and EPA staff-directed programs in the area of solid wastes had dwindled in size, as the agency responded to the concerns around hazardous wastes in crisis terms, forced to react to such major hazardous-waste-related events as the contamination of Love Canal in New York, and Times Beach in Missouri. With the inauguration of the Reagan administration and its anti-government-intervention bias, the federal solid waste effort literally disappeared from sight. Budget and personnel from solid waste programs were dramatically reduced, with budgets cut from $29 million in 1979 to $16 million in 1981 and then down to a minuscule $320,000 in 1982, while staff were reduced from 128 to 74 in 1981, and then 73 of those 74 positions eliminated the very next year.[24]

This redirection of RCRA framed the efforts around solid waste management during the late 1970s and early 1980s. On the one hand, RCRA played a direct role in limiting the landfill option at a time when a number of municipalities were beginning to seek additional landfill capacity. On the other hand, RCRA set the limits for federal intervention. This occurred through the dramatic reduction in EPA staff and funding, the absence of regulation and specific direction other than in broad, general terms (favoring sanitary landfills, resource recovery, and conservation-based programs as an overall conceptual approach), and the shift in EPA priorities toward hazardous waste. By establishing economic incentives to help develop statewide and regionally based solid waste management plans, RCRA also expanded jurisdictions, thus bringing in new players, and encouraging municipalities to redefine the problem within a larger geographic framework. As a result, by the early 1980s, at the moment when the impact of RCRA's Subtitle D was first starting to be felt, the focus of planning and management activity shifted away from the federal level to the states and the new jurisdictions established to deal with an issue that had become increasingly difficult and problematic for all parties concerned.

THE STATES INTERVENE: THE CALIFORNIA EXAMPLE

In the wake of the passage of RCRA, the efforts to develop statewide programs proceeded on an uneven and often piecemeal basis. While a number of states followed through on EPA guidelines and established management plans during the late 1970s and early 1980s, others progressed slowly, even as the issue became more politically volatile. EPA's incentive approach based on planning grants, furthermore, was eliminated when federal funding for the program was discontinued in 1983. At that time, not all states had initiated management plans or other statewide initiatives to handle solid wastes. This uneven level of activity on the state level was further compounded by the variation of the programs adopted. Even the most developed programs, such as the one that evolved in California, were limited by the complex of contending interest groups and jurisdictions that sought to influence the formation of policy.[25]

The state of California has often been perceived to be a leader in environmental legislation and planning, and that perception has applied in the area of solid waste as well. The California legislature passed its own solid waste management act in 1972, anticipating the approach embodied in RCRA enacted four years later. California's legislation established the state's Solid Waste Management Board (now the California Waste Management Board, or CWMB) to develop state policy for solid waste management and the state's resource recovery program. This board is responsible for setting policy and developing guidelines as well as conducting research into new management techniques and resource recovery technology. Additionally, the board has the power to review permits for land disposal sites and other solid waste facilities, including incineration plants, on the basis of their compliance with county solid waste management plans mandated by this legislation.[26]

Aside from the California Waste Management Board, a number of other state agencies are involved in some aspect of managing municipal solid waste. The California Energy Commission, for example, is responsible for licensing all energy-generating facilities, including waste-to-energy plants, that produce more than 50 megawatts of energy.[27] The Department of Health Services (DOHS) has ongoing responsibilities for the health impacts of solid wastes, including the review and evaluation of the characteristics of wastes and their methods for disposal. DOHS is also responsible for the designation of a waste as either hazardous or nonhazardous—a decision that can have a significant economic impact in terms of treatment and disposal.[28]

Air emissions from solid waste facilities are also subject to complex regulation. For one, the California Air Resources Board (ARB) has become increasingly involved in the regulation of landfills and incinerators, and has developed guidelines for their emissions and pollution control strategies. Under the ARB, there are 43 local air districts with responsibility for controlling air emissions from stationary sources such as incinerators. These regional bodies issue or deny permits for waste-to-energy facilities on the basis of plant emissions.[29]

These various jurisdictions also interact with other local and regional agencies that have been given the primary responsibility for the collection and disposal of wastes at the local or municipal level. Each California county, for example, is required to prepare a solid waste management plan that provides for the disposal of all wastes either originating in the county or brought into it from another location. The county plan in turn needs to specify possible sites for solid waste facilities such as landfills and incinerators. Any facility proposed by a municipality has to conform to the county plan. Thus, the county has an expanded role above and beyond the activities of the municipalities most directly concerned with the problem of disposing of their wastes.[30]

Most of the decisions made on the local level—whether by municipalities or counties—have been dictated primarily by technical and economic factors such as the tipping fees and the amount of disposal capacity. At the same time, political, environmental, and public health considerations, driven by public concern, have continued to play a substantial role, especially in limiting potential waste management options. Ultimately, no common strategy or specific mandate on how to manage solid wastes emerged in California during the 1970s and 1980s, although the emphasis on waste-to-energy became a favored (though often challenged) option for many jurisdictions.

The most substantial effort to devise a unified approach on the state level has been provided by the California Waste Management Board. When the board was first established, it conducted a survey of the state's landfills and developed overall standards for solid waste facilities in keeping with the concern about open dumps. By the early 1980s, as the landfill capacity problem became more severe, the California legislature directed the board to prepare a "comprehensive plan and implementation schedule for nonhazardous waste disposal in California." This plan was completed two years later and published in June 1985.[31]

The CWMB report declared that California was experiencing the beginnings of a waste management crisis that needed to be directly addressed. The state was generating 36 million tons of solid waste each year. Accounting for the state's 11 percent recycling rate, that meant that more than 30 million tons needed to be disposed of each year, almost all of which was being landfilled. More than 60 percent of the wastes were being generated in the southern part of the state, with the largest amount, 14 million tons per year, generated in Los Angeles County alone. Landfill space throughout the state and especially in Southern California was diminishing rapidly and becoming increasingly costly, with problems of air and water pollution and safety making it difficult to find new sites. The CWMB report estimated that the state would run out of landfill space by 1997, with the southern part of the state reaching capacity a year earlier.

The CWMB report characterized this overall situation as a "landfill crisis" and put forth a series of recommendations in order to accomplish "[the diversion of] waste from landfills." The key recommendation called for the construction of a substantial number of waste-to-energy facilities, particularly mass burn incinerators, as the strategy with "the greatest potential to reduce the volume of waste." To accomplish that goal, the CWMB's plan called for low-interest loans to incineration facilities and a revision of state laws to make it easier for such plants to bypass certain air quality restrictions. The plan also addressed the need to guarantee a specified volume of wastes for waste-to-energy plants and to facilitate plant financing. Most significantly, the board called for its empowerment as a "one-stop" permitting agency, bypassing the myriad of jurisdictions that slowed down and restricted the siting of new solid waste facilities.

With that kind of support, the CWMB report concluded that statewide waste-to-energy facilities could annually dispose of as much as 20 million tons, or more than 55 percent of California's solid wastes by the turn of the century.[32]

The CWMB report also mentioned recycling, composting, and reduction programs, areas in which the board had previously expressed interest, particularly when Jerry Brown was governor. But with its 1985 report, CWMB now suggested that these strategies were at best secondary to waste-to-energy. The report's recommendations for recycling, for example, were limited to the introduction of legislation to help finance development of new collection technologies to aid recycling efforts as well as the publicizing of recycling programs already initiated or under consideration.[33] Composting was discussed only briefly and dismissed, and reduction programs were not elaborated on at all, despite earlier studies commissioned by the CWMB during the Jerry Brown period, which suggested a variety of programs that could be pursued. The board interpreted its mandate to address only the issue of wastes *after* their generation and rejected consideration of any source reduction measures, such as packaging designs or product substitutions. The report, furthermore, was limited in its discussion of health and environmental impacts and failed to address the *cumulative* air quality, health, water, and transportation impacts resulting from the rapid and concentrated development of waste-to-energy facilities it now advocated.[34]

Publication of the CWMB report, presented as a management plan for the state, nevertheless failed to signify the emergence of a consistent waste management strategy throughout the state. For one, the advisory nature of the CWMB and its lack of authority to implement specific programs was compounded by the planning process at the county level. These county plans, developed during the 1970s and 1980s, did not necessarily share the CWMB's emphasis on waste-to-energy. Many plans focused on extending landfill capacity. Most had only limited data on the composition of the local waste stream, information critical in the design of successful waste management programs. While several counties discussed waste-to-energy as a disposal option, their estimates as to the significance of this strategy varied. At the municipal level, where local plans were also prepared, the approaches were even more distinct and varied. The city of Los Angeles, for example, developed plans to rely on incineration for the disposal of as much as 70 percent of its residential solid wastes, while the city of Berkeley enacted a provision preventing any consideration of incineration facilities for at least five years.[35]

The California approach toward solid waste management, though detailed and complex, ultimately failed to provide a clear and specific strategy to be implemented at the state and local levels, despite the preference and heavy emphasis on waste-to-energy by many of the jurisdictions, including the California Waste Management Board. Legislative efforts ranging from a bottle bill to incineration-related health and environmental issues came to be addressed on a piecemeal basis, often subject to powerful lobbying influences by industry forces. The effort to devise a "comprehensive" solid waste management plan offered little by way of direction, other than to recognize a landfill capacity problem, encourage the shift toward waste-to-energy, and largely dismiss the significance of alternative reduction and recycling strategies. The thrust around waste-to-energy, embodied in the CWMB report and supported by a number of local agencies, was less a strategy for integrated waste management than the push for a new, comprehensive disposal technology.[36]

By the late 1980s, many of the state's waste-to-energy advocates, including the California Waste Management Board, dozens of counties and municipalities, and the growing waste-to-energy industry in the state, had decided that the existence of extensive regulations and multiple permitting processes functioned as a *deterrent* to the "proper means of waste disposal" such as incineration and more sanitary landfills, as the California board report put it.[37] These groups preferred a single-stop planning authority, one that could implement a plan with waste-to-energy as the preferred, most logical and comprehensive option. They perceived the existing system as filled with roadblocks, allowing too much input and debate.

This perception was based significantly on the ability of a growing number of community groups, derided by incineration advocates as part of a "NIMBY" ("Not in my backyard") movement, to impact this decision-making process. Already by the late 1970s, the EPA declared that "public opposition to new facilities" had become "the major problem confronting efforts to improve solid waste management."[38] A decade later, these community groups had evolved into a strong and potent opposition force, consisting mostly, but not exclusively, of local residential groups. In the process, these groups had laid siege to nearly all of the proposed waste-to-energy projects in California and in several other areas of the country. They accomplished this in part by successfully engaging in the state's intricate planning and permitting process, despite the technical and often obscure nature of the proceedings.

By the late 1980s, the breadth and intensity of the community opposition had effectively slowed any rush to incineration in California that the CWMB and several other planning agencies and public officials had been hoping to stimulate. By 1989, of the 34 major waste-to-energy plants that had been proposed, 28 had either been terminated or put on hold within just a few years. Of the other six facilities, only *one* was operating, with a second in the test-burn phase soon to begin operation, and a third under construction. The other three proposed facilities were still subject to further review, their status potentially threatened. Moreover, no plant had received its final "permit to operate" from the appropriate air regulatory agency, including the only plant that had managed to come on line in Commerce, California.[39] "The NIMBYists are winning," lamented a *Waste Age* article about this new version of California's "disposal crisis." Ultimately, these new players had "caused a reversal," as the CWMB put it, of the major push by the planning agencies toward a waste-to-energy future in the state. Garbage policy in California and other states, couched by the public agencies for more than a decade in more neutral technical and administrative terms, had been forced to confront the biases located in the policy process.[40]

THE POLITICS OF CONTAINMENT

For the industry and the waste management bureaucracy, the role of the public has always posed problems, creating uncertainties in the formation of policy. The intervention of the public, expressed through community protests, and later, with the development of the environmental movement, through lobbying and litigation, was considered a significant factor in the evolution and eventual demise of an earlier generation of waste disposal strategies, including landfills. Siting issues were of particular concern, with much of the public opposition directed at impacts from a specific facility to the immediate neighborhood, whether in terms of air quality, groundwater contamination, or property values. And with the rise of community involvement and opposition in the waste area, particularly with respect to landfills, the ability to contain and/or redirect such political movements came to be recognized as a central component of the effective implementation of policy.

This focus on containment was reinforced with the take-off of the incineration industry and the evolution of the position of the EPA and many state and regional agencies in favoring waste-to-energy as the primary disposal option. With mass burn on the rise, the waste industry assumed a much higher profile in lobbying and ultimately shaping the agenda of the public agencies. As a result, a new industry-agency axis emerged in the solid waste arena. The revolving door that developed between public agencies and the waste-to-energy industry underlined this relationship. It included, among others, EPA staff, such as William Ruckelshaus, twice head of EPA, and Marcia Williams, the former head of EPA's solid waste program who joined Browning-Ferris Industries (BFI) as its CEO and director of environmental and regulatory affairs, respectively. Others include elected officials such as Alfred DelBello, and state waste management bureaucrats, such as Richard Shuff, a solid waste official with the Wisconsin Department of Natural Resources who became a district engineering manager for Waste Management. As a consequence, public officials and waste managers at the federal, state, and regional levels often became the visible, up-front proponents of the new burn technologies, attempting to influence the political process while answering to an increasingly volatile public.[41]

As the question of politics and public intervention became more dominant in the siting and permitting of incinerators and landfills, the focus of the industry-agency axis turned increasingly toward the community opposition, or "Fighting the NIMBY Syndrome," as a special section in *Waste Age* was called.[42] By the 1980s, opposition to landfills and incineration facilities emerged as a real economic and political threat to the implementation of the new industry-agency waste disposal strategy favoring incineration. Unlike the volunteer social movements of the late 1960s and early 1970s that offered a countervision centered on recycling as a kind of cultural or lifestyle alternative to landfills, the community and neighborhood groups that emerged a decade later organized themselves largely around individual solid waste facilities with a specific goal in mind—stop the project. Often impressive in their ability to mobilize supporters and focus their activities, the groups frequently crossed class and race boundaries in terms of participation. Some of the groups were largely black or Hispanic in composition, and were based in poor and working-class as well as middle-class neighborhoods. The intensity of their opposition was based primarily on the consideration of siting a facility too close to home, both literally and figuratively, a situation that in turn produced the charge of "NIMBYism."[43]

The use of the term "NIMBY" by incineration and landfill advocates provided a convenient means of explaining and potentially countering what they identified as a "self-interest" movement. The NIMBY concept had in reality become a catchall phrase over the years to describe any opposition to a large industrial facility by nearby residents. During the late 1960s and 1970s, for example, the rapid and massive opposition to nuclear power plants was criticized as a form of self-interest protest, that is, opposition by downwind residents. Community engagement around zoning and land use issues to protect neighborhoods and challenge specific rezoning or urban renewal plans were another form of this politics. Most importantly, the NIMBY phenomenon could be seen as a reflection of the single-issue political culture that prevailed through much of the history of twentieth-century America, where politics was segregated from ideas and ideologies and the outlets for public intervention in daily life were narrowly drawn. The "general interest" in this context became defined in terms of technical solutions to immediate problems, such as solid waste disposal, with little space for public input. Public opposition to projects couched in the general interest came to be defined as parochial, unable to comprehend the larger picture. The NIMBYists in this setting were accused of being a social disease, "a public health problem of the first order," a "recurring mental illness" that continued "to infect the public," according to one especially vitriolic statement in a waste management publication.[44]

During the 1970s, public intervention in the waste area was primarily directed at landfill sites, both at new facilities and increasingly, through the decade, at closing existing landfills. RCRA was passed in part because of the growing strength of community groups that had emerged to challenge specific dumps. During this period, the NIMBY issue was raised in relatively defensive terms by waste policy-makers who included appeals to civic duty and promises that "new and better disposal alternatives" would eventually alleviate the problem. "This implies commitments from the public and sometimes supporting difficult decisions," an EPA guide suggested, since "someone's county, city, town, and backyard" ultimately had to be selected for a facility.[45]

During the 1980s, the NIMBY problem was reinforced by the rise of the hazardous waste issue. The sharp and bitter conflicts that developed between community residents on the one side and public agencies, elected officials, and a wide variety of industries on the other heightened public mistrust of the policy process. It also linked more traditional "self-interest" concerns such as neighborhood preservation with health and environmental problems. Moreover, the absence of any existing technical solution—there was no developed hazardous waste management and treatment industry at the time of Love Canal—pushed the community opposition into the forefront of the search for alternatives.

The continuing growth of community opposition to solid waste facilities during the 1980s paralleled in some ways the movements that had emerged in the hazardous waste arena. The toxics issue and the solid waste issue brought forth some of the same responses and forms of protest, particularly with the take-off of the incineration option. This fusing of issues expanded the oppositional base, much to the consternation of incineration advocates who argued that the community movements had confused and misrepresented the two waste issues.[46]

The groups who felt most sensitive to the critique of NIMBYism were the nationally based, professionally staffed, environmental organizations. Most of these groups had largely steered clear of the solid waste issue at the local level during the 1970s, concentrating instead on federal environmental legislation (such as the hazardous waste sections of RCRA) and its implementation by EPA. The conventional environmental groups became most adept at the politics of negotiation, seeking to improve their lobbying capabilities while promoting alternatives such as source separation and recycling programs. EPA's "resource recovery" approach, linking studies and demonstration projects regarding recycling and reuse strategies with "energy recovery," was generally supported by the large environmental organizations that were sensitive to the charge that they only opposed and didn't propose solutions. But, by the late 1980s, in response to the surge of localized community opposition, the nationally based environmental groups finally began to include solid waste issues as part of their agenda, though keeping largely removed from the more insistent and action-oriented movements at the community level.[47]

In certain instances, environmental groups or leaders were solicited to support the waste industry directly. In 1987, Dean Buntrock, the chief executive officer of Waste Management, Inc. (WMI), was named to the board of directors of the National Wildlife Federation (NWF), the largest environmental group in the country. As the largest and at the same time most controversial landfill operator in the country, WMI itself was subject to a number of investigations of allegations for bid rigging, price fixing, and other antitrust violations, several of which had resulted in guilty verdicts for the company and its officers. WMI was also a special target of local activists who were directly challenging a number of WMI facilities. Several activist groups were outraged by the Buntrock appointment, arguing that it provided undesired credibility for his company. But NWF refused to heed the protests and declared that WMI was "conducting its business in a responsible manner," as NWF executive director Jay Hair wrote to the head of a Vermont-based community group. Hair, furthermore, suggested that he had come to this conclusion based on a review of materials provided by WMI itself.[48]

The lobbying of "moderate," "positive-oriented" environmental groups was just one component of the strategies to contain community opposition that emerged during the 1970s and 1980s. As part of that evolution, the incineration industry as well as federal, state, and local government officials began to turn increasingly to various consultants and "idea men" on how to handle the community opposition. A cottage industry of anti-NIMBY consulting groups emerged to offer advice and frame strategies for the many trash-burning facilities now proposed. The issue these consultants addressed—the question of public participation and intervention—was increasingly considered the key to a successful project.[49]

These consultants defined a number of overlapping themes and strategies. Included were the use of "advisory committees" consisting of members of the public; enlisting "responsible" environmental and civic organizations, or, as one consultant put it,[50] "more rational interest groups" such as the Sierra Club and the League of Women Voters, through the creation of a broader-based plan with both trash-burning *and* recycling elements; arguing that plants create jobs and much needed community revenues; using neutral-sounding terms such as "waste-to-energy," "sanitary landfill," and "resource recovery," as opposed to the more graphic terms such as "incineration," "dumps," and "trash-burning" plants; and focusing the site selection process on a neighborhood least likely to cause trouble. Some consultants also recommended what New Jersey incineration advocates called "environmental flying squads." Using friendly scientists and community relations specialists, these groups would enter resistant communities with their set of data and technical explanations to "explain the truth about an environmental hazard," as New Jersey governor Thomas Kean put it, and in the process attempt to overwhelm community opponents. All of these were just the more obvious components of a public participation and containment approach that became essential to the planning process for a new waste facility.[51]

This kind of "concerted public participation program," as the Cerrell Associates report called it,[52] complemented some of the harsher tactics employed by industry and public agencies to try to marginalize community opposition. Among other approaches, a number of waste companies decided to threaten and in several instances pursue libel and slander suits to attempt to intimidate the most vocal of the community leaders. These "intimidation" suits, as they came to be known, though initially effective in frightening local activists who had little experience with the court system, ultimately backfired as community residents rallied behind those who had been sued, established legal defense funds, gained greater sympathy among nonsupporters, and strengthened their ties to the regional and national networks of community activists that had sprung up throughout the country.[53]

The various efforts at containment, moreover, had failed to disrupt the momentum of community action and opposition. By the late 1980s, these local residential groups, most initially organized in the Northeast and Mid-Atlantic states, where the incineration boom had begun, had quickly spread to rural areas and small cities in the South as well as the major metropolises of the Southwest and West Coast. They had further consolidated their position by joining together in a series of national and regional networks. Groups such as the Citizen's Clearinghouse for Hazardous Wastes, the National Coalition Against Mass Burn Incineration, and the California Alliance in Defense of Residential Environments were able to provide information and resources to the hundreds of local groups that became involved in the waste issue, many of whom were little more than ad hoc neighborhood organizations formed to fight a specific facility. The networks became adept practitioners at what they considered the politics of democratic mobilization, responding to requests for assistance and developing organizational forums that reinforced the primacy of the grass roots, multiracial, and multiclass character of the movement.[54]

The core of the opposition, the neighborhood group, had become a fixture in the politics of garbage. Though some groups had sprung up in middle-class communities, much of the opposition tended to be located in poor and working-class neighborhoods, where most of the waste disposal sites were located, for reasons that often paralleled the recommendations provided by Cerrell Associates. Often, these groups were led by housewives and other residents who lacked professional expertise, and who were quickly obliged to master the technical jargon, complex regulations, and the arcane world of health risk analysis, among numerous other issues. They became their own self-taught experts, not only in terms of environmental issues such as air pollution or landfill leachate but with respect to a wide range of waste management issues as well.[55]

In several ways, these community groups fit the NIMBY profile: organized to block a specific facility, ad hoc in origin, focused on stopping a project rather than negotiating some modest changes, and thus refusing to accept the imperative of building such a facility. Yet, in crucial ways, these groups refused to follow the central precept of the NIMBY concept: "Not in *my* backyard." Attempts to pit different communities against each other around locational issues often backfired; the community groups in fact quickly coalesced rather than divided. "If we don't unite," one community leader told the *Los Angeles Times*, "somebody's going to get one of these in their backyard, and we don't want it in anyone's backyard."[56] Along the same lines, the newsletter of one of the network groups was entitled *Everyone's Backyard*.

Instead of organizing around a proposed project and then dispersing, these community groups became the leading edge of a new social movement struggling to redefine the waste issue. They became strong advocates of recycling and reuse programs on both the local and state levels and opponents of incineration as a waste disposal *strategy* rather than in terms of a particular siting decision. They were also successful in causing a shift in focus for many of the nationally based environmental organizations such as the Sierra Club in including solid waste on their agenda. Far from containing the opposition, the waste industry and the public agencies were realizing that their own favored waste management strategies were now at risk. The "disposal crisis" had become a crisis of politics as well as a problem of planning and implementation.

DEVISING NEW STRATEGIES

The question of politics had become, by the late 1980s, the issue most preoccupying the waste industry and the public agencies. "We believe public opposition will remain the toughest hurdle for this industry over the next several years," stated the Kidder, Peabody company in its April 1988 annual status report on the waste-to-energy industry.[57] "Uncertainty," the report concluded, "will continue to cause project cancellations, or at the very least, an extension of scheduled operating dates."[58]

Not unexpectedly, new, revised strategies began to emerge to offset or deflect the opposition concerns. In some instances, trash-burning advocates attempted their own form of political mobilization, creating proincineration coalitions, increasing their lobbying profile, and directly entering the political process by attempting to elect public officials sympathetic to the incineration option. Most attempts to counter the opposition, however, involved incorporating the language and at times the programmatic thrust of a recycling strategy, while still working at salvaging a waste-to-energy program, albeit with a lower profile and perhaps on a smaller scale.[59]

The key to this revised approach involved a trade-off strategy: increased recycling efforts *with* an acceptable waste-to-energy program. In Massachusetts, for example, leading public officials and waste managers, including Governor Michael Dukakis, moved quickly in 1986 and 1987 to devise an approach that would emphasize both recycling and waste-to-energy. From just a 5 percent recycling effort in 1987, some of which had resulted from the state's bottle deposit law, the Dukakis administration projected that, once an emergency-style effort was in place, such recycling could potentially lead to a 25 percent reduction of the waste stream. But then, as Dukakis put it rhetorically in one press conference, "Even if we meet a goal of 25 percent, what do we do with the rest of the rubbish?" The answer he and other officials projected was the construction of several waste-to-energy facilities then being proposed by the high profile and rapidly expanding waste-to-energy industry in that state. These projects, Dukakis and other public officials were well aware, had met with some of the most intense and widespread opposition in the country. By 1988, that opposition had pushed policy-makers to impose a one-year moratorium on new incinerators and to revise their waste management plan to increase recycling's potential contribution to as high as 40 percent of the waste stream.[60]

This new interest in recycling, furthermore, was being conceived as a kind of crash course in waste management without much effort toward increased public input or prioritization of programs. Even those states that offered what they characterized as an "integrated solid waste management program," such as New York State, provided little of substance by way of supporting such a concept. In the New York case, the Department of Environmental Conservation prepared, in 1987, a plan similar in intent to the 1985 California Waste Management Board plan.[61] This plan, according to New York governor Mario Cuomo, reflected "an intelligent hierarchy of management practices: waste reduction, recycling and reuse, incineration and landfilling. . . . We must become accustomed to thinking of these four strategies together," Cuomo told members of the New York State legislature, asserting that the state would "assume a more aggressive role" (see table 3.1).[62]

Table 3–1
WASTE MANAGEMENT HIERARCHY

SOURCE REDUCTION
REUSE
RECYCLING
INCINERATION (WITH MATERIALS RECOVERY—RDF)
INCINERATION (WITH ENERGY RECOVERY)
INCINERATION (WITHOUT ENERGY RECOVERY)
LANDFILL

In this hierarchy of preferred waste management strategies, the various alternatives are ranked in order of increasing negative impacts on the environment.

The New York plan, however, when translated into legislative initiatives and programs, emphasized much of what had already been proposed at the federal level: demonstration projects; waste exchanges; recycling by government agencies; facilitating household hazardous waste separation and collection; and facilitating the siting of "necessary regional solid waste facilities" such as incinerators. The limited programs proposed in the waste reduction area, moreover, such as a labeling requirement on consumer products, were put forth for future consideration rather than specific legislative initiative. Despite the rhetoric, the heart of the waste management strategy for both New York State and New York City continued to revolve around incineration, which had long represented their most costly and heavily subsidized option.[63]

The celebration of recycling strategies, furthermore, was set in a context that restricted them to a limited percentage of the waste stream. Recycling as a primary focus for government activity was dismissed as utopian and harmful to the development of this new, revised waste management perspective. "When recyclers wanted to get their foot in the door, they railed against relying on a single technology, waste-to-energy," stated one utility analyst. "They criticized us for putting all our eggs in one basket. They were right. Now they would have us put development of all other projects on hold while we try to make recycling work. There's just as much folly in that approach and we can't let them get away with it."[64]

If recycling were, in fact, to achieve a sizable reduction of the waste stream, say 50 to 70 percent or more, as several recycling advocates argue, then waste-to-energy would become increasingly impractical and economically unjustifiable. "A successful waste-to-energy facility," as two industry consultants argued, required among other conditions "sufficient quantity of waste suitable for processing."[65] Although public officials and industry executives were now arguing that incineration was compatible with recycling, a number of communities, such as Modesto, California and Detroit, Michigan, guaranteed delivery of all their waste to their proposed incinerator.[66] A number of mass burn plants, moreover, were conceived prior to this shift in strategy and had no recycling plan associated with their operation, though common sense should have dictated that it would have been wiser and easier to remove metals, for example, before they became part of the ash or went out the stack into the air. The new "integrated waste management strategy," now widely adopted by the industry-agency proincineration axis, essentially presented recycling as a *supplemental* option, as much a public relations device for reviving incineration as a serious waste management policy.[67]

The renewed emphasis on recycling gave rise to another strategy to deal with opposition movements: the "merchant plant" concept first developed by Wheelabrator Technologies and Waste Management, Inc. The key to the merchant plant strategy was that a waste-to-energy facility would be built and owned *without public funds* and therefore not be as dependent on the political process. Owning the land in advance of a siting decision would be one way to accomplish that goal, with already existing landfill sites a preferred location, as in the Wheelabrator/WMI arrangement. Though permits would still be required and opposition might still emerge, advocates of the merchant plant hoped that the existing tacit approval of landfill neighbors would blunt any new community opposition.[68] Despite these plans, however, community opposition had sufficiently coalesced and strengthened so that any waste-to-energy project, no matter the location, had become a project at risk.

By the late 1980s, solid waste policies had come to an impasse. The decline of landfills in the late 1970s and early 1980s had given way to a waste-to-energy boom that became the centerpiece of the new politics of waste. Though the boom became more problematic by the end of the decade, waste-to-energy advocates still pursued strategies to keep their option alive. The "merchant plant" concept suggested one strategy; a revised incineration plus recycling approach offered another.

The key problem for the waste-to-energy advocates, all agreed, remained the intense and widespread opposition of the community groups. Government intervention had provided only limited guidance; the predominant philosophy at the federal level, since the passage of the first solid waste legislation in the 1960s, was to let the private sector respond to the problem. This market-oriented approach was based on a cultural paradigm that assumed a technological solution to any problem, especially one as disagreeable as garbage. This position evolved over time from a focus on landfills to advocacy of a waste-to-energy future. But through the 1980s, local, state, and federal agencies were finding themselves obliged to respond to the myriad of issues such a future engendered, issues kept in the forefront by the growing oppositional movements. The issues these groups raised, predominantly those relating to the environmental and economic arenas, were continuing, as Kidder, Peabody noted, "to play a major role in this industry's development."[69] And the controversies that emerged in the process had become the centerpiece of the debate over the politics of garbage.

NOTES

1. "The EPIC of Upton Sinclair," *Nation*, October 31, 1934; *I, Candidate of California and How I Got Licked*, Upton Sinclair, Los Angeles End Poverty League, Los Angeles, 1935.
2. Interview with Craig Steele, Senior Account Executive, Cerrell Associates, 1988.
3. "Public Perceptions and Community Relations," Lawrence Chertoff and Diane Buxbaum, in *The Solid Waste Handbook: A Practical Guide*, ed. by William D. Robinson, New York, 1986.
4. "Political Difficulties Facing Waste-to-Energy Conversion Plant Siting," Cerrell Associates, J. Stephen Powell, Senior Associate, *Waste-to-Energy Technical Information Series*, chap. 3a, California Waste Management Board, Los Angeles, 1984; "What Do Citizens Want in Siting of Waste Management Facilities?" Lillie Craig Trimble, *Risk Analysis*, vol. 8, no. 3, 1988.
5. Solid Waste Disposal Act of 1965 (P.L. 89-272).
6. Clean Air Act Amendments and Solid Waste Disposal Act, Report No. 192, 89th Congress, 1st Session, Senate Committee on Public Works, May 14, 1965; "A History of Public Works in the United States: 1776–1976," ed. Ellis Armstrong, American Public Works Association, Chicago, 1976.
7. The Rivers and Harbors Act of 1899, 33 USC 407. See also *The History of the U.S. Army Corps of Engineers*, Army Corps of Engineers, EP 360-1-21, Washington, D.C., 1986; "The Refuse Act of 1899: Its Scope and Role in Control of Water Pollution," Diane D. Eames, *California Law Review*, November 1920, pp. 1444–1473.
8. "The New Federal Role in Solid Waste Management: The Resource Conservation and Recovery Act of 1976," William L. Kovacs and John F. Klucsik, *Columbia Journal of Environmental Law*, 3:205, March 1977.

9. Clean Air Act Amendments and Solid Waste Disposal Act, Report no. 899, 89th Congress, 1st Session, House Committee on Interstate and Foreign Commerce, August 31, 1965; *Federal Pollution Control Programs: Water, Air, Solid Wastes,* Stanley Degler, Washington, D.C., 1971.
10. Senate Report No. 192, 89th Congress, 1st Session, 1965; also "The New Federal Role in Solid Waste Management," William L. Kovacs and John F. Klucsik.
11. The Resource Recovery Act of 1970 (P.L. 91-512).
12. *Federal Pollution Control Programs,* Stanley Degler; Senate Report No. 192, 89th Congress, 1st Session, 1965.
13. *Resource Conservation and Recovery Activities: A Nationwide Survey,* 530/SW-142, Environmental Protection Agency, 1977; "The New Federal Role in Solid Waste Management," William L. Kovacs and John F. Klucsik.
14. *Waste Age* Tipping Fee Survey, November, 1979; House Report No. 1491, 94th Congress, 2nd Session, 12, 1976; *Resource Recovery and Waste Reduction,* 3rd Report to Congress, 530/SW-161, EPA, 1977.
15. *Resource Recovery and Waste Reduction,* 1st Report to Congress, 530/SW-118, EPA, 1974; 2nd Report to Congress, 530/SW-122, 1975; 3rd Report to Congress, 530/SW-161, 1977; 4th Report to Congress, 530/SW-600, 1978; also *Choices for Conservation: Final Report to Congress,* 530/SW-779, EPA, 1979.
16. Resource Conservation and Recovery Act of 1976 (P.L. 94-580); *Environmental Protection: Law and Policy,* Frederick R. Anderson et al., Boston and Toronto, 1984; interview with Susan Sakaki, 1988.
17. Resource Conservation and Recovery Act of 1976, §4002a; 42 USC § 6942 (a); see *Subtitle D Study, Phase One Report,* 530/SW-86-054, EPA, Office of Solid Waste, June 1987; "Legislation and Involved Agencies," William L. Kovacs, in *The Solid Waste Handbook,* 1986; *A Comprehensive Plan for Management of Nonhazardous Waste in California,* California Waste Management Board, June 1985, P.C.-15.
18. *An Environmental Assessment of Gas and Leachate Problems at Land Disposal Sites,* 530/SW-110.OF, EPA, 1973; *Subtitle D Study Phase One Report,* EPA, June 1987; "The New Federal Role in Solid Waste Management," William L. Kovacs and John F. Klucsik.
19. *Washington Post,* July 9, 1976; *American City and Country,* September 1976.
20. Congressional Record 122:S11061-11105, June 30, 1976; *Hazardous Waste in America,* Samuel Epstein, Lester O. Brown, and Carl Pope, Sierra Club Books, San Francisco, 1982; Toxic Substances Control Act, 1976 (P.L. 94-469), 90 Stat. 2003 USC 882601-2629.
21. Resource Conservation Recovery Act, Subtitle 40CFR July 1, 1985, parts 260-271; *Subtitle D Study Phase One Report,* EPA, June 1987.
22. "EPA and Recycling: An Interview with J. Winston Porter," *Resource Recycling,* vol. 7, no. 2, July 1988; interviews with Frank A. Smith, 1988; Susan Sakaki 1988.
23. *Hazardous Waste Sites,* Michael R. Greenberg and Richard F. Anderson, Center for Urban Policy Research, New Brunswick, New Jersey, 1984; "Solid Waste Management Alternatives," Elliot Zimmerman, Illinois Department of Energy and Natural Resources, Springfield, Illinois, 1988, p. 8.
24. EPA FY 1977 Budget—Region IX—002425-1977; EPA FY 1981 Budget—Region IX—002425-1981; see also "Rising from the Ashes: Our Trash Shouldn't Burn," Judy Christup, *Greenpeace,* vol. 13, no. 3, May–June 1988.
25. *Subtitle D Study: Phase 1 Report,* 530/SW/86-054 (NTIS PB 87-116810), EPA, October 1986.
26. The Nejedly-Z'Berg-Dills Solid Waste Management and Resource Recovery Act of 1972, Government Code Title 7.3, chapter 2, 66700 et seq.; interview with Chris Peck, Public Information Officer, California Waste Management Board, November 14, 1987; see also California Waste Management Board, *1987 Annual Report.*

27. Warren-Alquist State Energy Resources Conservation and Development Act, 1974, Public Resources Code 25000 et seq.; "Presentation on Air Quality Impacts of Waste-to-Energy Facilities in the South Coast Air Quality Management District," Sanford M. Weiss, Director of Engineering, October 16, 1986, p. 3.
28. California Health and Safety Code, Section 4520; also SB 2292 (Campbell), California Stats. (1984), Chap. 1160, Section 1–4.
29. *Air Pollution Control at Resource Recovery Facilities,* California Air Resources Board, 1984, pp. 15–18; also Weiss, supra.
30. Nejedly-Z'Berg Solid Waste Management and Resource Recovery Act of 1972; also California Waste Management Board, *1987 Annual Report.*
31. *A Comprehensive Plan for Management of Nonhazardous Waste in California,* California Waste Management Board, Sacramento, June 1985; also, Z'Berg-Kapiloff Solid Waste Control Act of 1976, 66795 et seq.
32. *Comprehensive Plan,* CWMB, pp. 12–16, 25, 30, A-1.
33. Nejedly-Z'Berg Solid Waste Management and Resource Recovery Act of 1972, Title 7.3, Cal. Govt. Code Chapter 2, Article 2, Section 66780-66784.
34. *Comprehensive Plan,* CWMB, pp. 16–19, 22–26.
35. Interview with Ariel Parkinson, 1988; Mary Lou Van Deventer, 1988; also City of Berkeley Construction and Building Code, 1988; interview with Terry Anderson, Department of Public Works, City of Berkeley, 1988.
36. *Comprehensive Plan,* CWMB, pp. 12–31; California Waste Management Board *1987 Annual Report;* interview with Ariel Parkinson, 1988.
37. Cited in "California Faces a Disposal Crisis," Hal Rubin, *Waste Age,* September 1987.
38. *Report to Congress on EPA Activities Under the Resource Conservation and Recovery Act for Fiscal Year, 1978,* 530/SW/755, EPA, p. I–6.
39. "The CWMB Waste-to-Energy Update," California Waste Management Board, Sacramento, April 1987; interview with Manuel Rubivar, South Coast Air Quality Management District, September 20, 1988.
40. "California Faces a Disposal Crisis," Hal Rubin.
41. *Waste Not* no. 4, Work on Waste USA, Canton, New York, April 26, 1988; " 'Bribes' Work in Wisconsin," Richard G. Shuff, *Waste Age,* March 1988; *Hazardous Waste News,* no. 103, Princeton, New Jersey, November 14, 1988.
42. "Is NIMBYism Here to Stay?" special section in *Waste Age,* March 1988.
43. "Where Will We Put All That Garbage," Faye Rice, *Fortune,* April 11, 1988; see also *Not in Our Back Yards,* Nicholas Freudenberg, New York, 1984.
44. "It's a National Mental Illness," H. Lanier Hickman, Jr., *Waste Age,* March 1988; "Not on *My* Block You Don't: Facility Siting and the Strategic Importance of Compensation," Michael O'Hara, *Public Policy,* vol. 25, no. 4, Fall 1977.
45. *Waste Alert: A Citizen's Guide to Public Participation in Waste Management,* Environmental Protection Agency, Washington, D.C., p. 25.
46. "Political Difficulties Facing Waste-to-Energy Conversion Plant Siting," Cerrell Associates; "California Faces a Disposal Crisis," Hal Rubin.
47. Interview with Penny Newman, 1988; "Environmental Lobbies Face Grass Roots Challenge," Robert Gottlieb, *Los Angeles Times,* October 19, 1988.
48. *Hazardous Waste News,* Environmental Research Foundation, Princeton, New Jersey, no. 27, June 1, 1987; no. 34, July 20, 1987; Jay Hair letter cited in no. 72, April 11, 1988.
49. "The Tussle for Public Support," Clinton C. Kemp, *Solid Waste and Power,* February 1987; "Using Citizens to Site Solid Waste Facilities," Terry A. Trumbell, *Public Works,* August 1988; "Get Citizens Involved in Siting—And Do It Early," Carolyn Konheim et al., *Waste Age,* March 1988; "Public Acceptance Strategies Memorandum," from Howard Coffin et al., to Melanie Thomas et al., Fanfare Communications, December 29, 1986.

50. "NIMBYism Just Is Not Viable," Eugene J. Wingerter, *Waste Age*, December 1987.
51. "Dealing with NIMBY," Thomas Kean, Keynote Address, Air Pollution Control Association Annual Meeting, June 1987, in *Journal of the Air Pollution Control Association*, vol. 37, no. 9, September 1987.
52. "Political Difficulties Facing Waste-to-Energy Conversion Plant Sitings," Cerrell Associates, p. 64.
53. "A Chilling Flurry of Lawsuits," *U.S. News and World Report*, May 23, 1988; "Legal Corner," Ron Simon in *Everyone's Backyard*, vol. 6, no. 3, Fall 1988.
54. Presentation by Penny Newman, UCLA, November 10, 1988; interview with Will Collette, 1988.
55. "Coping in the Age of NIMBY," William Glaberson, *New York Times*, June 19, 1988.
56. "Burning Unity: Grass Roots Organizations Merge Effort to Fight Trash-to-Energy Incinerator Plants in Southland," Bob Baker, *Los Angeles Times*, September 7, 1986.
57. *A Status Report on Resource Recovery as of December 31, 1987*, Kidder, Peabody & Co., New York, April 29, 1988.
58. Kidder, Peabody, ibid.
59. "WTE Developers: What Are They? Who Are They?," Mike Kilgore, *Solid Waste and Power*, August 1988; "Waste-to-Energy Incinerators Fuel Intense Lobbying Effort," Mark Gladstone, *Los Angeles Times*, December 7, 1986.
60. "Toward a System of Integrated Solid Waste Management," *The Commonwealth's Solid Waste Masterplan*, Draft, Executive Office of Environmental Affairs, Commonwealth of Massachusetts, December 20, 1988; "Massachusetts Fiddles as Capacity 'Burns' Away," W. Tod Cowen, *Waste Age*, December 1987.
61. *An Integrated Solid Waste Management Program*, Department of Environmental Conservation, State of New York, Albany, January 1988.
62. "Partnership in Solid Waste Management," "The State of the State," Address to the New York State Legislature, Governor Mario Cuomo, January 6, 1988.
63. "Environmental and Economic Analysis of Alternative Solid Waste Disposal Technologies: III. A Comparison of Different Estimates of the Risk Due to Emissions of Chlorinated Dioxins and Dibenzofurans from Proposed New York City Incinerators (Including a Critique of the Hart Report)," Center for the Biology of Natural Systems, December 1, 1984, Flushing, New York; "Top of the Heap," *Solid Waste and Power*, December 1988; *An Integrated Solid Waste Management Program*, Department of Environmental Conservation, State of New York, Albany, January 1988.
64. Cited in "WTE Development: A Zigzag Course," Harvey W. Gershman and Nancy M. Petersen, *Solid Waste and Power*, August 1988, p. 15.
65. "WTE Development: A Zigzag Course," Harvey W. Gershman and Nancy M. Petersen; *Comprehensive Plan . . .*, California Waste Management Board, pp. 154–56.
66. "Market Prospects for Refuse-to-Energy Development in California," Paul R. Peterson et al., paper presented at "Waste Tech '87," October 26–27, 1987, San Francisco, p. 5; *Coming Full Circle*, Environmental Defense Fund, p. 55.
67. "Is Mandated Recycling Possible?" Herschel Cutler, *Solid Waste and Power*, August 1988; "Solid Waste Industry Enters Recycling Arena," Jim Glenn, *Biocycle*, vol. 29, no. 4, April 1988; "Waste Disposal Quagmire."
68. Interview with Alfred DelBello, 1988.
69. *A Status Report on Resource Recovery as of December 31, 1987*, Kidder, Peabody & Co.; "How Can We Get on with the Job," Representative James J. Florio, *Solid Waste and Power*, June 1988, pp. 21–22.

PART II

THE ISSUES: HAZARDS AND RISKS

4

THE ENVIRONMENTAL ISSUES: EYE OF THE STORM

SOUTHERN CALIFORNIA'S FIRST AND FINEST

The tour guide, a smiling, affable, PR man with the Los Angeles County Sanitation Districts, began moving through the buildings, pointing to the new equipment and the different operating procedures, pleased to conduct one of the first tours of the plant since its opening. Located in the city of Commerce, in a heavily industrialized area in the south-central part of the county, where mostly poor Hispanics resided, this modest-sized, 420-ton-per-day mass burn facility was the first of its kind to go on line in Southern California. Though plans were afoot to build at least ten other plants in the region, several of them considerably larger than the Commerce facility, this new "state-of-the-art plant," as the PR people called it, was unique in some ways, though representative in other respects of many mass burn facilities then on drawingboards throughout the country.[1]

The plant had opened a few weeks earlier and was not yet in full commercial operation. It had been built by a contractor—not one of the leading firms—who had used various equipment from several different manufacturers to produce a kind of stitched-together facility, not unlike other plants built in this period. Despite the tendency toward concentration in the industry, the Commerce plant was not exceptional, given the great variation in the design specifics and component parts of these new trash-burning facilities. Commerce was primarily designed to burn commercial trash, everything from the large volume of office paper in the commercial waste stream to the huge shrouds of plastics that came from the myriad of daily products and processes used by commercial firms.[2]

Although built to burn nonhazardous wastes, the Commerce plant continually processed hazardous waste items that managed to find their way onto the tipping floor.

The hazardous waste issue is one that has continued to confound the industry. More than ten years earlier, the Resource Conservation and Recovery Act had set in motion a chain of regulatory actions and policy initiatives that sharply segregated solid and hazardous waste disposal activities. Different classes of landfills were established, as were different regulations governing management by type of waste. The costs, as for construction or tipping fees, were considerably higher for hazardous waste sites, and the methods and technologies of disposal were different as well.[3]

Solid waste managers frequently invoke the distinction between hazardous and nonhazardous waste, in part to allay fears about potential pollution problems from the mass burn process. Incineration of hazardous waste can cause explosions threatening plant workers and nearby residents as well as air and groundwater pollution. But the tour guide implied that hazardous wastes were a minimum problem at the Commerce facility. He confidently described their two-part program to detect and prevent hazardous wastes from entering the furnace. About one in every 15 trucks is pulled aside for a brief inspection for both hazardous waste and large, unburnable items such as refrigerators, which would then be taken to a landfill. The crane operator, seated fifty feet above the "tipping floor" where the trash was dumped, looks for and removes large items, such as 55-gallon drums, which might contain hazardous wastes. Since the crane handles 5,000 pounds at a time, the detection capability of the operator is limited.

While inspecting the ash bin in the rear of the plant, the tour group noticed several charred 55-gallon drums that had apparently escaped detection and had made their way through the complete cycle. This wasn't the first such incident in the brief life of the plant; the tour guide explained that a few days before, a huge engine block had escaped detection and been put into the incinerator, falling out of the bottom and onto and through the ash conveyor belt that transported the ash residue from the incinerator to the ash bin.

The problem of detection is one indication that hazardous wastes, especially in small amounts, are nearly impossible to avoid at incinerators. For one, about 1 percent of household wastes, as much as 55 to 60 grams weekly per household, contain a wide variety of hazardous substances, from paints and household pesticides to batteries, cleaning solvents, and cosmetics.[4] Solid waste landfills and incinerators are constantly subject to some form of illegal disposal of hazardous wastes, a practice that has grown as the cost of hazardous waste disposal has increased due to stricter regulations. A large-volume traffic in illegal dumping, in fact, has spread across the country, much of it controlled in one form or another by organized crime.[5]

This situation has been compounded by differing state regulations governing waste disposal facilities and the increasing tendency to transport wastes across state lines. A particularly devastating example of this problem took place in Akron, Ohio, where an explosion occurred at the local solid waste incinerator a few days before Christmas in 1984. Three workers at the plant were killed and seven more were hospitalized. The suspected cause of the explosion was the incineration of 2,500 gallons of wastes that had been taken to the plant by a New Jersey firm and that had been declared on delivery to be a combination of sawdust and used motor oil. This particular combination of materials was considered hazardous waste in New Jersey, but not in Ohio—the reason the wastes had been brought to Ohio for burning.[6]

While the problem of hazardous wastes remains troubling, the operational procedures regarding the ash can be even more problematic. At the Commerce plant, the door to the ash bin (where the burned drums resided) was open, and the ash was spread out over the parking lot and could be seen and touched throughout the entire plant. As with other plants, the points of collection for the two kinds of incinerator ash—the fly ash and the bottom ash—were separate from each other. The bottom ash from the furnace was collected about 50 yards from the point where the fly ash from the air pollution control equipment was collected. Both kinds of ash fell onto various parts of the plant's open-air conveyance system, where the ash was first mixed and then moved another fifty yards to be dumped into the combined ash bin.

The combining of the ash at all incinerator plants has become an economic necessity, given that the more toxic fly ash residue, with its brew of heavy metals and toxic compounds, has been clearly identified as a hazardous substance. Since the Commerce plant, like most plants, obtains even under ideal conditions a 70 to 90 percent reduction of garbage by volume and about a 60 to 75 percent reduction by weight, the amount of ash to be disposed of is considerable, ranging from 43 to 307 tons per day.[7] The more hazardous the waste stream that enters the incinerator and the more effective the air pollution control system, *the more hazardous the ash residue*. The mass burn incinerator in effect concentrates toxics in the waste stream. Despite the intent to segregate the issues of hazardous and solid wastes, the two are continually joined at every stage of this waste disposal process.

In the parking lot of the Commerce plant stood one of the 32 workers employed at the plant, whose job it was to shovel the ash in the lot onto a conveyor belt back into the ash room. The worker was leaning on his shovel, killing time. He was closest to the area where the fly ash was collected. Did he know what kind of toxics were present in the fly ash? Did anybody know? The worker, a lean, young man, said that nobody had told him that his work might be hazardous. He had been given a paper mask to wear, but, never having been told that he might be at risk, had not been using it. He shrugged off the questions, explaining that he had only begun work a few days earlier and would probably quit the next day.

As the tour broke up, it had become apparent that this "state-of-the-art" facility generates more questions than answers. Some hazardous waste was inevitably entering the plant; the question was, how much. The Commerce facility, furthermore, was located in one of the dirtier parts of the dirtiest air basin in the nation. Los Angeles, for example, exceeded the federal standard for ozone more often and by a greater margin than any other area of the country. It was also the only air basin to exceed the standard for nitrogen oxides, or NOx, a key constituent of photochemical smog.[8] The Commerce plant, furthermore, had been obliged to utilize a technology never before used on a trash incinerator in the United States, called Thermal De NOx, to reduce NOx emissions in the combustion process. Other sophisticated and expensive pollution control devices were also supposed to be added on to deal with other air contaminants.

During the tour, the guide was quick to claim that the Commerce plant was, of course, in full compliance with regulations, and represented a model to emulate, given its control technology. The Commerce plant, proclaimed the plant's boiler manufacturer in ads for industry trade publications, is "the cleanest waste-to-energy plant in the world."[9]

That claim, it ultimately became clear, left much to be desired. For more than a year and a half after the tour, the Commerce plant failed to receive its final "permit to operate" from the local air district because of poor emissions data submitted to the district. Then, when more accurate figures became available the following year, it turned out the plant had repeatedly violated air pollution standards, including serious violations of nitrogen oxides, more than triple the limit of sulfur dioxide emissions on an hourly basis, and the exceeding of carbon monoxide limits as well. Some of the much touted air control technology that had been required by the Air District, moreover, had still not been put in operation. As a result, the Air District denied the facility its full operating permits. The response of the plant manager, the plant's "state-of-the-art" presumption now in question, was to ask the Air Quality Management District to nevertheless grant the permit by lowering the emission limits![10]

These problems were not just limited to the excess air emissions and unwanted hazardous wastes. There was also the question of the ash, the most difficult and potentially the most costly environmental factor related to the burning of wastes. Should the ash be combined, and if so how should it be designated: hazardous or less hazardous and therefore not subject to the regulations governing hazardous wastes? Would the ash that was eventually to be buried somewhere create a new generation of Superfund sites? And at Commerce, what of the ash in the parking lot becoming a ubiquitous element in the environment of the plant and the surrounding neighborhood.

As the tour group pulled away in their cars to leave the plant, the questions seemed to linger in the lot, much like the ash that could be felt in the eyes and lungs of the visitors: both were an unwanted presence. What seemed inescapable, no matter how the answers were framed, was that the environmental factors had become a reality from which the incineration process could not really escape.

A GROWTH INDUSTRY

Looking out the tenth-floor window of the mid-Wilshire Los Angeles quarters of the Southern California Association of Governments one day in early March of 1987, you could see the bottom of the thick layer of smog that hung heavy over the Southern California basin. More than 30 people were in attendance—nearly all representatives of industry and government, from organizations such as Union Oil and Northrop, the EPA and the California Air Resources Board, the Automobile Association of Southern California and the South Coast Air Quality Management District. Only three people identified themselves as members of the public, though the subject at hand, the Transportation and Land Use Element of the 1988 Air Quality Management Plan, had significant repercussions for the future air quality of the region.[11]

The air issue was a volatile one for Southern California, as for a number of other communities—at least 96 areas, comprised of more than 150 million people living in 400 counties within 33 states plus the District of Columbia—who were not in compliance with one or more provisions of the 1970 Clean Air Act.[12] Though the Clean Air Act was ready to expire at the end of 1987, there was general recognition that sooner or later it would be reauthorized, likely strengthening rather than weakening its health-based standards and means for implementation. Just a few days before, in fact, EPA had issued a news release threatening to impose sanctions (as indeed mandated by the legislation) on areas that did not meet the federal standards. It was a situation that would be repeated on several other occasions, as deadlines would be extended and new threats made until sanctions were finally imposed on September 1, 1988.[13]

While there was concern in the room that sunny day in March, there was no appreciable panic. Preparation of the Air Quality Management Plan was another mandated provision of the Clean Air Act, and since Los

Angeles had the worst air quality in the country and past plans had done little to change that status, the consensus in the room was that it would indeed be prudent to begin to draft another new plan. The talk quickly turned to the EPA press release and speculation concerning the nature of possible sanctions. It was clear that one possible sanction, if the threats were ever to come to pass, would be a moratorium on the construction of major sources of air pollution. Everyone in the room kept asking, "Would EPA really do it this time?" referring to past threats and the growing pressure on EPA to "do something." "What would the economic effects be?" industry representatives wondered out loud, raising a frequent and often successful counter to the threat of action.

The EPA representative at the meeting assured his attentive listeners that most industries should have no fears about economic effects, since any moratorium, were it to occur, was likely to affect significantly only one new major source of air pollution planned for the Southern California region. This source: the ten or so mass burn incinerators on the drawingboards for a number of communities in Southern California, including the large LANCER project proposed for the city of Los Angeles.

In terms of air pollution, incineration is the number-one growth industry in the country, as well as Southern California, the EPA official noted. With no plans for new oil refineries or fossil-fuel-burning power plants, the more traditional major sources of air pollution, the big incinerator plants, though redesigned in the 1970s and 1980s in an attempt to reduce air emissions, had nevertheless thrust themselves onto center stage of the air quality debate. The sheer number and size of the facilities being proposed and scheduled to come on line in just a few short years had changed the dimension of the discussion about air impacts. Incinerators had become a new and significant concern of the *air* regulators, who were finding themselves, more often than not, in conflict with the *waste* regulators. The air officials, nervous about impending deadlines, seemed to seize on this growth industry to indicate some movement and show some willingness to act.

Once the EPA official had laid out his thoughts, there was an audible sigh of relief in the room. The focus of discussion changed at this point, not to waste, but to such things as cars, traffic, and smog. The environmentally related problems for incinerators, however, would not disappear, as concerns over air emissions joined a range of other environmental and health questions that increasingly placed in doubt the future momentum of the industry.

Air emissions, in some respects, represent the most heavily regulated consequence of incineration. In this country, air pollution issues are dealt with by an overlapping network of governmental agencies at the national, state, and regional levels, and by a multitude of complex and confusing statutes, programs, and rules. Despite the complexity, the regulatory system has over the years at best only provided uncertain protection from the growing hazards of increased levels of air pollution that have plagued urban/industrial societies for more than 150 years. By the late 1980s, this system, moreover, had failed to adequately address many of the pollutants emitted by incinerators.

Air pollution concerns have long been linked to the operation of trash-burning facilities. As far back as the Progressive Era, the debate between different technologies of collection and disposal included, at least as background issues, the problems of discharges and emissions. As the air quality problem in urban and industrial areas increased in magnitude in the post–World War II era, the substantial air emissions of incinerators became an area of concern, despite modest improvements in the technology. Regulations spawned by the Air Quality Act of 1967 directly addressed the emission problems and appeared to signal the death knell for this technology in its requirements for some control of such emissions.[14]

The contemporary era of air quality regulation was ushered in with the passage of the Clean Air Act of 1970 and its subsequent amendments in 1977.[15] The text of the Clean Air Act alone includes 118 pages of the United States Code, while its regulations require four volumes in the Code of Federal Regulations. The enforcement of the Clean Air Act has largely been passed on to the states who have in turn delegated varying degrees of planning and enforcement responsibility to regional air pollution control agencies such as the Air Quality Management Districts in California. State and regional agencies have issued additional volumes of their own regulations, compounding the complex and often confusing structure of air pollution regulation.[16]

Since passage of the Clean Air Act in 1970, a series of deadlines established for all areas of the country to meet the act's standards have not been attained. Meanwhile, the nation's air quality has continued to deteriorate. In 1988 alone, the number of cities whose air quality violated federal ozone standards increased by more than 40 percent, as 28 additional areas with 15 million residents fell out of compliance with the federal standards during the course of the year. These numbers threatened to get worse rather than better in the foreseeable future, with some analysts arguing that several of the worst areas might never attain compliance. Failure to meet the standards was supposed to result in the imposition of sanctions, which included a moratorium on the construction of new, major sources of air pollution, and the cancellation of federal funding for highway construction and sewage treatment facilities.[17]

The sanctions issue has generated its own uncertainties. In one 1988 internal memorandum, EPA administrators argued that they had the discretion not to impose sanctions on an area that appeared to have an acceptable plan of action, even if the area had not met its cleanup deadline.[18] This interpretation was strongly criticized by others, including community activists and environmental groups, who argued the law required the imposition of sanctions regardless of ongoing efforts. For members of Congress, the sanctions issue was less a substantive deadline than a prod to action for communities whose air problems continued to get worse rather than show significant improvement. Unlike the Clean Water Act, which had resulted in a demonstrable improvement in the conditions of certain lakes and streams (though by no means for all waterways), the Clean Air Act, with some notable exceptions (for example, lead emissions), had not meaningfully improved the condition of the air in most metropolitan areas.

Part of the problem has been the inability of both public agencies and industry to translate the complex regulatory system into effective monitoring and enforcement. The situation with regard to waste incinerators is revealing in this respect. Through the impact of the 1967 Air Quality Act legislation, a generation of incinerators, some of them making significant contributions to the deteriorating air quality in their air basins, were nearly entirely eliminated within a ten-year period.[19] Though some of the new incinerators designed in the 1980s had reduced emissions with advanced air pollution controls, both the level and nature of uncontrolled emissions remained unclear.

During the 1980s, a relatively small number of incinerators were operating; thus only limited operational information about their effectiveness in controlling air emissions was available. The vast number of projects were still, by the late 1980s, in the planning or construction stage. Air regulators found it difficult, if not impossible, to accurately estimate a plant's emissions before an incinerator was built. Project consultants and local air districts have attempted to estimate emissions based on other, "representative" facilities, but such comparisons present a number of problems. At any one incinerator, changes in the waste stream tend to make both the quantity and nature of emissions not only unique, but subject to rapid change, and, therefore, difficult to predict. Differences in the composition and quantity of the waste burned, the combustion processes, and the age and design of the equipment (particularly the air pollution control systems), all combine to produce emission rates that vary substantially.[20]

Even among existing facilities, the information is far from uniform. In 1986–87, for example, the EPA produced an emissions data base for all the existing solid waste incinerators in the United States, Canada, and Europe, and discovered that the variations in emissions for all pollutants were substantial, with the variation for one key pollutant, carbon monoxide, as much as 7,200 percent.[21] The reliability of the data was compromised when tests were performed by the companies proposing to build and operate the facilities, a frequent occurrence.

Surprisingly, there is no standardized mechanism to review emissions. That problem is compounded by the number of pollutants detected from incineration. The lack of standards has created yet another level of uncertainty about emission estimates. The procedures used produce at most only a set of informed guesses, with unreliable and possibly invalid figures. Ultimately, what remains clear is that this new generation of incinerators, despite changes in technology and pollution control devices, already contributes to the deterioration of the neighboring community's air quality. With the proposed massive expansion of such plants, the problem will become more critical. And it is the growth of this industry most especially that has made it a leading target in the struggle to reclaim the country's dirty air.

TRACKING THE AIR

The contemporary air pollution regulatory system has essentially divided air pollutants into two classifications: criteria pollutants, for which the EPA has set health-based standards, and noncriteria pollutants, which include all other airborne compounds that may threaten public health (see table 4.1).

Since 1970, EPA has set standards for seven criteria pollutants, all produced in varying and often substantial amounts by incinerator plants. These seven are: oxides of nitrogen (NOx); ozone (O_3), which is formed from a combination of NOx and hydrocarbons (HC); particulate matter (TSP or PM); carbon monoxide (CO); lead (Pb); and oxides of sulfur (SOx) (see table 4.2).[22]

Each of the criteria pollutants can contribute to a range of health problems, most of which are acute (short-term and reversible). Ozone, a colorless gas, results in eye irritation and damage to lung tissues. It reduces resistance to colds and pneumonia, contributes to asthma, bronchitis, and emphysema, and aggravates chronic heart disease as well. It can also damage crops and forests, and corrode materials such as rubber and paint. By the late 1980s, the EPA had begun to admit that the health effects of ozone were "more serious than envisioned," and that "existing standards may provide little or no margin of safety."[23]

Several of the other criteria pollutants produce serious, chronic effects; heart and lung difficulties are of particular significance. They also contribute to serious environmental problems, such as the formation of acid rain. These pollutants can also damage materials, such as statues and build-

Table 4–1
PARTIAL LISTING OF POLLUTANTS FROM RESOURCE RECOVERY FACILITIES

1. Criteria pollutants, governed by California or federal ambient air quality standards

Nitrogen dioxide	(NO_2)
Sulfur dioxide	(SO_2)
Carbon monoxide	(CO)
Particulate matter	(PM)
PM10	(Particulate less than 10 microns)
Lead	(Pb)
Nonmethane hydrocarbon	(HC)

2. Noncriteria pollutants, not governed by California or federal ambient air quality standards

METALS		ORGANICS	ACID GASES
Arsenic	Molybdenum	Benzo-a-anthracene (BaA)	Hydrogen chloride
Mercury	Selenium	Benzo-a-pyrene (BaP)	Hydrogen fluoride
Tin	Tellurium	Benzo-e-pyrene (BeP)	Sulfuric acid
Cesium	Thallium	Coronene (Cor)	
Strontium	Manganese	Fluoranthene (FA)	
Beryllium	Antimony	Polychlorinated dibenzo-p-dioxin (dioxins)	
Cadmium	Barium		
Chromium	Sodium	Polychlorinated dibenzofuran (furans)	
Nickel	Indium		
Zinc	Bismuth	Polyaromatic hydrocarbons (PAH)	
Copper	Vanadium		
Cobalt	Magnesium	Chlorinated benzene (CB)	
Boron	Aluminum	Chlorinated phenol (CP)	
Gold	Potassium	Polychlorinated biphenyls (PCB)	
Phosphorus			

SOURCE: *Air Pollution Control at Resource Recovery Facilities*, California Air Resources Board, May 1984.

ings, and harm vegetation, reducing crop yields in some instances. The impact on the timber industry is also severe, as underscored by the American Forestry Association assertion that "forest resource impacts from air pollution constitute a national threat."[24]

Most often, chronic effects are manifest by a significant time delay between exposure to pollutants and the onset of a particular disease. This has made it difficult to pinpoint a cause-and-effect relationship between the pollutants and the disease. But criteria pollutants also contribute to a wide range of acute disorders, including colds, bronchitis, emphysema, and other forms of lung disease that are easier to document. In general,

Table 4–2
SUMMARY OF MUNICIPAL WASTE COMBUSTION CRITERIA POLLUTANT EMISSION RANGES

	Range of Pollutant Emission Concentrations[a]		
	MASS BURN	STARVED AIR	RDF-FIRED
PM	6.18–1,530	142–303	176–533
	(0.003–0.669)	(0.062–0.132)	(0.077–0.233)
SO_2	0.040–401	61–87	54.3–188
NO_x	140–630	258–309	128–263
CO	18.5–1,350	33–67	217–1,580

SOURCE: "Emissions Data Base," *Municipal Waste Combustion Study*, EPA, 1987.

[a] All concentrations are in units of ppmdv corrected to 12 percent CO_2 except particulate matter concentrations, which are reported as mg/Nm^3 (gr/dscf) corrected to 12 percent CO_2.

acute effects have been studied more extensively and have been the focus of most regulatory efforts.[25]

Despite the standards set for criteria pollutants, the magnitude of the problems they pose has only increased in time. Such health problems, a U.S. Senate committee argued, are "serious and pervasive. There is no choice but to breathe the air, whether it is clean or polluted."[26] The problem is particularly pronounced in air basins that exceed these standards, standards that might already be inadequate in protecting against acute and chronic disorders.

The recent boom in the incineration industry has directly compounded the problems of these "nonattainment" areas. Of 220 planned incinerators, 206 (93.6 percent) are for areas that do not meet federal air quality standards. Though EPA estimated in 1987 that about 30 of those incinerators, or 13.6 percent, might be affected by the ban on the construction of major new sources, repeated extensions of the Clean Air Act have protected this small group from sanctions.[27] Even with the best air pollution control equipment, mass burn incinerators emit large amounts of criteria pollutants. Many are likely to emit more than 100 tons per year of any single criteria pollutant and would therefore be classified as a "major, new source" of air pollution, as defined by the EPA.[28] Nevertheless, such emissions could be legally permitted if the facility met certain other requirements, such as the emission offset rules and the use of best available control technology (BACT). Thus, under the complex Clean Air Act system, an incinerator could obtain a license to significantly pollute an already dirty, unhealthy, air basin. The projected large size and number of such facilities in these areas, furthermore, has been the main reason the EPA, some members of Congress, and opposition groups have characterized waste incineration as the number-one air pollution growth industry in the country.

There are, moreover, a number of other pollutants emitted by incinerators that can directly threaten public health but that are not regulated by the EPA or any other agency. These noncriteria pollutants generally are emitted in much smaller amounts than criteria pollutants, but they can be of considerable danger, particularly in terms of such serious, long-term health effects as cancer.

The noncriteria pollutants of most concern, primarily because of their extraordinary toxicity, are a variety of chlorinated organic compounds known collectively as polychlorinated dibenzo-p-dioxins (PCDD) and polychlorinated dibenzofurans (PCDF), or dioxins and furans. There are 75 different dioxins and 135 furans in existence, all of which are commonly referred to as simply "dioxin."[29]

"Molecule for molecule," said Donald Barnes, director of the EPA Science Advisory Board of dioxin, "this is the most potent carcinogen we've ever seen in the laboratory."[30] It was the discovery of dioxin on the streets of Times Beach, Missouri, that prompted the EPA to pay for the relocation of that entire community. Dioxin also played a large role in the contamination at Love Canal, and is the compound of concern in lawsuits over the exposure of Vietnam veterans to Agent Orange.[31] Dioxin and its related compounds have been conclusively proven to cause cancer in animals and are suspected of causing cancer and birth defects in humans. Some toxicologists contend that these compounds also damage the genetic, neurological, and immunological systems.[32]

The range of health-related problems potentially caused by dioxins is extraordinary. The preliminary results of a study of Vietnam veterans conducted between 1984 and 1988 by U.S. Air Force physicians, for example, linked dioxin with increases in cancer, birth defects, psychological damage, liver damage, cardiovascular deterioration, and degeneration of the endocrine system.[33] Liver damage can be particularly severe. In test animals, dioxin has been proved to cause lesions in the liver and thymus glands, while in humans, it has been shown to cause changes in liver enzyme levels, and, in some cases, enlargement of the liver. Dioxin is also capable of creating disturbances in the responses of the peripheral nervous system and has been shown to cause severe weight loss and chloracne, a disfiguring and persistent form of acne growth.[34]

The role of dioxin in causing cancer remains one of the most widely debated and contested issues. Some medical experts have described the great potential for dioxin to act as a promoter of cancer by increasing the carcinogenesis of other chemicals. A number of animal studies have concluded that dioxin has a direct role in inducing cancer. These studies were responsible in part for EPA's Carcinogen Assessment Group stating, in 1983, that dioxin should be regarded as both an inducer and promoter of cancer. But top administrators at EPA, as well as others in the scientific community, have argued that the evidence to date is still inconclusive. The continuing nature of the debate is partly responsible for the lack of any enforceable national standard for dioxin in the United States.[35]

Since 1977, when dioxin was first discovered at the Hempstead, New York, facility, dioxin has been discovered in varying levels in the emissions of *every single incinerator* tested to date.[36] Similar to its much debated role as a carcinogen, the origin of this mysterious compound in the incineration process also remains uncertain. Though dioxin per se is not likely to be found in the waste stream itself, several scientists have argued that certain products made from plastics, such as polyvinyl chloride, provide the necessary raw materials for the creation of dioxin. Other scientists have argued that dioxin results from the combustion of any product in the waste stream containing chlorine. Other studies, however, have called into question the relationship of the composition of the waste stream to the amount of dioxin created and whether the combustion process plays a role in its formation.[37]

One key theory suggests that dioxin is formed after combustion when the gases cool in the exhaust stack. If dioxin is created in the combustion chamber, as industry proponents claim, then efforts that improve combustion may very well reduce the amount of dioxin emitted. However, if dioxin is formed after the gases leave the combustion chamber, then the design of the combustion chamber would have little effect on the amount of dioxin created.[38]

There are disagreements as well regarding how much of the dioxin detected in the environment can be attributed to existing waste incinerators. One set of studies, though challenged by some, argues that waste incinerators are responsible for much of the dioxin in the environment where such incinerators are operational. A study by the California Air Resources Board concluded that of 15 potential sources of dioxin, waste incinerators were among the four top sources, and, of the four, the only potential source for which any construction was planned. The waste industry has countered those arguments, pointing to studies, such as one in *Waste Age*, an industry trade journal, which suggested that automobiles might be a greater source of emissions than incinerators.[39] "The risk from emissions [of dioxin]," an Ogden Martin executive told the Long Island newspaper *Newsday*, "is far less than from many other combustion sources that we deal with every day—automobiles, fireplaces, power plants. Those are the biggies."[40]

Perhaps the most crucial point that remains subject to debate concerns the estimate of how much dioxin is likely to be emitted from *proposed* incinerators. As with other pollutants, the procedures for estimating emissions can generate more questions than reliable numbers. There are no standardized tests employed by the industry or public agencies, and not even every plant tests for dioxin in the first place. Even within the same facility, and with tests conducted by the owner of that facility, results have varied enormously. Furthermore, with proposed facilities, estimates must rely on test results from an operating facility that is considered an equivalent plant. Yet those comparisons have been clearly inadequate. Often information has not been provided on what wastes were burned, nor have they always been based on plants with the same air pollution control technology. The same data, moreover, might yield different results when different methods for calculating the relative toxicity of the various types of dioxin are used.[41]

The wide range of uncertainties, discrepancies in numbers, and unresolved health issues have all tended to produce an uneven response from regulators, public agencies, and industry in establishing safe dioxin emission limits at any given plant. Though dioxin has in many ways come to symbolize for the public the issue of toxic compounds linked to the waste disposal system, it remains largely unregulated and continually debated. And while the air pollution control bureaucracy struggles to come to terms with the question of how safe is safe, another question, potentially touching the entire spectrum of emissions, looms possibly larger: is any of it safe?

Dioxin, it turns out, is not the only dangerous noncriteria air pollutant emitted by incinerators. Lead, mercury, and beryllium, the only metals for which the EPA has issued standards, are just three of many hazardous heavy metals that are continually being released into the air and found in the ash from incinerators. Others include cadmium, zinc, and arsenic.

As with dioxin, the amounts of these heavy metals emitted into the air varies from facility to facility, creating more difficulties for estimating potential emissions from unbuilt plants. The source for these metals has long been assumed to be common products discarded in the waste stream such as cadmium from flashlight batteries and lead and zinc from automobile batteries. Incineration neither creates nor destroys metals, which account for approximately 10 percent of all municipal solid waste. The same amount that enters an incinerator is eventually released, whether in ash from the furnace, residue from the air pollution control equipment, or directly into the air from the smokestack.[42]

The heavy metals emitted from waste incinerators can pose "severe health risks," according to the California Air Resources Board.[43] Most heavy metals are carcinogens. Many of them generate a wide range of other adverse health effects, including irritation of the lungs, which can cause respiratory problems such as fibrosis; ulcers of the nasal tract; digestive disorders; dermatitis; and a contribution to bacterial pneumonia. Lead, in particular, has been shown to cause damage to the gastrointestinal system, the liver, kidneys, and blood and central nervous system. Similar to lead, antimony and possibly other heavy metals are thought to impair the development of young infants within the range of levels generally emitted by incinerators. Moreover, children are especially threatened by these compounds.[44]

The release of heavy metals into the air by incinerators, even from those with the most advanced pollution control systems, might well be a more serious problem than the heavily publicized dioxin issue. Unlike dioxin, there exists a significant amount of evidence of the adverse health effects of metals on humans, eliminating the need to extrapolate from animal data. But similar to dioxin, heavy metal emissions tend to be ubiquitous, passing through the incinerator system from their multiple sources to either the emissions that disperse into the atmosphere or the ash that remains to be disposed.

Small amounts of numerous other toxic gases have been detected in incinerator emissions, many of which are products of incomplete combustion, created in the burning process. One key group of compounds are "trace organics," including such well-known toxics as PCBs, vinyl chloride, and various polyaromatic hydrocarbons. Other potentially harmful emissions from the incineration process include acid gases and bromated compounds. One of these compounds—benzo-a-pyrene (BaP)—which has been studied extensively, is considered both carcinogenic, mutagenic (that is, it damages genes), and teratogenic (it causes birth defects). There are indications, though not yet conclusive, that other members of this chemical family, also incinerator by-products, will produce effects similar to BaP (see table 4.1).[45]

The great uncertainty about the health effects of most of these toxic compounds is magnified by the need to rely primarily on extrapolating data from animal studies. Furthermore, very little is known about possible synergistic or interactive effects, which could cause the toxicity of an individual substance to be greater once it's combined with other substances.[46] Very few of these chemicals are regulated, and for almost all of the others, there are no guidelines to provide even targeted goals.

The air quality issue, for incinerators, remains far from resolved, despite the claim that the new "state-of-the-art" plants have laid it to rest. Pollutants are still emitted into the air 24 hours per day, 365 days per year, often in great amounts, and by processes not fully understood. Most plants have not tested for the full array of possible compounds, and those that have

have discovered greatly varying results. Incinerator plants, most of them located in dirty air basins, have at best minimum information on the nature of their waste stream, including the serious and widespread presence of the hazardous substances that become part of the burn. Predicting the pollutants that might enter the air and their effect on the community has remained a process rife with uncertainties and false claims.

DETERMINING RISK

The issue of uncertainty has been compounded by the use of risk analysis to provide guidance for policy-making. Since the early 1970s, when the Environmental Protection Agency was confronted with a range of decisions over how to implement such legislation as the Clean Air Act and the Safe Drinking Water Act, EPA has increasingly come to rely on techniques known as *risk assessment* and *risk management* to frame its choices. By basing policy choices on a scientifically derived, numerical value of increased risk, public officials hoped to avoid the more contentious and subjective task of choosing between potential dangers and presumed benefits, of what is an acceptable risk and what is not, a choice that would otherwise be considered political and not technical. Risk analysis became a way of removing environmental decisions from the political arena.

Risk analysis procedure, as it evolved during the 1980s, came to consist of two separate and distinct parts. The first step in this process, risk assessment, relied on a scientific review of the question at hand (for example, quantifying the predicted health risk of a particular facility such as a mass burn incinerator), a task often contracted out to consultants. EPA defined this review process as a tool "to analyze scientific evidence in order to evaluate the relationship between exposure to toxic substances and the potential occurrence of disease." This could be accomplished by "ascertaining the health effects and the potency of the substance in question," as well as "estimating the number of people exposed, their sensitivity, degree of exposure, route of exposure, and the presence of other hazards that might increase the risk."[47]

The outcome of this review procedure is the calculation of a numerical estimate of the amount of increased risk involved. This estimate is usually provided in two forms: as the number of deaths due to cancer from an exposure of 70 years at the point of maximum concentration *in excess* of what would be predicted without the exposure; and as the increased odds of one individual contracting cancer given that same scenario. Most regulatory agencies including EPA have used one excess cancer in a population of 1 million as the limit for acceptable risk, although there have been numerous occasions when such agencies have raised the figure to one in 100,000 or more by including other factors such as the cost of the pollution control technology in the eventual risk management decision. Moreover, by the late 1980s, a number of agencies were considering raising the acceptable risk number to one in 100,000 as the general rule of thumb for acceptable risk.[48]

The numerical estimate of risk has become the centerpiece of risk assessment. The calculation is at best limited and uneven, given the nature of the information involved and the various methods and problems of demonstrating a direct link between a hazard and its health impact. One official with the California Department of Health Services called the procedure "voodoo science" because of the large number of uncertainties and absence of information.[49] To offset potential criticism, EPA and other agencies have established guidelines to help base the numerical estimate on "conservative assumptions," where errors, if made, would be in the direction of protecting public health. In that context, many risk assessments use an "upper bound" value for increased risk that would be based on the greatest risk calculated from the most conservative assumptions, which can be the maximum of the reported range of increased risk or a single numerical estimate based on a "worst case scenario." More recently, however, the conservative approach has been challenged by some consultants in the field who argue that a "realistic" approach to the review process is more reasonable than the conservative bias, an approach that has appealed to industry in its claim that risks are often overblown. Regardless of the method, all risk assessments claim to produce an objective, scientifically justifiable estimate that protects public health and provides a basis for the second phase of the risk analysis process: risk management.[50]

Risk management is the decision phase of the process. It is based on an integration of the "results of risk estimation with engineering data and with social and economic factors to carry out the regulatory process of managing the risk," as defined by EPA.[51] Much of the risk management procedure involves trade-offs between a perceived risk and a presumed benefit. Part of that process has been answered by the risk assessment itself, which suggests a measurement of acceptable risk. Since this is couched in terms of scientific objectivity, despite the uncertainties in the review process and the numerous subjective assumptions made in the procedure, the numerical estimate develops a kind of life of its own, separate from the more "subjective," nonscientific input of community residents or other members of the public who might define "acceptable risk" in different terms.

When the risk assessment/risk management procedure has been utilized in the siting and permitting of incinerator plants, the question of how to define "acceptable risk" has invariably been raised. In the risk assessment phase, the calculations have primarily focused on air emissions, while problems associated with the ash have not been evaluated. Furthermore, the calculations themselves have varied widely and no single set of methods or numbers have been utilized for many key questions such as dioxin emissions.[52] Only a limited discussion occurs, if any at all, over which kinds of groups might be most at risk. Nor do most risk assessments provide a full comparison of alternative waste management options and their associated impacts. Such a discussion, if undertaken in a comprehensive, public forum, would not only influence the definition of what risk is acceptable and what is not, but it would place policy decisions in their appropriate setting: as public matters decided by the community and not technical questions settled by experts and rubber-stamped by public officials.

Given the ambiguity of the information and the problem of establishing a uniform procedure, most regulatory agencies have acknowledged the limitations of risk analysis. At the same time, however, risk analysis has developed into the primary tool for decision-making and is presented as "currently the best available technique," as three public agency analysts argued. Increasingly, in fact, risk analysis has incorporated techniques of cultural anthropology, the better to alleviate the fears and manipulate the responses of communities at risk.[53] This approach, which acknowledges both the uncertainties and subjectivity associated with the procedure, has been criticized as insuring that "an industrial status quo relatively free of socially enforced limits" would inevitably be sustained.[54] In this setting, permits tend to be issued and projects built despite the objections of the affected community and on the basis that the numerical estimate and the risk management outcome (often a contract stipulating an acceptable level of air emissions) provide a sufficient technical basis for proceeding with the facility.

In 1985, William Ruckelshaus, having stepped down as EPA chief for the second time in a decade to resume his career in the waste management industry, described risk analysis as "a kind of pretense." "To avoid paralysis of protective action that would result from waiting for 'definitive' data," Ruckelshaus concluded, "we assume that we have greater knowledge than scientists actually possess and make decisions based on those assumptions."[55] As this spur to action, risk analysis appears to elevate health and environmental questions in the determination of policy, while in reality serving as a mechanism to remove such issues from public input. The determination of risk, as uncertain as the regulatory process that spawned it, has become another way of justifying a project already favored by both industry proponents and policy-makers.

RELYING ON TECHNOLOGY

For the waste-to-energy industry and its public agency allies, the key to the concerns over health and environmental risks such as air pollution has been the location of an appropriate technological means to reduce the problem to acceptable levels. Recent improvements in combustion technology and more advanced air pollution control devices have been proclaimed as the answer to the air problem. Emissions that were once measured in micrograms (μg), a millionth of a gram, are now measured in nanograms (ng), a billionth of a gram. Even the emissions of dioxin, though still released into the air by all waste incinerators, have been significantly reduced since the early 1980s, in a few cases by as much as a factor of 1,000.[56]

The new technologies, however, failed to end the debate about the air issue. With dioxin, for example, much of the incineration industry and the regulatory community have focused on improving combustion as the most effective means of reducing dioxin emissions. A more complete combustion could be accomplished by maintaining a high temperature, upwards of 1,800 degrees Fahrenheit, among other procedures. But high temperatures also increase the emissions of NOx, which in some air basins, such as Southern California, necessitate additional controls for this pollutant.[57]

Among the best-known air pollution control technologies are the scrubber systems that are used to control emissions of sulfur dioxide (SO_2) and acid gases (HCI, HS). There are two kinds of scrubbers: "wet" and "dry." Wet scrubbers, in particular, though more effective at removing SO_2, create wastewater that must be treated, while also increasing the potential for contaminants to leach from the ash.[58] Dry scrubbers, which eliminate the need for wastewater treatment and are considered by some more reliable and effective, have generally become the favored choice for the newest incinerator facilities.

During the early and mid-1980s, with the significant rise of community opposition movements, the waste-to-energy industry intensified its search for more effective air pollution control technologies as the best means of allaying environmental fears and containing public opposition. The most effective strategy, proposed by EPA in 1987, was a combination dry scrubber, fabric filter/baghouse system that was capable of reducing the emissions of several pollutants, including SO_2 and particulates. Reduction of particulate emissions was especially significant since both dioxin and heavy metals adhere to particulates; thus, their emissions would be reduced as well. This combined system, with removal efficiencies for certain pollutants reported by the industry to be greater than 99 percent,[59] quickly established itself as the preferred technique for air pollution control. By 1989, even this two-part system, already disregarded by some European waste managers, came to be challenged by civil engineers in this country who argued that the technology would not operate at the same efficiency over its 20-year life-span.[60] Nevertheless, the dry scrubber, fabric filter/baghouse system has received the designation of best available control technology (BACT) by many, but not all, regional air quality agencies. The BACT designation comes into play in granting Clean Air Act permits, and this system will thus be required for many but not necessarily all new incinerator projects in the years to come, since BACT determinations can vary region by region.

Most significantly, the effective reduction of emissions can be accomplished without an exclusive reliance on technology. This can be achieved by utilizing in one or another form a source separation approach; that is, removing certain items from the waste stream before they become part of the burn. For example, a standard source separation program that removed paper, plastics, glass, and metal cans, including ferrous metals, would significantly reduce the level of NOx and heavy metal emissions. Similarly, with SO_2, the removal of certain organic compounds, such as mixed paper, rubber, corrugated boxes, and plastic, might, according to the California Air Resources Board, "produce the greatest reduction in SO_2 emissions," rather than the existing array of scrubbers and other techniques. The Air Board also concluded that the "removal of certain plastic products could lead to some reduction in chlorinated organic compounds, and total mass emissions generally."[61]

The reliance on technology also tends to obscure the more fundamental questions with respect to air emissions: at what level are they considered a problem, and thus require some form of intervention or regulation. The feasibility and especially the economics of control technologies too often tend to determine how the problem is defined. Regulations are created *after* the fact, that is, after a technology is selected, rather than regulations forcing the industry to reduce emissions *to the greatest extent possible,* given the uncertainties that exist. Though several criteria pollutants and certain toxic compounds such as dioxin have seen significant reductions, the control technologies themselves cannot provide an answer as such to the health-based question of whether such reductions are sufficient. This problem is reinforced by the elusive nature of the standards for many pollutants in many areas of the country. And while the waste industry and the public agencies consider technology to be the key, the problems engendered by the limits of inadequate and haphazard regulation still remain.

THE REGULATORY MORASS

The major regulatory mechanisms, built around the BACT system and a reliance on technology, have almost exclusively focused on the standards and regulatory procedures for criteria pollutants. Noncriteria pollutants, including air toxics like dioxin, have been subjected to far more limited regulation, primarily dependent on a program established through the Clean Air Act amendments of 1977.

This program, the National Emission Standard for Hazardous Air Pollutants (NESHAPS), administered by EPA, has been criticized as inadequate in its scope and performance. Since the passage of the Clean Air Act in 1971, the EPA has adopted standards for only eight compounds. Mercury, beryllium, arsenic, and viny chloride are the air toxics presently regulated by NESHAPS that are known to be emitted by incinerators. But standards for most of the numerous other air toxics associated with incineration, including dioxin, have yet to be set. The slow pace and unwillingness to set standards, unless an exacting criteria of certainty is met, indicate that standards for some air toxics might never be adopted. Given EPA's unwillingness to intervene in the standard setting for dioxin and other air toxics, it is likely that the agency will, once again, as it has on other matters, "leave the tough choices to the states," as one congressional committee noted.[62]

But the states, too, have been slow to control air toxics. In California, for example, a lengthy process of review that was adopted in 1983 still had not developed by 1989 any emission standard for any compound, more than five years after the process had been initiated. Nevertheless, the state's Air Resources Board decided to designate, on its own, dioxin as a toxic air contaminant. The board's Science Review Panel had made its recommendation to list dioxin on the basis of findings that pointed to dioxin as a known animal carcinogen and a potential human carcinogen. The panel argued, furthermore, that there was no known safe threshold for dioxin and that the current and planned waste-to-energy facilities in the state would "provide a high potential for emissions of dioxins into the air."[63] But despite the panel's conclusions, such an analysis did *not* result in the adoption of a statewide dioxin standard.

The California experience has been more the rule than the exception. By 1987, only eight states had adopted specific regulations for limiting dioxin emissions from incinerators, and of these eight, only five had specific, numerical standards. Most review systems in the country tend to be performed at the local level by local air pollution control districts, who estimate and assess the impacts of individual pollutants such as dioxin on a case-by-case basis, but without such assessments linked to specific standards.[64]

Similarly, by 1989, EPA, state agencies, and local districts had cumulatively issued just one emission standard for incinerators as a special class of polluters. This is particularly striking since, among existing operating plants across the country, 16 have no air pollution control equipment at all, and, of those in the planning stages, 18 are being designed in a like manner. The lack of any standardized testing procedures further compounds the problem. In the state of Florida, for example, where governmental agencies have increasingly turned to the trash-burning option, dioxin emissions have not been measured at any of the state's nine waste-to-energy plants. State officials have justified that decision by arguing that the costs of such measurements—about $100,000 per test—were too high. When Massachusetts finally began testing for dioxin in 1986, they discovered such high levels at one plant that they shut it down. The plant, however, had already been operational for more than two years.[65]

The lack of standards in the United States contrasts sharply with the situation in Europe and Japan. The Japanese government has developed a series of national incinerator emission standards, while allowing local governments the option to impose stricter standards, which most have done. The Japanese, furthermore, have developed a system of computer-controlled monitoring of the plants as well as an electronic communication system that keep them in continuous contact with the regulatory agencies. These systems not only indicate the level of pollutants going into the atmosphere, but help determine how well the control equipment and the furnace are working. This allows the regulatory agency to impose changes on the operation of the incinerator, including a complete shutdown if problems develop with the incinerator or if the ambient air conditions in the surrounding neighborhood deteriorate.[66]

The most publicized incinerator emission regulations, however, were developed in Sweden, where upwards of 55 percent of that country's wastes are incinerated. In 1985, the Swedish Environmental Protection Agency requested a temporary halt to the construction of new incinerator plants, citing, among other reasons, its discovery of elevated levels of dioxin in both fish and mother's milk. When the moratorium was lifted the following year, new, more stringent emission figures were established, reducing the emissions levels of several pollutants, including dioxin, by about 90 percent. Existing plants were also required to be retrofitted within five years to meet their new levels.[67]

The Swedish regulations created immediate problems for the U.S. industry. None of the operating incinerator plants in the United States, and almost none of the numerous proposed plants, were able to meet such standards. As a result, industry groups and public agency officials began discounting the Swedish decision, arguing that these were *guidelines*, and should not be interpreted as *standards*. "Rather, they should be seen as goals to be attained as the technology of the waste-to-energy industry advances," one analyst argued in *Waste Age*. "If the guidelines are imposed as standards," the analyst warned, "the waste-to-energy technology will not be a viable option for solid waste disposal."[68]

ASH: THE CROSS-MEDIA EFFECT

Over the years, identifying the environmental hazards of waste disposal systems has become something of a shell game. When wastes were disposed of in the ocean or in rivers and streams, downstream communities discovered that they had become the recipients of this transfer of pollutants. Land disposal methods, established to minimize the pollution of surface waters, were eventually found to contaminate not only the land, but, in many instances, the groundwater basin underlying a particular landfill site, and consequently the community's drinking water supply. Landfill gas emissions, with potentially explosive levels of methane gas, were yet another example of the kinds of problems that have been generated when pollutants are transferred from one medium such as land or water to another medium such as air.

The perpetuation of this pattern, known as "cross-media pollution,"[69] has become the most striking environmental consequence of the new generation of waste-to-energy plants. These plants were designed to address the problem of solid waste disposal. Once operating, they were confronted with a second, onerous problem of air emissions. But, in the process, the redesign of these plants to reduce air emissions created a potentially more serious and intractable problem—the ash residue—which threatens the very viability of this disposal option. Incineration, in fact, should more appropriately be identified as a waste *processing* rather than a waste *disposal* technology, since, as two Environmental Defense Fund scientists have argued, "its products [ash residue] pose substantial management and disposal problems of their own."[70] In fact, the Japanese, with their more developed conception of solid waste, describe incineration as an "interim treatment."[71] The sudden emergence in the late 1980s of ash as a health issue, furthermore, created what Marcia Williams, the then EPA director of the Office of Solid Waste, characterized as a "crisis" at the agency, forcing it to develop a quick response to what could well be incineration's most difficult and contentious issue.[72]

The key to the ash problem is that burning trash concentrates toxic substances located in the waste stream and any other toxics, such as dioxin, that are formed during the incineration process, while opening up additional pathways of exposure. Recent studies have made it increasingly clear that the more efficient the air pollution control equipment, the more toxic the ash becomes. These studies already indicate that at least part of the ash formed meets the definition of hazardous waste. Moreover, this particular question of how to define ash has significant economic as well as environmental consequences, since the cost of ash disposal is dependent on where it goes (that is, to a hazardous waste landfill or to some other disposal site). Classifying ash as hazardous, waste officials have recognized, could determine that a particular project is no longer viable.[73]

The two kinds of ash produced by incinerators, fly ash and bottom ash, are defined by the point at which they are collected. Bottom ash is taken from the bottom of the furnace, while fly ash is collected from the air pollution control system. The fly ash, the more toxic of the two residues, can be disposed of separately or combined with the bottom ash for joint disposal. If the fly ash is not captured by the control system, it leaves the plant through the stack and falls out on the land or surface waters.[74]

Among U.S. plants, ash residue has generally been about one-third of the weight of the waste stream. Of this amount, about 10 percent is fly ash. Thus, for each 100 tons of waste that is burned, about 3 tons of fly ash and 27 tons of bottom ash will remain. Currently, most of the ash from operating plants is buried in sanitary or non-hazardous-waste landfills.[75]

U.S. incinerators, moreover, generate far more ash than many of their European and Japanese counterparts. In Japan, incinerators that burn over 200 tons of waste per day are required to produce no more than 5 percent ash, and several of the Japanese incinerators produce even less. This is largely due to the widespread practice in Japan of separating noncombustible material from incinerator-bound waste. To prevent leaching from landfills, some of the ash is combined with cement and asphalt or is melted down and made into briquettes and then disposed of in landfills.[76]

Incinerator ash raises serious questions about health risks. It invariably contains several heavy metals known to be toxic to humans. Among the most dangerous are lead, nickel, cadmium, chromium-6, mercury, and arsenic. In addition to toxic metals, fly ash also contains detectable concentrations of toxic organics, including dioxin.[77]

The toxics in ash pose a threat to human health by either leaching into the ground or escaping into the air. When ash is buried in landfills, the more soluble heavy metals are especially likely to leach into the groundwater. Moreover, the lime used in incinerator scrubber systems increases the alkalinity of the ash, and actually accelerates leaching caused by such natural phenomena as erosion and runoff from rainfall. Some ash escapes into the surrounding air and is a threat to anyone who comes into contact with it in the collection, transportation, or landfilling process.[78]

Just how toxic is the ash? In November 1987, EPA released a report that concluded that ash from trash-burning plants contained significant levels of lead, cadmium, dioxin, and other toxic substances. A week earlier, the Environmental Defense Fund released a study that found that the regulatory limit defining a hazardous waste for lead was exceeded by the averages of all samples taken at incinerator plants for both the fly ash and the bottom ash, and the combination of the two. The EDF study also determined that the average of all samples for the fly ash exceeded the regulatory limit for cadmium (see table 4.3).[79]

Table 4–3

SUMMARY OF AVAILABLE EXTRACTION PROCEDURE TOXICITY TEST DATA FOR LEAD AND CADMIUM FROM MUNICIPAL SOLID WASTE INCINERATOR ASH

	LEAD	CADMIUM	EITHER
Fly Ash (19 facilities)			
Number of samples analyzed	87	85	87
Number of samples over EP limit	83	83	87
Percent of samples over EP limit	95%	98%	100%
Average of all samples (mg/L)	<u>23.0</u>[a]	<u>28.4</u>	—
Bottom Ash (10 facilities)			
Number of samples analyzed	245	210	245
Number of samples over EP limit	93	4	94
Percent of samples over EP limit	38%	2%	38%
Average of all samples (mg/L)	<u>6.7</u>	0.19	—
Combined Ash (26 facilities)			
Number of samples analyzed	366	272	366
Number of samples over EP limit	171	54	176
Percent of samples over EP limit	47%	20%	48%
Average of all samples (mg/L)	<u>7.6</u>	0.68	—

SOURCE: Environmental Defense Fund data base, 1989. A full list of references is available upon request.

Note: Due to the large number of individual samples analyzed from certain facilities, the aggregate data tend to be skewed and overly dependent on the quality of ash from those few dominating facilities. Caution should therefore be exercised in drawing conclusions about overall exceedance rates.

[a] Underlined values exceed EP limits defining a hazardous waste:
lead: 5.0 milligrams per liter (mg/L)
cadmium: 1.0 mg/L

The release of these studies and the indication that the ash residue at even the newest incineration plants has high levels of toxics have made the problem of ash disposal far more difficult than anticipated. Fearing potential liability, landfill operators in this country have become increasingly reluctant to accept incinerator ash, a problem that will be compounded when more plants come on line by the early 1990s.[80] Attempts to dump ash in developing countries such as Haiti, Sri Lanka, and Panama have also brought worldwide attention to the issue. The dumping of ash from Philadelphia on the small island of Kassa, four miles from the mainland capital of the African country of Guinea-Bissau in January 1988, for example, touched off an international furor when it was discovered that the ash had immediately damaged plant life and potentially posed a health threat to the island residents. With concerns growing both in this country and abroad, where the ash should be disposed of has become the most contentious of the waste-to-energy issues.[81]

The Resource Conservation and Recovery Act of 1976 initially established the system governing the disposal of hazardous wastes. Designed to address both hazardous waste and solid waste issues, RCRA also provided that a facility burning municipal solid waste would not be considered to be managing hazardous waste and thus not be subject to RCRA's Subtitle C section covering hazardous wastes. This provision effectively granted an exemption to solid waste incinerators from RCRA controls.[82]

But is the ash also exempt? This controversial question became a fierce point of contention between incineration's supporters and its critics, with EPA forced to choose sides. Previously, EPA had not listed ash as a hazardous waste, thus allowing plant operators themselves to determine whether it showed hazardous characteristics.

Under EPA regulations developed through the 1980s, the plant operator was the one who decided whether to consider ash hazardous. Until recent challenges, plant operators simply assumed that the ash was not hazardous, and therefore needed no special treatment or even monitoring. Even more egregiously, of more than 200 proposed incinerator plants, not one facility, as of 1988, had prepared an assessment of health risks posed by the ash.[83]

This casual interpretation of the EPA regulations came under widespread attack when the ash toxicity studies were released. In March 1987, the Environmental Defense Fund, as a means of forcing plant operators to change their procedures, sent letters with material on the hazardous nature of the ash to the managers of more than 100 incinerator plants. By doing so, EDF argued that plant operators were from that point obliged to test their ash regularly to determine whether it needed to be disposed of as a hazardous waste. Many plant operators, however, continued to rely on their understanding that the language of RCRA exempted the ash.[84]

With such high stakes involved, the lobbying on ash classification became intense. EPA's own position in the course of this lobbying changed considerably. In a March 1987 memorandum, EPA administrators argued that "although the law on this point has been the subject of debate among EPA and the regulated and environmental communities, EPA's position has been and remains that fly ash and bottom ash which are determined to be hazardous must be managed as hazardous wastes." Six months later, however, J. Winston Porter, the assistant administrator of EPA, stated that "ash is legally exempt from treatment as hazardous waste in most cases." A much anticipated review and ruling on the issue by EPA has been delayed, with former EPA chief Lee Thomas arguing in 1988 that there was "no hurry to get it out," despite the mood of crisis that had prevailed within the agency over the issue. "Literally, there is a change [in plans] every day," concluded one agency source.[85]

The exemption issue is compounded by controversies over the testing procedures for ash. Like the procedures for estimating air emissions, testing the toxicity of wastes is an area in which scientific disagreements reflect policy disputes. For its tests on ash, EPA has relied on the "extraction procedure," or EP toxicity test, which determines the concentrations of metals that are likely to leach out of the waste when it is placed in a typical solid waste landfill. The EP test, however, was intended to predict the maximum concentrations of toxic constituents at any *one* time, rather than average or cumulative releases over a period of time, which results in the detection of only the most hazardous substances. Criticism of the EP test led some at EPA to propose an alternative procedure, the "toxicity characteristic leaching procedure" (TCLP), which further compounded the uncertainty over ash regulation, with attention shifting from the scientific to the political arena.[86]

As with air emissions, differences in the waste stream, in combustion technology, and in air pollution controls can all cause the characteristics of the ash to vary.[87] Despite these variations, however, there have been consistently high levels of heavy metals, especially lead and cadmium, exceeding the limits for defining a hazardous waste. Dioxin is also present in almost all samples. The lead content, even of combined ash (which, due to dilution, is less toxic than fly ash alone), has been shown to be particularly toxic, tens of thousands of times higher than the existing drinking water standard for lead.[88]

Though the EPA has delayed its response to ash regulation, a few states have attempted to pursue the issue on their own. The state of Washington, in particular, decided to test ash from seven incinerators around the country in order to determine how to designate ash. After the tests, conducted by the state Department of Ecology, were reviewed and analyzed, the state decided to issue regulations classifying the ash as a dangerous or extremely hazardous waste, thus requiring more stringent landfilling procedures.[89]

As the issue of the designation of the ash intensified, the incineration industry began to develop its own fallback position: burial of the ash in special landfills, known as monofills or "ashfills," designed to contain only ash. In terms of costs and technology, these ashfills would fall somewhere between a hazardous waste and "sanitary" (non-hazardous-waste) landfill. Some designs called for two liners and a leachate collection system, as opposed to the single liner for sanitary landfills and the three liners required for hazardous waste facilities. The costs would fall between those of the two existing categories, though some industry analysts argue that since the costs of sanitary landfills are rising so rapidly, the additional costs for the ashfill would not be significantly higher.[90]

The industry, furthermore, has hoped to avoid a hazardous waste designation for the more toxic fly ash, and therefore also to hold down costs by diluting fly ash with bottom ash and then sending the combined ash out for burial. Combining the ash can require additional equipment and costs, such as at the Commerce plant, but those new plants that have been designed to facilitate the combination process can be modified easily if it were necessary to collect the ash separately.[91]

The move toward the ashfill represents in some ways an apotheosis of the environmental problems for incinerators. The initial, limited data that have already been gathered on the handful of existing ashfill sites suggest serious problems are in the making. For example, one EPA study of three ashfills found that the drinking water standard for lead was exceeded for eight of nine tests, and the average value at the three sites exceeded the lead standard more than twelvefold. Another review of a New York State ashfill, monitored during its first year of operation, when leachate should be considerably lower than in future years, found that pollutant levels in the leachate already exceeded, in some cases dramatically, a number of drinking water standards for various substances.[92]

The increasing use of ashfills, furthermore, is likely to present a more exaggerated form of cross-media pollution. By creating a new class of landfill sites, a decision based essentially on economic considerations, an entire new generation of potentially dangerous landfills may be established at a time when there is growing sentiment for yet more restrictions on new landfills. Since all landfills eventually leak, a position that a number of scientists and even EPA and industry officials have adopted, then a leaking ashfill only compounds a problem that the rush to incineration was designed to address.[93]

* * *

Mass burn incineration, a new technology put forth as a solution to the range of waste disposal problems, including environmental ones, had become, by the late 1980s, a technology under siege, with environmental issues a central concern. The unresolved debates around ash and air emissions created a climate of uncertainty for which the industry, public agency supporters, including the EPA, and even several established environmental groups all sought to address and resolve. These different constituencies had come to the conclusion that clarity of the regulatory framework could ultimately lead to new technological breakthroughs to reduce hazards as well as the legitimation of this disposal option. EPA stood at the forefront of these efforts, having adopted a position by the late 1980s of strong support for incinerators as "a viable waste management alternative for many communities."[94] Several environmental lobbyists, critical of the absence of regulation, felt that an effective compromise around standards could slow down the incineration bandwagon and establish a more limited but still important place for incineration within an overall solid waste management plan.

The waste-to-energy industry also agreed that regulation was now important and that lack of action by Congress was holding up new facilities. By the 100th Congress, a range of bills were introduced dealing specifically with the ash problem. This legislation, however, stalled in Congress, with community opposition compounding the problems of negotiation and trade-offs.[95]

Despite the search for consensus, the level of opposition to incineration continued to impact the industry, with environmental factors framing the mistrust of this new technology. Once again, the analogy with nuclear power is compelling. The nuclear industry proclaimed, in the 1950s and 1960s, that this unproven technology provided a dramatic improvement, in environmental terms, over such competitors as coal (with its high air emissions) and hydroelectricity (with its destruction of natural environments). But nuclear power's developing problems, from low-dose radiation, serious safety problems, and, most importantly, safe disposal sites for its waste residue, have belied that claim.[96]

The incineration advocates, similarly, portrayed their newly developing technology as preferable, in environmental terms, to landfills, their main waste disposal competition. Yet incineration has also demonstrated that it, too, poses a series of environmental risks, from the continuing and protracted problem of air emissions to the growing realization that incineration's waste disposal problem—the ash residue—might ultimately turn into its own environmental nightmare.

These technologies have been promoted, moreover, not so much for their environmental properties, as for their comparative economics. Not that such plants were cheaper initially; the argument was made that with the growth of the industry and the economies of scale such plants represented, there would ultimately emerge a picture of clear economic benefits.

As these industries did grow, however, and more and more plants came on line, the issue of risk became paramount. The enormous number of *scientific* uncertainties, and the disturbing data that started to become available, all of which pointed to new environmental risks, offered just one set of dilemmas. To these had to be added the wide range of *economic* uncertainties, and a series of numbers and estimates that translated into substantial financial risks. And while environmental questions have raised the problem of public acceptance, the economic questions go to the heart of whether this waste disposal option will be able to survive, let alone thrive as the industry once anticipated it undoubtedly would.

NOTES

1. The plant tour took place April 18, 1987, and included one of the authors, Louis Blumberg.
2. Presentation by Don Avila, Los Angeles County Sanitation Department, April 18, 1987.
3. *Environmental Protection: Law and Policy,* Frederick R. Anderson, et al., Toronto and Boston, 1984; "Legislation and Involved Agencies," William L. Kovacs, in *The Solid Waste Handbook,* ed. William D. Robinson, 1986.
4. *Characterization of Household Hazardous Waste from Marin County, California and New Orleans, Louisiana,* W. L. Rathje et al., Environmental Monitoring Systems Laboratory, Office of Research and Development, EPA, July 1987; "Hazardous Waste Found in Household Garbage," *Chemical and Engineering News,* vol. 65, no. 41, October 12, 1987.
5. *Organized Crime Involvement in the Waste Hauling Industry,* from Chairman Maurice D. Hinchey to the New York State Assembly Environmental Conservation Committee, Albany, New York, July 24, 1986.
6. *New York Times,* December 21, 1984; *NCRA News,* Northern California Recycling Association, vol. 1, no. 10, March 1985.
7. "Air Emissions Tests at Commerce Refuse-to-Energy Facility: May 26–June 5, 1987," Energy Systems Associates, prepared for the County Sanitation Districts of Los Angeles County, Whittier, California, July 1987.
8. "National Air Quality and Emissions Trends," 450/4-88-001, EPA, February 1988, Research Triangle Park, North Carolina, table 4–6.
9. See "The Cleanest Waste-to-Energy Plant in the World?" Foster Wheeler ad, *Waste Age,* January 1989; same ad in *Solid Waste and Power,* February 1989.
10. Interview with Manuel Rubivar, Southern California Air Quality Management District, Engineering Division, September 20, 1988; "Trash-to-Energy Plant Under Fire for Air Violations," Rick Holguin, *Los Angeles Times,* February 1, 1989.
11. One of the authors, Louis Blumberg, was in attendance at the meeting.
12. "U.S. Cities Flunking Ozone Standards Increase by 40% in 1 Year," Douglas Jehl, *Los Angeles Times,* February 17, 1989; Federal Register, vol. 53, no. 108, June 6, 1988, pp. 20722–20734.

13. "Environmental News," EPA, Region IX, Regional Administrator, April 7, 1987; Federal Register, vol. 53, no. 173, September 7, 1988, pp. 34500–34507.
14. 1967 Air Quality Act (P.L. 90-148), 81 Stat. 485; "The Place of Incineration in Research Recovery of Solid Waste," J. H. Fernandes and R. C. Shenk, *Combustion*, October 1974; for an account of early efforts to control air pollution, see *Vanishing Air*, John C. Esposito et al., New York, 1970.
15. Clean Air Act of 1970, Clean Air Act Amendments of 1977, codified together as 42 U.S.C.A. 7401 et seq.
16. *Environmental Protection: Law and Policy,* Frederick R. Anderson, Daniel R. Mandelker, A. Dan Tarlock, Little, Brown and Company, Boston, Toronto, 1984, pp. 135–36.
17. Gerald A. Emison, Director, Office of Air Quality Planning and Standards, U.S. EPA, Internal Memorandum, April 29, 1988, p. 1.; *Los Angeles Times,* February 17, 1989; United States Senate, Report of the Committee on Environment and Public Works, staff report to accompany S. 1894, the Clean Air Standards Attainment Act of 1987, November 20, 1987, U.S. Government Printing Office, Washington, D.C.
18. Letter from Judith Ayres, EPA Regional Administrator, Region 9, to Jananne Shapless, Chair, California Air Resources Board, April 3, 1987; Gerald A. Emison, supra, at "Questions and Answers . . . ," p. 3.
19. *COPPE Quarterly,* Council on Plastics and Packaging in the Environment, vol. 1, no. 2, Fall 1987.
20. Science Advisory Board, U.S. EPA; "Evaluation of Scientific Issues Related to Municipal Waste Combustion"; Report of the Environmental Effects, Transport, and Fate Committee; April 1988.
21. "Emissions Data Base for Municipal Waste Combustors, Review Draft, *Municipal Waste Combustion Study,* EPA, January 7, 1987, chapters 2 and 7; *Final Report to Congress,* EPA 530/SW-87-021 a-i.
22. California Air Resources Board (CARB), *Air Pollution Control at Resource Recovery Facilities,* May 24, 1984, pp. 4, 27, 74–180, 191–204; hereafter referred to as CARB, 1984.
23. *Ozone and Smog: A Threat to Your Community,* Bruce Jordan et al., Office of Air Quality Planning and Standards, U.S. EPA; June 1986, p. 3. *An Environmental Review of Incineration Technologies,* Neil Seldman, Institute for Local Self-Reliance, Washington, D.C., October 1986.
24. Cited in U.S. Senate, Committee on Environment . . . , supra at p. 3.; see also "Crop and Tree Damage Are Tied to Air Pollution," Philip Shabecoff, *New York Times,* September 11, 1988.
25. *Control Measures and Concepts: An Issue Paper,* Southern California Association of Governments, Los Angeles, CA, February 24, 1987, p. 9.
26. U.S. Senate, Committee on Environment . . . , supra at p. 3.
27. Internal Memorandum, October 17, 1987, Barry Gilbert, Office of Air Quality Planning and Standards, EPA, Research Triangle Park, North Carolina.
28. Interview with David Jesson, EPA Region IX, October 14, 1988.
29. Dioxin has been called the most toxic substance ever synthesized by man. This conclusion is derived from the work of A. Poland and E. Glover as published in *Molecular Pharmacology,* vol. 13, 1977, pp. 924–38. The toxicity of dioxin is discussed in CARB, supra, at 198.
30. Cited in "Toxic Fear, Anger Billow from Town's Two Mills," Ronald B. Taylor, *Los Angeles Times,* December 5, 1988, part 1, p. 3.
31. *National Dioxin Study,* Office of Solid Waste and Emergency Response," EPA 530/SW-87-025, August 1987, pp. I-1, I-2.
32. *Public Hearing to Consider the Adoption of a Regulatory Amendment Identifying*

Chlorinated Dioxins and Dibenzofurans As Toxic Air Contaminants (AIDTAC), Staff Report, California Air Resources Board, February 1986; see also presentation by Dr. John Froines, UCLA School of Public Health, January 1987; AIDTAC, part B, "Health Effects of Chlorinated Dioxins and Dibenzofurans," California Department of Health Services, p. I-1.

33. *United States Air Force Personnel and Exposure to Herbicide Orange: Interim Report for Period March 1984–February 1988,* Richard A. Albanese, United States Air Force, Brooks Air Force Base, Texas, February 1988.
34. "Morphology of Lesions Produced by Dioxins," R. D. Kimbrough in *Human and Environmental Risks of Chlorinated Dioxins and Related Compounds,* New York, 1983; "A Consideration of the Mechanism of Action of 2, 3, 7, 8- Tetrachlorodibenzo-p-dioxin and Related Halogenated Aromatic Hydrocarbons," A. Poland et al. in *Human and Environmental Risks of Chlorinated Dioxins and Related Compounds.*
35. *Dioxin Strategy,* EPA, Office of Water Regulations and Standards and Office of Solid Waste and Emergency Response, in Conjunction with Dioxin Management Task Force, Washington, D.C., November 28, 1983; AIDTAC, part B, supra at p. 1–4.
36. "A Data Base of Dioxin and Furan Emissions from Municipal Refuse Incinerators," Milton Beychok, *Atmospheric Environment,* vol. 21, no. 1, 1987, pp. 29–36; *Dioxin Strategy,* EPA, 1983.
37. "Assessment of Potential Health Impacts Associated with Predicted Emissions of . . . Dioxins and . . . Furans from the Brooklyn Navy Yard Resource Recovery Facility," Fred C. Hart Associates, New York, August 17, 1984, Executive Summary, pp. 6–8; see also Statement of Ogden Martin Co. to California Senate Committee on Toxics, July 31, 1986; also CARB, 1984, supra at pp. 204, 215.
38. "Environmental and Economic Analyses of Alternative Municipal Solid Waste Disposal Technologies," Barry Commoner et al., Center for the Biology of Natural Systems, Flushing, New York, December 1, 1984, parts 2 and 3, pp. 26–30; "Environmental Levels and Health Effects of Chlorinated Dioxins and Furans," Barry Commoner et al., presentation at the AAAS Annual Meeting, Philadelphia, May 28, 1986; see also L. Stieglitz, G. Zwick, H. Deck, and H. Vogg, 1988; "On the De Novo Synthesis of PCDD/PCDF in the Municipal Waste Incineration Process"; Seventh Annual Symposium on Chlorinated Dioxins and Related Compounds; October 4–9, 1987, Las Vegas, Nevada; Technical Support Document, part A: *A Review of Chlorinated Dioxin and Dibenzofuran Sources, Emissions, and Public Exposure,* February 1986; AIDTAC-A, supra, p. 5.
39. *Waste Age,* Carolyn Konheim, pp. 69–76, November 1986; "Environmental Fate of Combustion-Generated Polychlorinated Dioxins and Furans," J. M. Zurzwa and R. A. Hites, *Environmental Science and Technology,* vol. 18, no. 6, 1984.
40. *The Rush to Burn: America's Garbage Gamble, Newsday* reprint, December 13, 1987, Long Island, New York.
41. Interview with Milton Beychok, 1987; "Data Base of Dioxin and Furan Emissions from Municipal Refuse Incinerators"; *Atmospheric Environment,* vol. 1, no. 1, 1987, pp. 29–36; "Methods of Inferring Total Potency of a Mixture of PCDD/PCDF," appendix B. Also "Interim Risk Assessment Procedures for Mixtures of Chlorinated Dibenzodioxins and Dibenzofurans," Chlorinated Dioxin Workgroup Position Document, EPA, April 1985.
42. *An Environmental Review of Incineration Technologies,* Neil Seldman, Institute for Local Self-Reliance.
43. CARB, 1984, supra at p. 182.
44. CARB, 1984, supra at p. 185; also interview with Paul Papanek, Chief, Toxics Epidemiology Division, Los Angeles County Health Services Department, April 30, 1987.
45. CARB, 1984, supra at p. 4; also Environment Canada, "National Incinerator Testing and

Evaluation Program," September 1985; John Froines, UCLA presentation; *Brominated and Brominated/Chlorinated Dibenzodioxins and Dibenzofurans,* Hans-Rudolf Buser, Swiss Federal Research Station, presented at the Sixth International Symposium on Dioxin, Fukukoa, Japan, September 16–19, 1986.

46. California League of Women Voters, *Gray Areas in Black and White, Hazardous Materials Management in California,* Alison Fuller, principal author, 1986, p. 7.
47. "Assessing Risks from Health Hazards: An Imperfect Science," Dale Hattis and David Kennedy, *Technology Review,* May/June 1986; League of Women Voters, 1986, p. 12.
48. *Proposed Rule 1401,* South Coast Air Quality Management District, March 26, 1987.
49. League of Women Voters, supra at p. 7.
50. Interview with Robert Valdez, UCLA School of Public Health, 1987; *Health Risk Assessment of the LANCER Project,* Dr. Alan H. Smith, Health Risk Associates, Draft, April 17, 1987; "Science Policy Choices and the Estimation of Cancer Risk Associated with Exposure to TCDD," Michael Gough, *Risk Analysis,* vol. 8, no. 3, 1988, pp. 337–42; "Risk Assessment in Environmental Policy Making," Milton Russell and Michael Gruber, *Science,* April 17, 1987.
51. *Regulating Carcinogenic Air Pollutants: Current and Potential Future Approaches,* J. Grisinger et al., Planning Division, South Coast Air Quality Management District, March 1987; "Incorporating Risk Assessment and Benefit-Cost Analysis in Environmental Management," Robert W. Rycroft et al., *Risk Analysis,* vol. 8, no. 3, 1988, pp. 415–20.
52. "Risk of Municipal Solid Waste Incineration: An Environmental Perspective," Richard A. Denison and Ellen K. Silbergeld, *Risk Analysis,* vol. 8, no. 3, 1988, pp. 343–355; *Comprehensive Management of Municipal Solid Waste Incineration: Understanding the Risks,* Richard A. Denison and Ellen K. Silbergeld, Oak Ridge National Labs, 1988; "Dioxin in the Agricultural Food Chain," Jeffrey B. Stevens and Elizabeth N. Gerbec, *Risk Analysis,* vol. 8, no. 3, 1988, pp. 329–335.
53. "Perceived Risk, Real Risk: Social Science and the Art of Probabilistic Risk Assessment," William R. Freudenberg, *Science,* vol. 242, October 7, 1988, pp. 44–49; "Risk Communication: Who's Educating Whom?" Amy K. Wolfe, *Practicing Anthropology,* vol. 10, no. 3–4, 1988, pp. 13–14; "Culture and the Common Management of Global Risks," Luther P. Gerlach and Steve Rayner, *Practicing Anthropology,* vol. 10, no. 3–4, 1988, pp. 15–18; *Regulating Carcinogenic Air Pollutants,* J. Grisinger et al., South Coast Air Quality Management District; *Risk and Culture: The Selection of Technical and Environmental Dangers,* Mary Douglas and Adam Wildavsky, Berkeley, 1982.
54. "Risk: Another Name for Danger," Langdon Winner, *Science for the People,* May/June 1986.
55. "Risk, Science, and Democracy," William D. Ruckelshaus, *Issues in Science and Technology,* Spring 1985, p. 26; "Assessing Risks from Health Hazards: An Imperfect Science," Dale Hattis and David Kennedy, *Technology Review,* May/June 1986.
56. *Municipal Waste Combustion Study,* EPA, supra at table 7-34.
57. Interview with Robert Pease, South Coast Air Quality Management District, Engineering Division, April 1987.
58. "The Hazards of Ash and Fundamental Objectives of Ash Management," Richard Denison, Environmental Defense Fund, January 27, 1988, Washington, D.C.; "Swedish Incinerator Ash Study Heightens Concerns Over Heavy Toxic Metal Leaching," Richard Denison, translation pp. 1–4 of *Residues from Waste Incineration: Chemical and Physical Properties,* Swedish Geotechnical Institute, June 1986, Report No. 172, Linkoping, Sweden.
59. *The National Incinerator Testing and Evaluation Program,* D. J. Hay et al., presented to the 79th Annual Meeting of the Air Pollution Control Association, June 22–27, 1986; also, CARB, supra at pp. 138–148.

60. "A Critical Review of EPA's Plan to Establish a Dry Scrubber Technology Standard for MSW Incinerators," Craig S. Volland, cited in *Hazardous Waste News*, no. 118, February 28, 1989.
61. CARB, 1984, supra at pp. 82, 119, and 215.
62. U.S. Senate, Committee on Environment and Public Works, November 20, 1987, supra at p. 6; also Opening Statement, Representative Henry A. Waxman, Chairman, Subcommittee on Health and the Environment, Hearings on H.R. 2576, June 11, 1985.; Statement of Representative Tim Wirth, Colorado, Subcommittee on Health and the Environment, Hearings on H.R. 2576, June 11, 1985.
63. Dr. Emil M. Mrak, Chairman, Scientific Review Panel, *Report on Chlorinated Dioxins and Dibenzofurans*, April 16, 1986, p. 1.
64. "Waste Age Survey," *Waste Age*, April 1987.
65. *The Rush to Burn: America's Garbage Gamble, Newsday.*
66. *Garbage Management in Japan: Leading the Way,* Allen Hershkowitz and Eugene Salerni, Inform, 1987, p. 7.
67. Press Release, Sunsdsvall Municipal Energy Authority, June 17, 1986, reproduced in Seldman, 1986, supra at pp. 101–3; see also "Determination of PCDDs and PCDFs in Incinerator Samples and Pyrolitic Products," in *Chlorinated Dioxins and Dibenzofurans in Perspective,* Christopher Rappe et al., Chelsea, Michigan, 1987, chap. 6.
68. "Why the Swedish Moratorium Was Ended," Edward Repa, *Waste Age,* November 1986.
69. *Environmental Planning and Decision Making,* Leonard Ortolano, New York, 1984, pp. 27, 125; *The Closing Circle,* Barry Commoner, New York, 1971.
70. "Risks of Municipal Solid Waste (MSW) Incineration: An Environmental Perspective," Richard A. Denison, presented at the Annual Meeting of the American Association for the Advancement of Science, San Francisco, December 17, 1988.
71. *Garbage Management in Japan,* Allen Hershkowitz and Eugene Salerni, p. 67.
72. Interview with Marcia Williams, Director of the Office of Solid Waste; U.S. EPA, 1987; *Waste Age,* February, 1987, at p. 20.
73. "Current Legislation Could Regulate Waste-to-Energy Ash," Roger D. Feldman and Howard L. Sharfstein, *Solid Waste and Power,* June 1988; letter from Delwin A. Biagi, Director, Los Angeles Bureau of Sanitation to David J. Lau, California Department of Health Services, Toxic Substances Control Division, November 20, 1984.
74. Ellen K. Silbergeld, Senior Scientist, Environmental Defense Fund, "Testimony Before the City Council of Philadelphia on the Subject of Municipal Waste Incineration," p. 8.; see also *Waste-to-Energy Facilities,* National League of Cities, 1986, p. 50.
75. J. Winston Porter, Assistant Administrator of the Office of Solid Waste and Emergency Response, USEPA, testimony before the Subcommittee on Transportation and Hazardous Materials of the Committee on Energy and Commerce, U.S. House of Representatives, March 19, 1987, p. 1. (The 30 percent figure is for dry ash. Wet ash can be as much as 45 percent by weight of the incoming waste.) See also Environmental Defense Fund, Testimony before the Subcommittee on Transportation and Hazardous Materials of the Committee on Energy and Commerce, U.S. House of Representatives, March 19, 1987, p. 2.
76. Hershkowitz and Salerni, 1987, supra, p. 76.
77. Silbergeld, supra, p. 9; Wakimoto and Tatsukawa, "Polychlorinated Dibenzo-p-dioxins and Dibenzofurans in Fly Ash and Cinders Collected from Several Municipal Incinerators in Japan," *Environmental Health Perspectives* 159, 1985; "Incinerated Garbage Ash Is Found to Have Several Toxic Substances," Philip Shabecoff, *New York Times,* November 26, 1987.
78. Mark M. Bashor, Ph.D., Director, Office of Health Assessment, Center for Disease Control, in letter to Gordon Milbourn, Office of Inspector General, EPA, September 28, 1987.

79. In 1986, EPA conducted a literature search to determine the results of toxicity testing of ash from municipal waste incinerators. EPA found results from only 14 incinerators. Among those tested were eleven samples of fly ash of which nine failed the EP test, and seven samples of bottom ash, of which one failed. Of 15 samples of combined ash, four failed the test. The agency, however, believes that the data are insufficient to justify rules requiring routine testing of ash.

    The Environmental Defense Fund conducted its own literature search of 31 incinerator ash tests and found that all of them showed high levels of heavy metals, especially lead and cadmium. (Source: Testimony Before the Subcommittee on Transportation Tourism, and Hazardous Materials, U.S. House of Representatives, Environmental Defense Fund, March 19, 1987.)

    The state of Washington's Department of Ecology tested ash from seven incinerators around the United States in order to determine the designation of ash as a hazardous waste. In five tests where fly ash was collected separately, the state designated four as extremely hazardous, and one as dangerous waste. In those same incinerators, the bottom ash also tested as hazardous waste, one as extremely hazardous, and four as dangerous wastes. In the remaining two incinerators, the bottom and fly ash were combined. The state designated the combined ash as a dangerous waste. (Source: James C. Knudson, Solid Waste Engineer, "Study of Municipal Incineration Residue and Its Designation as a Dangerous Waste," State of Washington, August 1986.)

    See also "Environmental News," Office of Public Affairs, U.S. EPA, March 12, 1987, p. 1.
80. Interview with Cliff Jessberger, President, American Ref-Fuel, *Waste Age,* April 1987, p. 88.
81. "After Two Years at Sea, Ship Dumps U.S. Ash," *New York Times,* November 10, 1988; Greenpeace press release, June 13, 1988.
82. Resource Conservation Recovery Act Subtitle C, Section 3001 (i).
83. "Risks of Municipal Solid Waste Incineration: An Environmental Perspective," Richard A. Denison and Ellen Silbergeld, *Risk Analysis,* 1988.
84. "Memorandum to Interested Parties," Dr. Richard A. Denison, Environmental Defense Fund, November 17, 1987, p. 1.
85. EPA Press Release, Solid Waste Task Force, Robin Woods, September 22, 1988; Feldman and Sharfstein, supra at p. 15.
86. Interview with Marcia Williams, 1987; *Comprehensive Management of Municipal Solid Waste Incineration,* Richard A. Denison and Ellen Sillbergeld; "The Confusion and Questions about Ash," Edward Repa, *Waste Age,* September 1987.
87. *Waste Age,* September, 1987, ibid, at p. 89.
88. Environmental Defense Fund, *The Hazards of Ash and Fundamental Objectives of Ash Management,* January 27, 1988, p. 1.; "Municipal Solid Waste Composition and the Behavior of Metals in Incinerator Ashes," T. L. Clapp et al., *Environmental Progress,* February 1988.
89. *Study of Municipal Incineration Residue and Its Designation as Hazardous Waste,* J. C. Knudson, Department of Ecology, State of Washington, August 1986.
90. *Waste Age,* September 1987.
91. Interview with John Phillips, Vice President of Marketing, Ogden Projects, March 18, 1987.
92. "Risks of Municipal Solid Waste Incineration: An Environmental Perspective," Richard A. Denison and Ellen K. Silbergeld, *Risk Analysis,* 1988; *Comprehensive Management of Municipal Solid Waste Incineration,* Richard A. Denison and Ellen Silbergeld.
93. *The Prevalence of Subsurface Migration of Hazardous Chemical Substances at Selected Industrial Waste Land Disposal Sites,* D. Miller et al., EPA 530-SW-634, 1977; "Design-

ing and Maintaining Landfill Caps for the Long Haul," David I. Johnson and Glenn R. Dudderar, *Journal of Research Management and Technology,* vol. 16, April 1988.
94. EPA Solid Waste Task Force, Robin Woods, Press Release, September 22, 1988.
95. "Current Legislation Could Regulate WTE Ash," Feldman and Sharfster, *Solid Waste and Power;* "How We Can Get on With the Job," Representative James J. Florio, *Solid Waste and Power,* June 1988.
96. *Forevermore: Nuclear Waste in America,* Donald L. Barlett and James B. Steele, New York, 1985.

5

THE ECONOMIC FACTORS: RISKS AND GAMBLES

LET THEM EAT ASH

It was a tense moment for the Butler, New Jersey, company, one of the 30 or so waste-to-energy developers in the country active in the New England area. Vicon Recovery Systems had, in fact, established five different corporate entities to run its various waste management operations, including a 240-ton-per-day mass burn incinerator in Rutland, Vermont, and landfills in nearby Sunderland and Bristol. At one point, Vicon had as many as 11 different proposals pending, hoping to obtain a substantial market share in an industry that projected exponential growth over the next several decades.[1]

The Vicon representative knew it was crucial to obtain backing at that July 1988 public hearing for its proposed ashfill for the residue generated at the Rutland plant. It had been a long haul for the company. The Rutland incinerator idea was first proposed in the early 1980s, when the company enjoyed a kind of honeymoon period with elected officials. It would be a "state-of-the-art" facility, company officials had proclaimed, financed by the Industrial Bank of Japan, and would hopefully serve as a model for future facilities.

But from the moment it opened in 1987, the Vicon plant had been beset with problems. When Vicon first approached Vermont officials, it quoted a low tipping fee of $16.50 (making it competitive with some landfills) and a reliable source of income from its energy contract with a local utility. In 1984, the utility, Central Vermont Public Service, had agreed to pay $12.02 per kilowatt hour for a long-term contract of 26 years for the electricity generated at Rutland. But, in 1987, as the incinerator neared completion with a price tag of $35 million, well over budget, the Vermont Public Service Board vetoed the rate as excessive, and Vicon had to settle for a short-term contract with a rate of only $3.06.[2]

The financial problems stemming from the electricity contract were further compounded when the Vermont Health Department, following earlier warnings that Vicon had ignored, required the installation of additional air pollution control equipment. This requirement for an expensive scrubber system was made necessary by Vicon's significant contributions to Rutland's nightly air inversions, which created the worst air quality in the state.

Even more contentious was the issue of ash disposal. Vicon had anticipated burying the ash at its own landfill in Sunderland, near the banks of the Battenkill River, a popular trout stream. But Sunderland residents mobilized against the plan, and upwards of half the adult population of the community attended meetings to fight Vicon. Eventually, the residents obtained a court order to stop Vicon from completing the ashfill. Beset by "precarious financing, poor planning, and opposition by environmentalists," as the local Rutland paper put it, the Vicon facility, by the time it opened, had become an economic basket case, with the tipping fee for district towns jumping to $45 a ton, and nondistrict towns paying even more.[3]

Now this tense public hearing was the company's last chance to recoup. Without the ashfill, the company needed to transport the ash long distances across state lines, an economic burden that had become too costly for the Rutland plant. The Vicon representative was a bit frantic that day, attempting to convince his skeptical audience that the ash was not a major problem. At one point, he picked up a bucket, explaining to his astonished listeners that it was full of Rutland ash. He then dumped the ash on the floor, declaring that since it was not toxic and there were no fumes, that the ash was indeed harmless.[4]

The Vicon executive concluded his remarks, and Rutland residents and officials left the room, wondering what the gesture portended. They soon had their answer. The next month, on August 23, 1988, Vicon Recovery Systems filed for bankruptcy, seeking relief under Chapter 7 of the federal bankruptcy laws for those corporations intending to liquidate their assets. The Vicon action stunned Vermont officials, some of whom had promoted the incineration concept and convinced local Rutland County towns to close their landfills and get on this new waste disposal bandwagon. In nearby New Hampshire, residents of the city of Manchester, appalled at the Vicon situation, and having themselves turned down a Vicon proposal for a 560-tpd incinerator *just two days before Vicon declared bankruptcy,* passed, by a two-to-one margin in a local referendum, a recycling plan that would explicitly remove mass burn incineration from Manchester's waste management future.[5] But in Rutland County, Vermont, the economic fate of Vicon and its overpriced incinerator had become unhappily intertwined with that region's own waste management options. The economics of incineration, once so promising, had become a local albatross.

AN UNCERTAIN BEGINNING

For local communities, confronted with the growing volume of trash and increasing difficulties in disposing of it, garbage, during the 1980s, had become an issue of economic survival. Overall solid waste collection and disposal costs were quickly becoming a major component of the total budget of municipalities and other government entities. The cost of garbage disposal in Philadelphia, for example, actually began to exceed the cost for fire protection, rising from $8.5 million in fiscal year 1983 to $40 million in fiscal year 1988. Even the grants, loan programs, and other forms of financial assistance from states and the federal government had been far too limited to offset the increase in unit costs of the two main options—landfills and incinerators—that have predominated in recent years. Landfill costs, though historically cheap, climbed rapidly, while the price tag for the big trash-burning plants also continued to go up. Waste management, both in terms of public expenditures and private revenues, had become, by the end of the decade, a big, albeit troubled, business.[6]

The rise of landfills a generation earlier had been driven by their low costs, their simplicity, and their relatively easy access to the points of collection. Landfill tipping fees continued to compare favorably, even through the 1970s and 1980s, with other disposal methods, including the first wave of waste-to-energy plants—despite an assortment of favorable loan programs and grants.[7]

By this period, however, increased attention to environmental and health-related problems began to have direct economic consequence for the landfill option, in terms of both increased regulation and problems with new sites. Landfills either closed for failing to meet new standards or reached capacity sooner than anticipated. As early as 1974, a National League of Cities survey projected that half the existing urban landfills would become filled, or deemed obsolete, within five years.[8] Two years later, the passage of the Resource Conservation and Recovery Act placed a series of new constraints on land disposal sites. Between 1979 and 1986 alone, partly as a result of the RCRA standards affecting open dumps, 3,500 landfills closed or reached capacity, while only a handful of new landfills were permitted. Newly sited landfills, moreover, were situated at increasingly distant locations, often outside the existing boundaries of a particular municipality. This, in turn, increased transportation costs and tipping fees. This growing scarcity factor, influenced both by urban growth and residential opposition to nearby sites, significantly increased the overall costs of landfilling.[9]

As late as 1984, however, the average tipping fees at landfill sites around the country had increased only marginally in comparison with other disposal alternatives, especially waste-to-energy facilities. During the 1970s and early 1980s, the shift away from landfills represented less an economic imperative than a political reality: it was becoming increasingly difficult to build new landfills, while old ones were filling up or closing down. In certain areas of the country, particularly the Northeast and Mid-Atlantic regions, landfill costs did increase enormously: Boston, for example, witnessed a 300 percent increase in landfill costs between 1983 and 1986, while Philadelphia's costs increased by 500 percent in that same period. In 1988, New York City doubled its fee at the Fresh Kills landfill on Staten Island, forcing a number of haulers to seek distant though less expensive sites. On a national basis, however, tipping fees for landfills increased by only 24 percent during those years.[10]

During the 1970s, the first waste-to-energy facilities were proposed with the expectation that they would become economically competitive with landfills as land siting problems increased. However, the early waste-to-energy efforts also assumed the sale of energy as an additional and crucial source of income, despite the fact that the first plants lacked plans to market the energy. As early as 1975, an EPA survey revealed that waste-to-energy developers and their financial backers anticipated that the energy generated and sold would eventually account for over half of the revenues. "Without revenue from energy products to offset processing costs," the EPA noted, "resource recovery plants cannot be economically viable."[11]

In economic terms, the 1970s represented, then, an uncertain beginning for waste-to-energy, despite the growing landfill troubles and the increasing interest of municipalities in considering such a facility for their community. The developing waste-to-energy industry tended to be cautious about its prospects, with lower estimates of future trash-burning facilities and the amount of waste they could dispose than government-related estimates.[12]

It quickly became apparent to the private developers that nearly all the new waste-to-energy facilities were experiencing problems, some quite serious. Garbage burning was still a high-risk venture requiring a sizable return on investment, upwards of 15 to 20 percent of capital invested. Most companies consequently favored a significant degree of public rather than private financing, given the capital required and the unproven nature of the technology. The companies were also concerned that the development of any plant proceed with a firm set of commitments, from the ability to secure a sufficient quantity of waste (to avoid underutilization), a clear resolution of jurisdiction for permitting and other regulations, a stable and favorable contract for the sale of the energy, and an overall financing package.[13]

Besides the need for energy revenues, the availability of sufficient financing remained paramount for the success of trash-burning plants. The variety of federal and state subsidies, such as demonstration grants, offered at most only supplemental financial assistance. The availability of capital, interest rate and tax factors, the bond markets, and the assignment of risk, all affected project feasibility. According to the 1975 EPA study of the industry, these "financial aspects of resource recovery [had developed as] the *most critical* limiters to implementation" of this new technology.[14]

These financial factors became the most overt and recognizable reasons for the slow and uneven development of waste-to-energy. Interest rates remained high through the 1970s, raising the cost of capital and increasing the overall cost of a project. The bond houses, furthermore, tended to view the industry in mixed terms, reflecting the unproven nature of the technology, the large capital requirements, and the unresolved legislative and jurisdictional issues, which made it difficult at times to obtain credit.

In 1976, in fact, a National Commission on Supplies and Shortages recommended against federal funding of waste-to-energy, citing hidden costs and a lack of cost-effectiveness. "To subsidize the recovery of energy, but not of materials, from solid waste," the commission warned, "would be to compound pressures for the inefficient use of resources."[15]

As a result of these problems, many projects remained on the drawingboards. Facilities that did come on line during the 1970s experienced operational difficulties, which created even more wariness about the track record of the technologies involved. Moreover, despite the primary role of the water-wall combustion technology, no single system emerged as *the* cost-effective and dominant waste-to-energy method, creating additional uncertainty regarding the option as a whole (see table 5.1).[16]

Financing methods also varied enormously, with revenue bonds, private capital, industrial revenue bonds, grant money, general obligation bonds, and even pollution control bonds explored at one point or another. The assortment of companies initially expressing interest in the trash-burning business remained cautious at best, concerned about who would pay for the facilities, how such payments would be made, and who would share the risks. "Industry is not willing to invest more private money to solve a public problem," the EPA warned.[17] With all trash-burning technologies accounting for only 1 percent of waste disposal by the end of the decade, waste-to-energy remained an unsure alternative with an unclear future.

Table 5–1
MUNICIPAL WASTE COMBUSTION FACILITIES CURRENTLY SHUT DOWN

LOCATION	TYPE[a]	DESIGN CAPACITY, TPD	SHUT-DOWN DATE	FUTURE STATUS
Sullivan Springs, AR	MI	16	NA	Unknown
Huntsville, AL	MI	50	1985	Shut down due to conveyer problems
Crossville, TN	MI	60	NA	Unknown
Genessee Township, MN	MI	100	NA	Unknown
Ansonia, CT	MB/OF	150	1984	Municipal waste is sent to Windham, CT
Braintree, MA	MB/OF	384	1983	Permanently shut down
Oyster Bay, NY	MB/OF	500	NA	Unknown
Oceanside, NY	MB/OF	750	NA	New incinerators ordered, negotiating redevelopment of project
Merrick, NY	MB/OF	NA	NA	Unknown
Monroe County, NY	RDF/D	2,000	1984	Currently preparing RFP for alternative use
Hempstead, NY	RDF/D	2,000	1981	Unknown

SOURCE: *Municipal Waste Combustion Study,* EPA.

[a] RDF/D—refuse-derived fuel fired in a dedicated boiler on site.
MI—modular combustor.
MB/OF—mass burn facility with overfeed stoker incinerator/boiler.

A NEW STAGE

The take-off of the trash-burning industry in the 1980s never successfully addressed all the concerns of the previous decade. But a series of events at the beginning of the decade, including upheavals in the energy arena, passage of RCRA, and changing governmental perspectives, produced a 180-degree change in outlook. The tentative discussions about possible trash-burning facilities suddenly turned into a flood of requests for proposals, new authorizations, and what turned out to be a breathtaking commitment of public capital. Wall Street also dramatically increased its financial stake in this new industry, as a number of bond houses sought out participation, hoping to make a major killing through a rapid expansion of the industry. "A resource recovery plant will go up in virtually every city," a Paine Webber vice president proclaimed in 1986 to the *Wall Street Journal*, an attitude shared by many investment houses and financial analysts.[18]

This change of perspective had several causes, though none was fundamentally due to major improvements in technology or reductions in cost. In fact, between 1983 and 1986, at the height of industry optimism, the average nationwide tipping fees for waste-to-energy facilities increased by more than 100 percent, *more than four* times the increase in landfill tipping fees. Even in California, New York, and New Jersey, where costs of landfills had increased more substantially, the tipping fees for new incineration and RDF plants remained unfavorable in comparison (see figure 5.2).[19]

Figure 5–2
CHANGE IN WASTE DISPOSAL TIPPING FEES

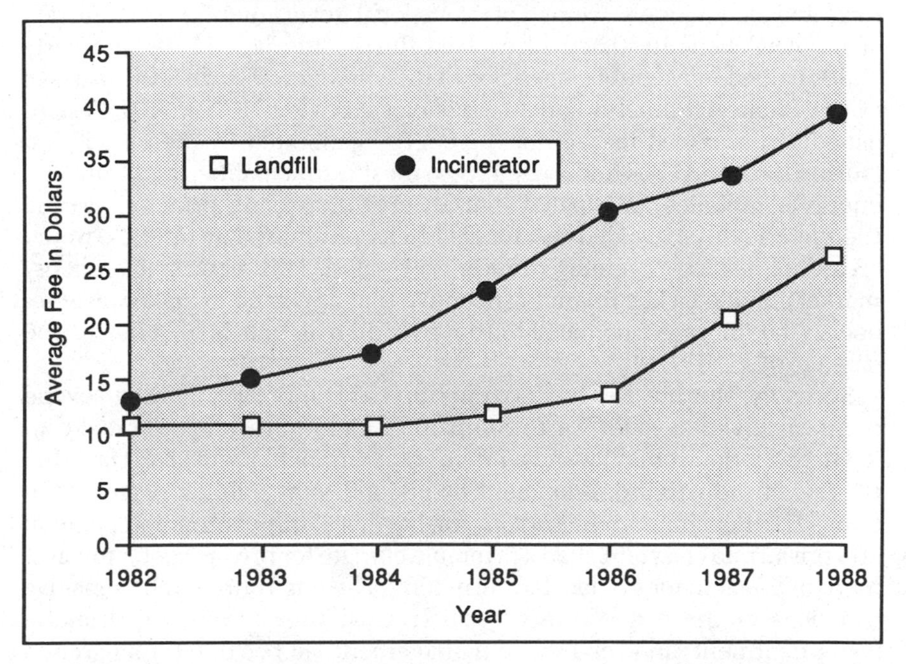

The trash-burning boom then was less a matter of comparative economics than of developing political and regulatory realities and an optimism over future cost comparisons between the two key disposal methods. The most significant changes, from the industry perspective, were induced by the implementation of RCRA. When signed into law in October

1976, RCRA's provisions and timetables for open dumps and landfills seemed relatively distant. But once implementation began on the state level and it sank in that a large number of landfill sites would be forced to close, the law's significance became apparent.[20]

The environmental conflicts around *all* landfills had also reached a crescendo, as public recognition and concern about groundwater contamination literally exploded in full force in the early 1980s. Problems of contaminated groundwater resulting from leaking landfills had already been recognized in EPA's reports to Congress during the early and mid-1970s and had been highlighted in some key studies during that period.[21] But active monitoring and widespread publicity about the extent of contamination only reached full force several years later, when wells in such places as California, New York, Wisconsin, Massachusetts, and Florida were discovered with trace amounts of volatile organic compounds.[22]

With the warnings from the 1970s turning into the frantic pursuit of new disposal options in the crisis-ridden 1980s, waste-to-energy, financial risks and all, suddenly seemed viable to local governments. Two new sets of financial benefits that had evolved during the late 1970s and early 1980s helped move public agencies in this direction. For one, the passage of the Public Utilities Regulatory Policies Act (PURPA) in 1978 established a guaranteed market for the energy generated by these facilities. During the late 1970s and early 1980s, most municipalities and waste-to-energy companies anticipated that an ever-increasing price of oil (and therefore a higher avoided cost for electricity purchased by utilities) would create increasing revenues from energy sales. This new, stable energy market provided a key financial incentive that appeared to make waste-to-energy far more economically attractive than it had been prior to the passage of PURPA.[23]

Secondly, the big Reagan tax cuts in 1981 and 1982 offered several financial advantages for waste companies, including an accelerated capital depreciation schedule and a range of tax financing provisions favoring the use of industrial development bonds and joint public-private operations. The Reagan administration, furthermore, in pushing deregulation and privatization, established a favorable climate for private sector involvement in the activities of local government. The environmentally sensitive functions of these governments, such as sewage treatment, drinking water treatment, and solid waste management, all became fertile ground for private sector intervention. The waste-to-energy boom, in fact, became part of a larger shift toward privatization among service- and engineering- and construction-based companies during the early 1980s.[24]

Once the boom began to unfold, the industry caution of the 1970s turned into the heady optimism of the 1980s, a mood that continued even when some of the key financial advantages became more problematic within a few years. The substantial decline in energy prices plus successful utility resistance to implementing PURPA deflated the hopes of significant revenues from energy sales. The Tax Reform Act of 1986, furthermore, undercut several key tax advantages. Nevertheless, through the mid- and late 1980s, even as these new constraints developed, the enthusiasm of both the industry and local governments failed to diminish. The overwhelming force of the decline in landfill space and the mounting wastes fueled the boom. "The surge to build resource recovery plants is being driven by the disposal crisis, not tax incentives," one analyst remarked of the changes in the tax code. Despite "mixed trends" such as the huge drop in the price of oil, "the intermediate and long-term outlook," industry consultants proclaimed, was "still positive."[25]

With orders for new plants increasing, and the financial exposure of both local governments and industry participants deepening through the 1980s, the troubled economics of incineration from the previous decade appeared to be a passing phase. With the new orders, the potential market for waste-to-energy seemed to explode. As a result, the more immediate considerations about costs and financial risk became less prominent for the industry. At the same time, the prevailing wisdom about long-term economic factors, far from discouraging a shift to trash burning, suggested that local governments move away from landfills and toward waste-to-energy, especially mass burn incineration, which was being touted as the cost-effective strategy for the future. Yet this presumption was not based on hard data. This was particularly striking given the difficult history of waste-to-energy plants in the 1970s, whose final costs had been "vastly underestimated, with revenues well below [initial] projections," as two industry analysts had put it.[26] The incineration boom of the 1980s had become less an economic imperative than a political calculation based on faith in a rosy economic future.

THE COSTS OF INCINERATION

Though much has been written in recent years about the waste-to-energy boom, surprisingly little analysis can be found by evaluating and comparing the costs of facilities, either mass burn incinerators or resource-derived fuel (RDF) facilities. The size of the facilities has varied significantly. Increasingly, however, communities have pursued, for purposes of economies of scale, larger plants with at least a 1,000-tons-per-day (tpd) capacity. While big cities such as Los Angeles, New York, and Kansas City have been able to pursue such undertakings on their own, smaller-sized communities have needed to combine efforts, often on a county-wide basis, to build the favored, large-scale units and meet the considerable financing requirements for construction and operation. No matter the size of the governmental entity, waste-to-energy plants, far more than landfills, are destined to become "one of the most complex and expensive public works projects a community may ever undertake."[27]

Total capital costs for waste-to-energy projects can vary considerably, from a 50-ton-per-day plant in Osceola, Arkansas, built for $1.2 million, to a 3,000-tpd facility in Hempstead, New York, with capital costs well over $350 million.[28] Besides size variation, there are significant differences in technologies, pollution control measures, and ownership structures, among other features, all of which impact cost. As a result, comparisons between facilities are difficult to make and can be misleading. For one, there are no effective comparisons between those plants built during the 1970s and the larger, often technically more complex facilities currently being built. Standardized measures have not been developed, and sorting through the maze of technological differences to establish a method of comparison has yet to occur.

There is sufficient information, however, to provide a narrower though crucial point of comparison: namely, by focusing on large plants with greater than 1,000-tpd capacity that utilize a mass burn system and water-wall furnace technology. These large-scale systems are favored by most industry proponents. By 1987, however, there were still only nine such plants on line and operational, all but one constructed within the previous four years. These nine plants included facilities in Massachusetts, Maryland, Oklahoma, New York, and Florida.

Despite the small numbers, these types of plants are projected to become the dominant industry form in the near or medium term. In 1987, at least 47 other mass burn plants, according to one EPA study, were in various stages of planning, with several scheduled to be completed before 1990. Other analysts forecast that upwards of 100 plants with a 1,000 or more tpd capacity could be built within the decade.[29] But it has been on the basis of the limited operational history of the existing plants that most of the economic claims are made about the future of the whole industry. In that light, a detailed analysis of the existing 1,000-plus tpd mass burn plants can provide insight into industry economics.

The few existing current studies on the economics of waste-to-energy have analyzed the tipping fees—the per ton charge for disposal—for each plant as the basis for comparison. These fees tend to be the most easily identifiable and accessible information. Most projects, however, have not included the same set of costs in calculating their specific tipping fee, which, in turn, seriously compromises any comparison based solely on these fees.[30]

Instead of relying exclusively on the tipping fees, a more comprehensive analysis of cost comparisons should also include capacity factors and construction costs. In the UCLA study comparing six of the nine operational incineration plants with greater than 1,000-tpd capacity, where such information was made available, factors evaluated included the total capital cost of each project (in 1986 dollars), divided by the *actual* throughput of the facility—the number of tons per day the plant actually processes. Per ton capital costs show considerable range, from $128,600 for the Baltimore Refuse Energy System Company (BRESCO) facility to $176,294 per ton for the proposed Los Angeles facility (LANCER, or Los Angeles City Energy Recovery Project), which has also been included in this comparison. The differences can only be partially attributable to economies of scale; other factors such as the cost of capital are reflected in the significant variations in per ton costs, even for plants of equivalent size, such as LANCER and the North Andover, Maryland, facility.[31]

The variation in construction costs between plants has also been analyzed for purposes of comparison. This was done by dividing the total construction cost by the design capacity of the plant. The range of variation between the six plants in this instance is much smaller, with the Tulsa facility at the lowest ($67,538) cost per ton and the Tampa, Florida, plant at the highest ($86,183) cost per ton. This smaller variation is largely due to the fact that construction costs tend to be dominated by equipment and materials, which vary less than labor costs. Labor costs, according to a *Waste Age* estimate, comprise only about 25 percent of total construction costs, making the large, mass burn incineration plant highly capital intensive.[32]

It is the tipping fee, in fact, which exhibits the greatest variation. The $55-per-ton fee charged at North Andover, Massachusetts, for example, is almost four times the $15-per-ton fee charged at the 2,000-tpd facility in Niagara Falls, New York. Tipping fees are ordinarily based on the debt service costs and operation and maintenance fees, minus the revenue obtained from energy sales and other sources. All these can vary from year to year.[33]

Part of the variation in costs reflects the variability of a number of factors, including the development costs and the financing and ownership structure of a facility. Of all the cost components, project development costs have been least available. A number of cities do not keep accurate data on expenditures undertaken prior to the start of construction, and they are seldom included in total project costs. Where projects involve a public agency/private vendor relationship, as in most proposed plants, the exclusion of development costs becomes a public subsidy to the project.

According to the California Waste Management Board, project development costs generally range from 2 to 5 percent of the total capital costs and will be expended over two to four years. Among such costs are fees for engineering, financial, and legal consultants; land acquisition costs; and informational expenses.[34] These costs may be capitalized into the project after the financing has been arranged, or written off by the local government entity. Such arrangements lower the listed project cost, thus making it appear more attractive in developing community support.

One of the key factors affecting the financing of a facility concerns its ownership and the other components of the relationship between the governmental agency (or agencies) and the private company selected to build and/or operate the facility. The ownership structure selected can implicitly if not explicitly influence the type of financing to be pursued. Most communities in turn base financial decisions on issues of cost and risk rather than on broader waste management objectives.

During the 1970s, the range and severity of technical and operational problems experienced by the new waste-to-energy plants caused most local governments to explore shifting the risk of potential failure to private owners. Local governments came to view private ownership as a "safer" option, though the transfer of risk ultimately remained an uncertain process. Thus, contracts involving private ownership of an incinerator plant assigned responsibility to the private vendor for assuming cost overruns and any unanticipated operation and maintenance problems. Certain areas, though, such as the cost of compliance with any new environmental regulations, were usually not included.[35]

The various tax provisions of the early 1980s also favored the private ownership model. One waste management guide, *The Solid Waste Handbook,* suggested that this variable could provide the margin between a profitable and unsuccessful waste-to-energy facility. "The majority of [these] projects," one of the essays in the handbook noted, "are not economically feasible when judged by the standard of comparable landfilling costs . . . without the use of private ownership financing. Private ownership financing can significantly reduce the financial cost of an energy resource facility by either passing on a major portion of the tax benefits through lower financing charges or enhancing the profitability of the builder, thus allowing lower Operations and Maintenance fees."[36]

The Reagan tax initiatives as a whole could amount to as much as 25 to 35 percent of total capital costs and were worth several times both the estimated residual value and the cost of financing the plant over the terms of the lease. These terms favored the private owner, and could also be attractive to the local government when the private vendor passed along some of the tax savings in the form of either increased equity contributions (thus lowering the total amount to be financed), or through lowered service fees. Thus, the private ownership option could reduce costs in the short run while shifting some degree of risk to the private owner.[37]

The private ownership model, however, had certain limitations, even prior to the passage of the new tax law of 1986, which reduced a number of these benefits. In several of the arrangements first put together in the early and mid-1980s, the private ownership structure represented some loss of control over the facility for the local government. Subsequent fears caused public agencies to stipulate the precise obligations of each party, thus narrowing the scope of private control. The length of time in pursuing and developing a facility has also been a factor in limiting the private ownership model. Most local governments are under significant time constraints when a decision has been made to build an incineration plant, and often greater time is required to work out the complex arrangements associated with private ownership.[38]

Perhaps most importantly, the private ownership model can lead to a situation in which a public agency may finance a facility that it will ultimately not own. The average life of incineration plants has been estimated at 30 to 40 years. Repayment of the bonds and the fulfillment of the initial contractual agreement often involve 20 years. The service agreement part of the contract would then have to be renegotiated, with the local governments likely to pay higher costs. Thus, private ownership, while producing some important short-term benefits, could have serious long-term disadvantages that could ultimately constitute a substantial form of public subsidy not apparent at the time a contract was signed.

A few of the local governments pursuing mass burn incineration have turned instead to a public ownership structure, in which a contract is signed with a private company to design, build, and often operate the plant under a full service contract. Of the six responding large incineration plants constructed prior to 1987, two, both Florida-based, are publicly owned. The private company involved in these instances has guaranteed to operate the facility at a specified performance level, while receiving a net service fee from the local government and a share of the energy revenues. In the Tampa, Florida, 1,000-tpd plant, the private company, Waste Management, receives 10 percent of the energy revenues and 90 percent of the recovered metals.[39]

Under a public ownership structure, two financing methods have emerged.[40] The first involves the issuance of general obligation (g.o.) bonds, which use the full faith and credit of the local government entity as security for the debt. Since the taxing power of the municipality or other local governmental entity is pledged, g.o. bonds generally have the highest rating. This involves the simplest method of financing and the lowest cost of issuance, but it involves the highest risk to the community and also requires voter approval. Given the risk factor and political calculus involved, general obligation bonds have been used sparingly, and only for smaller waste-to-energy projects.

Most communities have instead preferred a method of financing in which the bonds are backed not by the assets of the government entity, but by the income to be generated by the facility. Among both existing and proposed facilities opting for public ownership, this use of revenue bonds has been the most common method of financing.

With these plants, the guarantee of the revenues typically comes from the sale of energy and the tipping fee charged for disposal. With revenue bonds, there is a higher risk to bondholders and thus higher interest rates than with general obligation bonds. With many of the facilities, moreover, the simple pledge of project revenues may not be sufficient to assure investors that the debt service (the annual interest payments) will, in fact, be paid, a factor more important to financial analysts than whether the incinerator will actually work. Given the recent fluctuations in energy prices, these concerns have been intensified, as the revenues derived from the sale of electricity have generally dropped well below the expectations of the early 1980s. As with the Vicon situation, this has resulted in a big jump in tipping fees and/or necessitated pledges of further guarantees on the part of the local government.[41]

The Bay County, Florida, incinerator, for example, which processes 510 tons of waste per day, found that the big drop in energy prices forced a major funding shift, resulting in significant subsidies for the vendor, in this case Westinghouse. The project was originally designed to be privately owned, but the changes in the tax laws obliged the county "to start the project as if it were publicly owned," as one industry analyst put it, through the issuance of $60 million in revenue bonds. But when the price of electricity fell, the county had to scramble to find a way to supplement the plant's revenue. They did this by convincing county residents to vote for a half-cent sales tax to pay for running the plant, *while eliminating the tipping fees*. And since the plant, as with nearly all incinerators, was oversized to account for expected future growth of the waste stream, the county residents were now locked into a system that encouraged as much of the waste as possible to be sent to the Westinghouse plant, thus undercutting any potential recycling project. Meanwhile, once the tax situation had settled, Westinghouse sold the plant to Ford Motor Credit, which was still able to use the tax benefits from owning the plant. Ford purchased the plant for $15 million and assumed $35 million in obligations, with the rest of the revenue bond money stipulated for public works improvements around the plant and other landfill improvements.[42]

By far the most common method for financing large incineration plants has been the private ownership/vendor equity model. The private contractor/owner of the facility typically contributes between 20 and 35 percent of the project cost in equity, with the balance financed through the issuance of tax-exempt revenue bonds, usually industrial development bonds, or IDBs. In order to reduce the extent of the local government's financial exposure and liability, as well as to better coordinate the variety of governmental agencies involved, a Joint Powers Agency (JPA) is often created that is given the power to issue the bonds and to represent the interests of the local governments.

With this private owner/vendor model, the local government pays a negotiated service fee to the owner that covers the debt repayment as well as the costs of disposal. In return, the private owner guarantees to operate the facility at specified performance levels and to share the energy revenues through a negotiated formula.

The advantages for the private owner are considerable. They include full ownership of the facility after 20 years as well as significant tax breaks. In return for these advantages, the municipality or other governmental entity might negotiate to increase the size of the vendor's equity payment or to reduce its service contract payments. Consequently, in the short term, this method often leads to the lowest per ton charge of any of the various financing options. But it also remains the most complex method, given the greater number of negotiated payments and greater legal requirements needed to insure the tax exemption.

Selection of the private owner/vendor model often means that only a handful of firms are capable of bidding on such a project. Only the largest companies seeking to become vendors will have the financial capability to offer substantial equity payments or have a sufficient tax "appetite" involved in owning a large plant. This form of ownership has thus played a considerable role in increasing the market share of the biggest players in the industry.[43]

The variations in the form of ownership and financing methods ultimately have a significant impact upon the economics and feasibility of a project. A hypothetical case study by the California Waste Management Board calculated the economic effect of the various financing methods (g.o. bonds, revenue bonds, and IDBs). The study, abridged and reproduced here as table 5.3, shows substantial differences in the total savings to a community, termed net present value. This widely fluctuating outcome is the difference between what a community might currently pay for disposal, in this scenario $12 per ton at a landfill, versus the costs of a proposed incinerator. Though the California Waste Management Board concluded that the IDB method might be the most cost-effective when all factors are considered, it noted that each of the numbers can vary considerably once a specific project and a specific community's circumstances are taken into account.[44]

Using the case study, by calculating what appears at first to be only a minor change in any one of the specific negotiated agreements in the private ownership/vendor equity model, a significant change of the total

Table 5–3
COMPARISON OF FINANCING METHODS

Comparison Factors	*Public Ownership*		*Private Ownership*
	GENERAL OBLIGATION BOND	REVENUE BOND	EQUITY MODEL IDB
Bonds required ($000's)	$55,048	$66,746	$47,093
Interest rate (%)	10.0%	11.0%	11.0%
Debt coverage factor	1.00	1.50	1.25
Debt service reserve	No	Yes	Yes
Net present value ($000's)	$21,727	($14,935)	$13,679
Risk to community	High	Medium	Low
Complexity of financing	Low	Medium	High
Breakeven year vs. landfill	5	13	7

SOURCE: *Waste to Energy,* California Waste Management Board, p. 108.

project costs would result. For example, by decreasing the vendor's equity contribution by just 5 percent, the net present value would drop by 25 percent. Determining the specific equity contribution involves a complex process of negotiation between the vendor and the local government and there is no set standard. The amount is essentially up for grabs.

Another key component of the agreement, the debt service coverage factor (the "safety margin" banks require to assure repayment), can also significantly impact project economics. As an example, increasing the coverage factor from 1.25 to 1.35 (not an unusual increase) will bring about a 31 percent decrease in the value of the total savings from the project. Furthermore, just a 1 percent increase in the bond interest rate (where such changes occur frequently) can also produce a 35 percent decrease in a project's net present value. Even under the most favorable ownership and financing options, the ability to predict project costs and estimated values for incinerators remains highly limited.[45]

REDUCED SUBSIDIES AND HIDDEN COSTS

By 1986, the large mass burn incinerator, particularly one based on private ownership and financed by IDBs, had become a favorite in corporate boardrooms, on Wall Street, and at City Hall. The force of the boom, however, was already tempered by problems, from the Tax Reform Act to broad public opposition that brought to the fore questions regarding public subsidies, hidden costs, and extended risks. Though the boom continued, the economic picture had become far more uncertain.

The 1986 tax law threatened to unravel much of the structure of incineration economics and to force a reevaluation of project feasibility. Though the debates over the tax legislation focused little on waste-to-energy (or, for that matter, on various other privatization initiatives), several of its provisions have begun to have far-reaching impact.[46]

Among the changes was a state volume cap on "private activity" bonds, such as IDBs. Prior to the 1986 law, all privately owned facilities relied upon these tax-exempt industrial development bonds (IDBs), which allowed a municipality to finance any private project that fulfilled a "public purpose." Prior to 1984, in fact, there had been no limit to the total amount of debt a state or a local government could issue with these bonds. The interest paid and the loss of tax revenues were, in effect, public subsidies at both the state and federal levels.

An "initial volume" cap was first imposed by the Deficit Reduction Act in 1984, limiting states to a total of $150 per capita, or $200 million, whichever was greater. The 1986 tax law reduced that limit to $75 per capita in 1987, and $50 per capita beginning in 1988. The cap on a state's allocation of these funds meant that incineration facilities would compete with other public works and government-initiated activities such as sewage treatment, mass transit, student loans, and qualified mortgage bonds. Only a municipally owned facility (as opposed to the private ownership/vendor equity model) would be exempt from such a cap.[47]

The new tax law also created changes in the regulations governing the uses of the IDBs themselves. Private activity bonds were structured so that they could only be used to finance the "valueless" part of a project. For example, electrical generating equipment, which might constitute 10 to 15 percent of total project costs, was technically excluded from the qualifying costs applicable to this method of financing. Historically, local governments were able to circumvent this provision through what was known as the "bad money rule," which stipulated that 90 percent of the bond had to be used for qualifying costs, but allowed 10 percent to be spent on nonqualifying costs such as the electrical equipment. But under the new tax law, the bad money rule share was reduced to only 5 percent of the total bond proceeds. In addition, no more than 2 percent of the bond proceeds could be used to pay for the costs of issuing the bond. Since underwriter fees and bond counsel costs generally exceed this 2 percent limit, the difference had to be made up through taxable sources. That 2 percent, furthermore, counts toward the 5 percent total allowed by the bad money rule.

The new law also took away many of the specific tax advantages that had made such projects (and methods of financing) so attractive to private firms. For example, the law eliminated the investment tax credit that previously subsidized 10 percent of construction costs. For a typical 1,500-

ton-per-day facility, this could constitute as much as a $13 million to $15 million loss.

The elimination of the accelerated depreciation schedule was another key factor. Under the old tax laws modified by the Reagan administration, capital costs could be depreciated within five years. The new law extended depreciation to ten years and required the "straight line" method, thus eliminating the larger write-offs companies had been able to take in the early years of a project. While the specific financial impact of this change depends on the tax position of the individual firm, these losses, which could be considerable, reduce the savings that firms have typically passed on to their public partner in whole or in part under private ownership structures.

Finally, the minimum corporate tax rate was reduced under the 1986 law from 46 percent to 34 percent. While broadly supported by corporate interests, this provision also in effect further reduced the benefits from any remaining tax deductions. The combination of these changes, then, marked a significant reduction in the financial advantages offered by private ownership, the method favored by the larger companies who had emerged as the dominant players in the rapidly expanding incineration industry.

The new tax law, however, did provide for a "grandfathered" or transition status for those projects commencing on or before August 15, 1986. Under the new law, if $200,000 had been spent by that date, then the remaining costs of the project, including any further development costs, would qualify under the old provisions. Thus, a significant number of projects were initiated in the latter part of 1985 and early 1986 while the tax law with its grandfather clause was still being debated.

The 1986 tax law has begun to change both the structure of financing and the costs of incineration. When a private ownership structure is still pursued, it is likely that new forms of taxable debt need to be undertaken. Both the loss of the tax breaks and the need to develop more expensive forms of financing significantly push up costs, reducing much of the estimated $10- to $15-per-ton savings associated with IDBs.[48] And, as private ownership loses this financial advantage, there is likely to be greater interest in the more costly municipally owned project. "Changes in the tax code," argued one analyst, "push the resource recovery industry away from privatization and towards public financing deals."[49] The director of policy analysis for the U.S. Conference of Mayors in fact argued that the 1986 law would create a demarcation point between older projects already built or covered by the transition rule, which were largely privately owned, and new facilities planned and built after 1986, which would be predominantly publicly owned.[50]

The move toward public ownership was also creating the basis for more competition involving a greater number of smaller firms in contrast to the relatively few, large firms that had been bidding for the private ownership arrangements. At the same time, the tax law, with its reduced tax advantages and increased costs, was also contributing to a general slowdown of new plant orders, which in turn was also producing a shake-out of the market, with some of the smaller and overextended firms like Vicon facing bankruptcy and/or becoming prime takeover targets. Vicon itself, after declaring bankruptcy, became a takeover target of one of the industry giants, Ogden Martin. Even strong incineration advocates, such as Jonathan Kizer of the Institute for Resource Recovery, the main industry trade organization, admitted that the tax law, among other factors, had slowed the industry. "There is no denying," Kizer declared in 1988, "that orders peaked several years ago." This change had been mainly driven by "the rush to finance from the over-exaggerated situation around the IDBs," as Kizer put it. The subsequent slowdown had clearly become, for Kizer and others in the industry, a "cause for concern."[51]

Perhaps more than any other factor influencing the economics of large incineration projects is their enormous sensitivity to the price of energy and the policies affecting the sale of their electricity. Since energy revenues are used to offset the costs and service fees that a local government pays, any decline in the sale price of electricity generated at a waste-to-energy plant can significantly boost the per ton service fee (the sensitivity of tipping fees to energy prices is shown in figure 5.4).

Theoretically, an incinerator can sell the energy it produces to a utility on a "take-or-pay" basis based upon the avoided cost formula as designated in PURPA. But since PURPA was established, and particularly since there has been a drop in the price of oil, nonutility energy generators such as incinerators have been forced to negotiate these contracts on an individual basis and contract terms have varied widely. The "avoided cost" concept has been subject to varying interpretations, given the development of excess capacity by utilities and other factors that have made the purchase of additional energy less attractive to utilities. "Pricing at the 'avoided cost,' " one industry analyst argued, "may not always be sufficient to sustain a viable solid waste facility."[52]

The fall in energy prices significantly affected waste-to-energy facilities, including the six 1,000-plus tpd mass burn incineration facilities referred to earlier. Within less than six years, electricity rates dropped several cents per kilowatt hour, to as low as 1 and 2 cents/kwh in certain communities. At the same time, tipping fees went up, due in part to the change in price of electricity, with energy price fluctuations most dramatic in the early and mid-1980s. In all six plants surveyed, the fall in energy

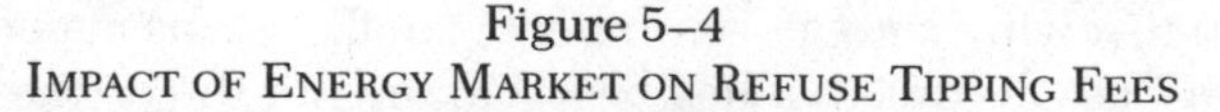

Figure 5–4
IMPACT OF ENERGY MARKET ON REFUSE TIPPING FEES

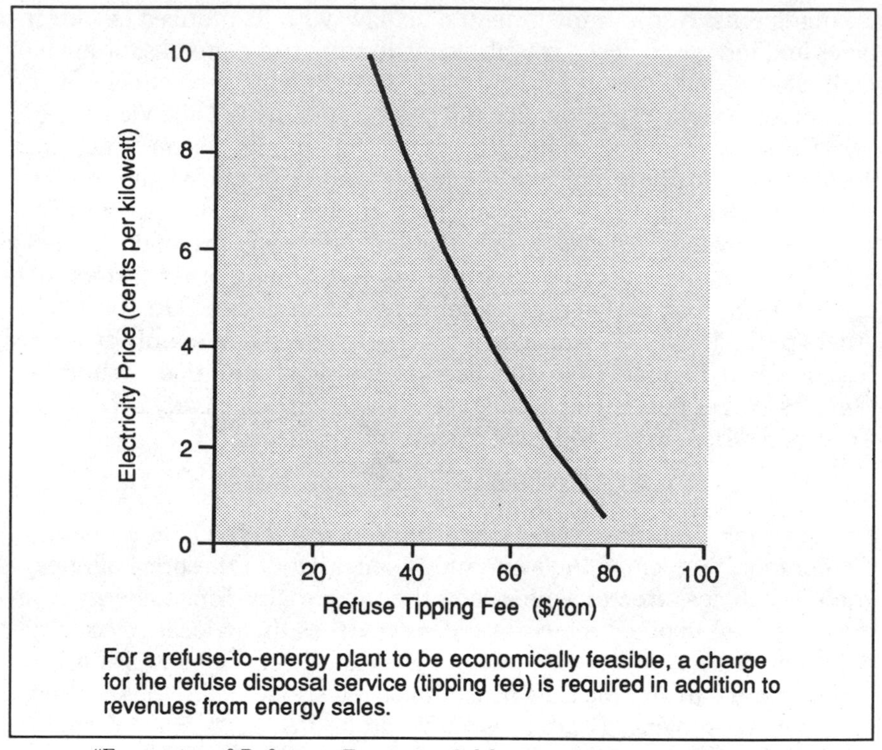

SOURCE: "Economics of Refuse to Energy in California," California Energy Commission, 1988.

prices led to significantly reduced energy revenues and consequently substantial increases in the overall cost per ton. At the Westchester County, New York, facility, for example, the county had guaranteed the owner/vendor a minimum price of 6 cents per kwh, but, as a result of the drop in prices, ended up paying an additional $30 million annually to the company, effectively doubling their net service fee.[53]

Other problems with the energy-related purchase agreements have developed from the uncertain status of PURPA-derived avoided cost contracts. While many of the waste-to-energy projects have relied upon PURPA's "take-or-pay" contracts with electrical utilities for an assured—and profitable—market for their energy, the utility industry has been able to successfully undermine the procedure through court actions, appeals to state utility commissions, and a hard bargaining posture with independent energy generators. This has further squeezed incinerator projects that depend on the ability to sell the generated electricity to be economically viable. By 1988, the dramatic changes in the energy market had not only undermined PURPA, but led the Reagan administration's Federal Power Commission to attempt to "deregulate" the business, in effect, eliminating PURPA's guaranteed market.[54]

This situation has been further compounded by the problem of excess capacity, which has caused some state utility commissions to look unfavorably on the generation of new energy. In California, for example, the state's Energy Commission, which has jurisdiction over waste-to-energy facilities generating 50 megawatts or more, has consistently maintained since the mid-1980s that "prospects for development [of incinerators] do not look favorable at this time," citing the situation around excess capacity as one crucial factor. Not surprisingly, this position is strongly backed by the utilities.[55]

A consequence of this price drop and energy glut (a kind of reverse energy crisis for incinerator projects) has been a search for new forms of public subsidy. In several instances, the energy produced has been offered to a publicly-owned utility (often required to qualify for tax-exempt financing) at a higher rate than it would pay on the open market. In effect, consumer electric bills increase to subsidize the local government's revenues from the incinerator. The private vendor benefits from this indirect subsidy through the percentage of the energy revenues that it collects in the negotiated contract. These subsidies, however, are not factored into the full estimated costs. This creates a hidden cost that one study estimated increased the overall project cost by as much as 20 percent. Once again, these factors have been seldom included in public debates.[56]

The issue of hidden or disguised costs is critical in evaluating the economics of a particular facility. All projects, for example, require some site development. Simple clearing and preparation of a site generally constitutes around 3 percent of the total construction costs and is ordinarily included as part of the fixed construction contract. But a variety of other costs relate to site development. Water and sewer lines, for example, almost always need to be expanded to accommodate the greatly increased flows to and from incinerators. The surrounding transportation arteries may need to be expanded to handle truck traffic. Most of these costs, however, have simply been absorbed by the local government and not calculated into project costs, thus becoming another form of hidden public subsidy.[57]

By the late 1980s, it was becoming clear that the shaky economic feasibility of incineration had been made possible only through public subsidies, many of them hidden from the public. These have included tax advantages, tax-exempt bond financing, electricity sales, and various costs absorbed by government and not incorporated into project costs. Many of these subsidies are being reduced or eliminated altogether while others are undergoing greater public scrutiny. The volatile nature of these subsidies, moreover, is undermining incineration's claim of cost-effectiveness, a claim already eroded by the number of costs not recognized at the time project contracts were signed. The economics of incineration has become, in the wake of these changing costs and hidden subsidies, an economics of uncertainty.

THE QUESTION OF RISK

Perhaps the most important area of uncertainty in evaluating the viability of incineration is the question of risk. "Disposal of wastes is a community responsibility," one waste management analyst wrote, "and the community is ultimately at risk should a waste-to-energy project fail to meet its economic and/or technical expectations."[58] These concerns led a number of local governments in the 1970s and early 1980s to try to shift the burden of risk to the private owner/vendor. Yet, as some local governments discovered, there were no "iron-clad contracts in a field as new—and still in the process of evolution—as resource recovery."[59]

This situation was forcefully demonstrated in the case of the Dade County, Florida, 1,800-tpd incineration project. Dade County originally negotiated a contract with a private owner/vendor, but the two parties soon came to a disagreement over the contract terms. When the private company brought suit, most of the significant clauses in the contract were changed in arbitration, with the tipping fee increased 2,000 percent and the profit-sharing arrangements fundamentally altered in favor of the private company.[60]

Both public and private ownership entail a wide variety of risks. Engineering and construction risks, for example, are either absorbed by the local government or passed along to the vendor through a construction contract with a fixed price and minimum acceptance standards. Although these kinds of contracts generally cover cost overruns during construction and guarantee that the plant perform to design specifications, unanticipated large-scale breakdowns can present a different set of problems.

In one instance, a community college in Lassen County, California, contracted for a small (96 tpd) mass burn incinerator to handle the college's solid waste as well as serve as a training facility for students. The plant, however, closed down because of design failures less than six months after it came on line. Since the college had issued bonds backed by the community college district, both the college and the district were forced into bankruptcy, sending the case into the courts.[61]

There are also operational risks, including the failure of a facility to perform up to design specifications, environmental standards, and energy generation levels. These also tend to be covered in a private ownership contract. Responsibility for system performance levels is murkier in the case of public ownership. In both cases, however, while the costs of equipment problems occurring during the life of the contract are usually picked up by the vendor, they are still likely to be passed through to the local government through operations and maintenance fees. Corrosion problems with many of the existing plants suggest in fact that this could be a significant cost factor affecting any project over time.[62]

There are financial risks, furthermore, regarding key elements of the revenue stream for waste-to-energy facilities, such as the sale of energy, tipping fees, and the sale of recovered materials. While energy production has generally been a private or vendor responsibility, the risk surrounding the price of energy has been assumed by the public. One key industry executive, Signal president John Sullivan, warned, in 1987, of several potential risks for vendors, including the failure to meet a perceived target for energy sales that had been assumed prior to construction of a plant and used by all parties to structure a plant's financing. According to the Signal president, just a 10 percent drop in that targeted figure could result in a revenue decline of as much as $1.5 million each year. Sullivan predicted that "fully half of the plants that are now under construction" would miss their target, an assessment he derived from Signal's own experiences. As a result, Sullivan concluded, "sound vendor pricing may start to disappear in the near future."[63]

For incinerators, moreover, effectiveness has depended upon their ability to operate continuously, which means that they rely upon a steady supply of waste, generally provided on a "put-or-pay" basis. If the local government for any reason (strikes, successful recycling or waste reduction programs, unanticipated shortfalls resulting, for example, from economic declines) cannot provide the required wastes, it nevertheless often remains responsible for the continued payment of the service fees to the operator. For example, New Jersey's first county-run incinerator incurred a $1.5 million deficit in its first six months of operation, in part because of New Jersey's new mandatory recycling law that reduced the amount of garbage available for burning. At the same time, a vendor could also be at risk if the plant becomes incapable, on the basis of any operational failure, to process the waste contracted for.[64]

Some of the most crucial yet rarely stipulated areas of risk for these projects are the myriad of environmental issues. Most contracts require the vendor to construct a plant that meets the environmental regulations that apply at the time the contracts are signed. But any costs needed to bring a facility into compliance with more stringent regulations subsequently adopted, such as pending air emission standards or ash disposal requirements, will invariably be borne by the public.

Disposal of the ash has clearly become the most important of these problems. This unresolved issue poses one of the greatest financial risks for a community. Estimates for disposal of ash, should it need to be disposed of as a hazardous waste, range from about $80 to $200 a ton, and, given the small number of and greater hauling distances for Class 1 hazardous waste landfills, this figure could easily increase in the near term. Most incineration projects have assumed that the ash can be disposed of legally in a sanitary landfill, where disposal costs for ash, as of 1987, averaged $12.79 per ton. The increased costs of disposal as hazardous waste, in fact, could be greater than the *total costs of construction* for many existing facilities. And with the debate over ash still far from resolved, and the strong preliminary evidence indicating that at the very least the fly ash and, in some cases, both the combined and bottom ash fail to meet key existing standards, the financial risk to the community posed by future disposal requirements hangs heavy over *all* incineration projects.[65]

The areas of greatest uncertainty regarding financial risk are essentially those that must be faced by the local community. The problems of the waste-to-energy plants of the 1970s—poor performance, financing concerns, inadequate revenue stream—have been magnified and redrawn as the risks of the 1980s. "One of the things I fear is a repetition of the events we experienced in this industry in the late '60s and early '70s," Alfred DelBello, then still head of Signal's waste-to-energy subsidiary, said in a 1986 trade publication interview. "A lot of manufacturers had very dramatic ideas. On paper they looked like they would work but just about every one of them went out of business. Bankers were walking away from projects and that created a very bad taste in the mouths of a lot of public officials."[66]

Despite the high premium placed on the sophistication of the current technology, the track record of the most recent facilities also leaves much to be desired. Problems that have emerged during this period, appear, in fact, to be endemic to the technology, such as corrosion of the boiler tubes, in part from burning certain plastics, "slagging" or clogging of the grate from burning glass, and superheating from the unpredictable, inconsistent nature of the waste stream. Corrosion problems, for example, forced a Hartford, Connecticut, incinerator to shut down *two days after it opened*, with preliminary repair estimates of $1 million.[67] Moreover, not one of the large incineration plants in the United States has gone through a single 20-year cycle, making it even more difficult to judge the long-term risks. Waste-to-energy, once championed by both the industry and many local governments as the "least risk" alternative to landfills, is offering, instead, a risk-filled and controversial future.[68]

THE COST OF PUBLIC OPPOSITION

Incineration, particularly in certain areas of the country such as California and New England, has become a system in trouble. With each new proposed facility, yet another new community group spontaneously forms to challenge it, then often links up with a larger movement opposing both incinerators and landfills. "The fight against plans to build garbage incinerators," one environmental commentator noted, "is creating a new generation of community activists and coalitions, much as the anti-nuclear movement did in the '70s."[69] By 1986, community groups opposed to mass burn incineration had organized their first national conference, and the successful event signaled the extent to which these activist networks were expanding. Some key antilandfill groups, moreover, such as the Citizen's Clearinghouse for Hazardous Wastes and Work on Waste (formerly the National Campaign Against Mass Burn Incineration), not only became effective critics of existing facilities and the developing plans of the incineration industry, but also became advocates for a range of solid waste strategies, including waste reduction and recycling.[70]

These networks of community activists have also become an economic factor through their increasing ability to either stop waste-to-energy plants or, at the least, slow their construction schedules. Throughout the country, in places like California, Massachusetts, New York, Connecticut, Pennsylvania, and Vermont, the rapid-fire arrangements of new authorizations and contracts for the big incinerator plants have been almost as quickly brought to a halt in the wake of the fierce community protests.

The costs of public opposition are real and varied. Community groups have developed the capability and resources to pursue legal intervention and to force revised procedures to offset public concerns. The analogy with nuclear power is again striking, as a much anticipated and favored technology has become subject to criticism and strong public opposition. Ultimately, with nuclear power, the costs of public opposition, combined with the enormous known—and unanticipated—environmental and operational costs, undermined whatever competitive edge the heavily subsidized nuclear industry had ever enjoyed. Similarly, with incineration, the costs of public opposition have added to the uncertainties, potentially placing it, too, beyond a competitive edge.

Public opposition, of course, is more than just an economic factor. Advocates of incineration have long been concerned that a decision to burn, despite the high technology involved, could be unpopular, given the health and environmental concerns. The battle that has emerged, sharper in focus and involving far more players than the conflicts over trash burning from earlier periods, has ultimately developed into the struggle over the future direction of the waste issue itself.

NOTES

1. *Rutland Herald,* August 23, 1988; August 24, 1988; also *Waste Not,* the Weekly Reporter for Rational Resource Management, no. 19, August 30, 1988, Canton, New York; "Garbage Incineration—Beating the Ash Attack," Theresa Freeman, *Everyone's Backyard,* Citizen's Clearinghouse for Hazardous Wastes, vol. 6, no. 4, Winter 1988, Arlington, Virginia.
2. "Where Will We Put All That Garbage?" Faye Rice, *Fortune,* April 11, 1988; *Rutland Herald,* August 24, 1988.
3. "A Bankruptcy Closes Sites for Waste in Vermont," Sally Johnson, *New York Times,* August 30, 1988; *Rutland Herald,* August 24, 1988.
4. Interview with Theresa Freeman, 1988.
5. *Waste Not,* no. 23, September 27, 1988; no. 29, November 8, 1988.
6. "Resource Recovery: Fact or Fiction?" Eugene L. Pollock, National Solid Waste Management Association *Reports,* July 1979; also interview with Alfred DelBello, 1988; "The Philadelphia Story: Recycling in an Urban Environment," Alfred Dezz, Philadelphia Recycling Office, *Solid Waste and Power,* vol. 3, no. 1, February 1989.
7. *Resource Recovery and Waste Reduction,* 4th Report to Congress, SW-600, EPA, 1977.
8. *Cities and the Nation's Disposal Crisis,* National League of Cities and U.S. Conference of Mayors, Washington, D.C., March 1973.
9. "Garbage," Richard Asinof, *In These Times,* February 11–17, 1987; "The 1985 Tipping Fee Survey," Charles A. Johnson and C. L. Pettit, *Waste Age,* March 1986.
10. "The NSWMA 1986 Tip Fee Survey," Charles Johnson and C. L. Pettit, *Waste Age,* March 1987; "Garbage Is One Thing, But Garbage from New York? Forget It," J. C. Barden, *New York Times,* February 12, 1989; "Cost Comparison of LANCER and Landfilling," InterCity Memo 0610-00019, from Keith Comrie, City Administrative Officer to the Finance and Revenue Committee, Los Angeles City Council, January 26, 1987; "Vermont Waste Law Sends Rates Flying," *New York Times,* August 9, 1987. Rates in Vermont increased 300 percent.
11. *The Resource Recovery Industry: A Survey of the Industry and Its Capacity,* SW-501c, Environmental Protection Agency, 1976.
12. Surveys cited in *The Resource Recovery Industry,* EPA, 1976, ibid.
13. *Potential for Resource Recovery in the United States: A Cost/Benefit Analysis of Resource Recovery in the Major Metropolitan Areas,* W. E. Franklin, Franklin Associates, Prairie Village, Kansas, March 1975.
14. *The Resource Recovery Industry,* EPA, 1976.
15. *Government and the Nation's Resources,* Report of the National Commission on Supplies and Shortages, December 1976, Washington, D.C., pp. 158–60.
16. "Resource Recovery from Municipal Waste," David B. Sussman and Stephen J. Levy, SW-789, EPA, 1979; "Paving the Way," Gary Burch, *Solid Waste and Power,* February 1987.
17. *The Resource Recovery Industry,* EPA, 1976.
18. "Burning Trash Is Becoming Big Business," *Wall Street Journal,* October 13, 1986; *Waste-to-Energy Facilities,* Alan Beals, National League of Cities, June 1986, p. 1.

19. "The 1986 Tip Fee Survey," Charles Johnson and C. L. Pettit, *Waste Age,* March 1987; "Refuse-to-Energy Development in California," Ann Peterson, *California County,* May 1988, pp. 32–34.
20. *EPA Activities Under the Resource Conservation and Recovery Act of 1976,* Fiscal Year 1978, EPA 530/SW-775, March 1979.
21. *Resource Recovery and Source Reduction: Second Report to Congress,* Environmental Protection Agency, EPA 530/SW-122, March 1974. Also Third Report to Congress, EPA 530/SW-161, 1975; Fourth Report to Congress, EPA 530/SW-600, 1977.
22. *Federal and State Efforts to Protect Groundwater,* U.S. General Accounting Office, GAO/RCED 84-80, February 21, 1984; *Contamination of Groundwater by Toxic Chemicals,* Council on Environmental Quality, Washington, D.C., 1981.
23. Interview with John Phillips, Ogden Vice President, 1987.
24. "Privatization: Trend of the Future," *Journal of the American Water Works Association,* February 1986; "New Reagan Privatization Concept: Industry Input Needed," Frank Sellers, *The Privatization Review,* vol. 2, no. 3, Summer 1986.
25. "Tax Code Will Not Hurt Market," Susan Darcy, *World Wastes,* December 1986; *A Status Report on Resource Recovery as of December 31, 1987,* Kidder, Peabody & Co., New York, 1988.
26. "Waste Disposal/Resource Recovery Plant Costs," W. D. Robinson and Sergio E. Martinez, in *The Solid Waste Handbook,* ed. W. D. Robinson, 1986, p. 107.
27. "Testimony on Bill 1005" (re: South Philadelphia incinerator), Nancy Hutter, Assistant Staff Attorney, City of Philadelphia, January 28, 1987.
28. "Resource Recovery Activities: Report on Semi-Annual Survey," *City Currents,* U.S. Conference of Mayors, October 1986.
29. *Municipal Solid Waste Combustion Study,* EPA 530/SW-87/021, 1987; "The 1987 Waste Age Survey," *Waste Age,* November 1987.
30. "The 1985 Tipping Fee Survey," Charles Johnson and C. L. Pettit, *Waste Age,* March 1986.
31. Information for the comparative study of the six plants is based on personal communications and materials from the plant operators, gathered by Terry Bills, member of the UCLA Urban Planning Comprehensive Project group.
32. "News Breaks," *Waste Age,* January 1987, p. 6.
33. "Waste-to-Energy Plant Questionnaire," North Andover RESCO Plant, Department of Resource Economics, University of Rhode Island; "The 1986 Tipping Fee Survey," Charles Johnson and C. L. Pettit, *Waste Age,* March 1987.
34. *Waste-to-Energy,* Technical Information Series, California Waste Management Board, Sacramento, June 1983, chap. 2, "Financing Resource Recovery Facilities in California: A Primer," Touche Ross & Co. and Brown, Vence, and Associates, p. 29.
35. "The New Tax Law: How It Will Affect Financing and Ownership of W-T-E Projects," Charles Samuels, *Solid Waste and Power,* February 1987.
36. "Economics and Financing of Resource Recovery Projects," Warren Gregory et al., in *The Solid Waste Handbook,* ed. W. D. Robinson, 1986.
37. "Who Should Own the Plant?" John Schopfer and George Graham, Jr., *Waste Age,* January 1987; "Economics and Financing of Resource Recovery Projects," Warren Gregory et al.; *Waste-to-Energy,* California Waste Management Board, chap. 2.
38. "Waste-to-Energy: Cash from Trash," Barry Mannis and Frank Prezelski, *Equity Research: Industry Comment,* a Publication of Shearson Lehman Brothers, New York, June 3, 1986.
39. "$83,375,000 Pinellas County, Florida, Solid Waste and Electric Revenue Bonds, Series 1983" (Resource Recovery System), *Official Statement,* December 20, 1983; "Pinellas County, Florida: Refuse-to-Energy Facility," Signal Environmental Systems, n.d.

40. On the various financing strategies, see "Financing Solid Waste-to-Energy Facilities," Wesley Hough and David C. Sturtevant, *Governmental Finance,* June 1984; City of Indianapolis, "Final Report on Resource Recovery Financing Alternatives," Wesley Hough and David Sturtevant, Government Finance Research Center, May 25, 1984; "Solid Waste-to-Energy Financing," John E. Petersen and David C. Sturtevant, Waste-to-Energy Conference, Seattle, Washington, August 23, 1984; "Financing Resource Recovery Projects," Maria Monet, *World Wastes,* vol. 28, no. 6, June 1985.
41. *New York Times,* August 30, 1988; *Economics of Refuse-to-Energy in California,* California Energy Commission, Sacramento, 1988.
42. "WTE Developers: What Are They? Who Are They?" Mike Kilgore, *Solid Waste and Power,* August 1988.
43. "Final Report on Resource Recovery Financing Alternatives," Wesley Hough and David Sturtevant, May 1984.
44. *Waste-to-Energy,* California Waste Management Board, chap. 2, pp. 71–108.
45. *Waste-to-Energy,* pp. 109–119, ibid.
46. For background on the 1986 Tax Reform Act, see "Tax Reform's Impact," Wendy Franklin and Stephen E. Howard, *Waste Age,* November 1986; "Dramatic Changes in Financing Waste-to-Energy," Philip M. Chen, *World Wastes,* January 1987; "The New Tax Law," Charles Samuels, *Solid Waste and Power,* February 1987; "Tax Code Will Not Hurt Market," Susan Darcy, *World Wastes,* December 1986.
47. "Dramatic Changes in Financing Waste-to-Energy," Philip M. Chen, *World Wastes,* January 1987.
48. "Tax Reform's Impact," Wendy Howard and Stephen E. Howard, *Waste Age,* November 1986; "Interview with Alfred DelBello," *Forbes,* July 1, 1985, p. 59.
49. "Tax Code Will Not Hurt Market," Susan Darcy; "New Trends for Resource Recovery Financing," Steven J. Allard, *The Privatization Report,* July 1988; "The New Tax Law," Charles Samuels.
50. Comments of Dave Gattor, Director of Policy Analysis, U.S. Conference of Mayors, cited in "Tax Code Will Not Hurt Market," Susan Darcy.
51. Interview with Jonathan Kizer, 1988.
52. "Resource Recovery Ratings Approach," *Journal of Resource Management and Technology,* vol. 13, no. 3, November 1984; *Waste-to-Energy,* California Waste Management Board, 1983; *Economics of Refuse-to-Energy in California,* California Energy Commission, 1988.
53. Letter from Calvin E. Weber, Deputy Commissioner, Solid Waste Management, Westchester County, to Terry Bills, March 16, 1987; also "$157,390,000 County of Westchester Industrial Development Agency," Resource Recovery Revenue Bonds, Westchester RESCO Company project, series A, October 1, 1982.
54. "Top of the Heap," *Solid Waste and Power,* April 1988, vol. 2, no. 2; and June 1988, vol. 2, no. 3.
55. "Market Prospects for Refuse-to-Energy Development in California," Paul R. Peterson et al., presented at "Waste Tech '87," October 26–27, 1987, San Francisco; *Economics of Refuse-to-Energy in California,* California Energy Commission, Sacramento, 1988.
56. *The Rush to Burn: America's Garbage Gamble, Newsday.*
57. "How to Pick the Right Location," Kevin J. Murray, *Solid Waste and Power,* February 1987; "Waste Disposal/Resource Recovery Plant," W. D. Robinson and Sergio Martinez, in *The Solid Waste Handbook,* ed. W. D. Robinson, 1986.
58. "Assessing Waste-to-Energy Project Risks," R. S. Madenburg, in *The Solid Waste Handbook,* 1986, p. 133.
59. "So You Think You Have an Iron-Clad Contract?" Deanne Siemer, *Waste Age,* November 1984.

60. "So You Think You Have an Iron-Clad Contract?," Deanne Siemer; *Titans of Trash: Big Profits, Big Problems,* Special Report, December 6–10, 1987, *Fort Lauderdale News/Sun Sentinel.*
61. "Waste-to-Energy: Anatomy of a Failure," Thomas Reilly and Linda Morse, *World Wastes,* January 1987.
62. *The Rush to Burn: America's Garbage Gamble, Newsday.*
63. Cited in "Industry Leader Predicts Industry Shake-Out," *Waste Age,* May 1987.
64. "Lacking Garbage, A Jersey Incinerator Is Losing Money," Robert Hanley, *New York Times,* January 23, 1989; *Waste-to-Energy,* chap. 2, California Waste Management Board, 1983.
65. "The Confusion and Questions About Ash," Edward Repa, *Waste Age,* September 1987.
66. "Exclusive Interview—Alfred DelBello," William M. Wolpin, *World Wastes,* February 1986.
67. "Top of the Heap," *Solid Waste and Power,* December 1988, p. 9; see also *Waste Not,* no. 30, November 15, 1988.
68. Part of the reason why such limited analysis has taken place is that the only data available on facility problems are self-reported; for example, no independent audits have been conducted of the six existing 1,000-plus tpd mass burn plants; see also "Resource Recovery Ratings Approach," *Journal of Resource Management and Technology,* vol. 13, no. 3, November 1984.
69. See, "Garbage," Richard Asinof, *In These Times,* February 11–17, 1987.
70. "Burning Unity: Grass Roots Organizations Merge Efforts to Fight Trash-to-Energy Incineration Plants in Southland," Kevin Roderick, *Los Angeles Times,* September 7, 1986; also interviews with Will Collette, 1988; Penny Newman, 1988.

PART III

THE CASE STUDY: POLITICS TO THE FORE

6

THE LIMITS OF THE INCINERATION STRATEGY: THE LANCER EXPERIENCE

LANCER'S ROLE

The room was jammed with reporters, television cameras, City Hall officials, and a host of other interested parties come to witness this latest episode of the LANCER saga.[1] The Los Angeles City Energy Recovery Project, commonly known as LANCER, was a project heralded by city and state officials and by waste managers everywhere. It was seen as a model for local governments, regulators, and the entire industry—a superior, environmentally sound, cost-effective solution to the problem of rapidly diminishing landfill capacity. The first of three planned 1,600-ton-per-day mass burn incinerators, LANCER was considered the key to the city's solid waste disposal plan into the next century, the "21st Century Solid Waste Management Solution."[2] Now the mayor, Tom Bradley, and LANCER's leading advocate on the City Council, Gilbert Lindsay, were about to make a pronouncement on the project's fate. It was almost ten years to the day since the city's waste bureaucracy, the Bureau of Sanitation (BOS), had first begun to conceive of an incineration future for their city.

Los Angeles's incineration strategy first emerged during the late 1970s, when the city was enduring a fierce conflict over existing and proposed landfill sites, some of which were located in the relatively pristine Santa Monica Mountains in the western section of the city. LANCER's history in some ways traced the recent—and spectacular—emergence of incineration and its subsequent troubles. The three proposed mass burn plants, to be located in three different areas of this vast, spread-out city, would eventually, it was hoped, burn as much as 70 percent of the municipal solid wastes collected each day, and in the process generate electricity for sale.[3]

The history of the LANCER project illustrates the unfolding of both the apparent strengths and eventual weaknesses of the mass burn strategy. It elucidated the controversies surrounding the technology: the continuing search for public acceptance, the importance of sound project economics, and the sharp conflicts around health and environmental impacts. LANCER also involved issues that were central to the political and social landscape of Los Angeles and the structure of its planning and decision-making. It highlighted the divisions and alliances between neighborhoods and political constituencies, and some of the particular environmental hazards that had become endemic to the region. It also brought the discussion of equity in the siting of environmentally controversial facilities into the public discourse over waste issues. The singular focus on both Los Angeles and California as testing grounds for new technologies and strategies gave additional weight to the project as a signpost for the future direction of municipal solid waste disposal. With major interest by the incineration industry, an active role for the investment community, strong commitment from the key political players, and strong opposition within the community, the political and financial stakes regarding LANCER were substantial.

LANCER was especially significant as part of the first wave of new-generation large mass burn incinerators, and was considered a model facility with respect to size and technology. LANCER was, of course, only one of the 11 incinerators proposed during the mid-1980s for Southern California, as well as one of 34 facilities proposed throughout the state.[4] But it figured prominently in the California Waste Management Board's advocacy of the shift toward waste-to-energy and was talked about within industry circles in a way that reflected the industry's change in mood from caution to optimism. The bruising battle between the top two industry leaders, Ogden and Signal/Wheelabrator, over who would obtain the first LANCER contract, was also indicative of the importance that LANCER held, not just as a source of potential profits but as a model plant. And the battle to stop LANCER had also become a test of the ability of the growing number of neighborhood community organizations that had formed to oppose this and other projects to slow down the incineration bandwagon. For many of the key players, LANCER had indeed become the symbol of the future direction of solid waste disposal, and the enormous amount of financial and political capital dependent on that future.

For Tom Bradley and Gilbert Lindsay, LANCER had become a political albatross. The mayor's long-standing and effective political alliance of inner-city blacks and Hispanics, West Side environmentalists, and key industrial interests was being severely strained in the wake of the battles over the project. Lindsay, in whose district LANCER was to be built, was a 27-year veteran of the council who had long been a favorite of the downtown business establishment, part of whose district he represented. The black councilman had managed to contain voter sentiment among his poor black and Hispanic constituents with a combination of wardlike politics (unusual in the diffuse and media-oriented political scene of the region) and a policy that rewarded his powerful friends in the high-stakes world of corporate lobbying, who were also generous Lindsay contributors. But Lindsay was finally starting to feel the heat over the LANCER situation.[5] When Bradley finally made up his mind about LANCER, he knew he had to bring in Lindsay, for only then would the position be seen as irrevocable. When the announcement was made and the room filled with speculation about its significance, the conflicts over LANCER, it appeared, had finally reached their climax. But the battles that LANCER had engendered over how to deal with the city's wastes, it was also clear, had only just begun.

FROM SWINE FEEDING TO LANDFILLS

Up through World War II, Los Angeles, as most other communities, had relied on an assortment of ways to dispose of its solid wastes. These included open dumps, various forms of incineration, and an active program of swine feeding to handle the organic and food wastes in the waste stream. Each of these methods had come under increasing criticism in the postwar years, at the same time that the region experienced its most intense and extended population and industrial boom to date. In just a little more than five years, Los Angeles was transformed during the 1940s from the ninth largest to the third largest city in the country.[6] This extraordinary growth compounded the waste issue, as the growth rate of the waste stream, tied also to the nationwide increase in per capita wastes, paralleled and even exceeded the rise in population and industrial activity. By the 1950s and early 1960s, waste collection and disposal issues came to figure prominently in the deliberations of both city and county government and in the politics of the period.

The rapid emergence of air pollution as a serious regional hazard during the 1940s and early 1950s was of special significance for the waste issue. Backyard incinerators designed to burn household trash were a major contributor to the smog problem in Southern California. By the 1950s, more than 1.5 million small or household incinerators were in use throughout Los Angeles County, which, along with the area's 2.5 million motor vehicles, 15,000 industries (many of which generated large amounts of air emissions), and 5 million residents, had created an air quality problem of major proportions. In 1954 alone, a series of smog episodes had reinforced Los Angeles's growing reputation as having the dirtiest air in the country.[7]

Local and state officials were at first slow to respond to the issue, fearful of interrupting Los Angeles's phenomenal growth. In 1948, five years after Los Angeles's catastrophic "Black Monday" air episode, the state legislature empowered counties to establish air pollution control districts (APCDs) to regulate air quality in their designated areas. These agencies, however, had only limited success in their efforts to tackle the smog problem, including the troublesome residential and industrial incinerators. In September 1955, the County Board of Supervisors finally passed an ordinance that amended the fire code and restricted the burning of combustible rubbish in backyard incinerators to specific hours of the day. Within two years, 45 cities in the county had established uniform procedures to enforce regulations regarding backyard incineration, which, according to air district officials, was responsible for emitting more than 500 tons of pollutants into the air each day. Residents caught firing up their single-chamber incinerators, or "Smokey Joes," as they were popularly called, could be fined $500 and imprisoned up to six months.[8]

With the rapid elimination of backyard incineration, the stress on the waste collection and disposal systems increased. The situation was compounded by the decline of swine feeding, which had been responsible for the disposal of most of Los Angeles's organic wastes. For several decades, the city had been able to sell some of its garbage to a group of hog raisers who ran a large swine farm about 50 miles from the city's loading site. At the peak of its operation, in the mid-1940s, it was the largest operation of its kind in the country, with upwards of 9 million pounds of pork sold annually.[9]

For a time, the private contractors who ran the operation paid a substantial fee to the *city* for the use of the garbage. But as business declined as a result of changes in the food business and a rapid increase in the use of garbage grinders, the payment structure was eventually reversed, until the city was obliged to pay the *farm* for disposal of its garbage. During the 1950s, swine feeding declined even further after public health procedures were instituted to address hog-related disease problems. By 1961, Los Angeles, one of the last holdouts in the country, finally discontinued its swine-feeding disposal practice as well.[10]

Nineteen sixty-one, in fact, became a pivotal year in setting the direction for the city's waste management options. Prior to then, city residents had engaged in various forms of source separation, reuse, and recycling activities. Materials shortages during World War II had reinforced these actions, and an active market for reused metals, tin, glass containers, and newspapers flourished. The city's collection and disposal system complemented these activities. Trucks collected different components of the waste stream on different days, which then traveled to different disposal sites, primarily open dumps and incinerators.

By 1961, after the elimination of swine feeding and the banning of the backyard incinerator, the city's trash collection system was organized around three separate trips: garbage, food, and organic wastes, one day; combustible trash on another day; and noncombustible trash on the third. For the collection of tin cans and other recyclables, the city also contracted with a private company that made a modest profit reselling the recyclables to industry.[11]

During the late 1950s, this system of collection and disposal had come increasingly under attack. It emerged as a major issue in the city's 1961 mayoral election campaign between the incumbent mayor Norris Poulson and his successful challenger Sam Yorty. Charges of poor service, organized crime connections, and mismanagement had plagued the city's activities and those of the private contractors. Poulson and Yorty traded charges during the 1961 campaign about the waste collection/disposal issue, especially the alleged organized crime connections. Yorty, however, as the challenger, was better placed to transform the issue into a populist appeal against the city's elites, calling them unresponsive to the needs of residents for a more "efficient," one-stop solution to the problem.[12]

While Yorty's victory helped shift the direction of solid waste management in Los Angeles, new approaches had already been developed by the waste bureaucracy at both the city and county level, which contributed to the demise of the source separation program. The key to this change was the emergence of the sanitary landfill, which soon came to be the region's dominant mode of disposal.

During the late 1950s and early 1960s, both the county's Sanitation Department and the city's Bureau of Sanitation began to aggressively pursue the sanitary landfill option. Their actions were reinforced in 1961 by the new mayor's decision to eliminate the source separation system and instead rely on a single container method of collection. This single collection and disposal method, Yorty argued, would save the city money and provide a more contemporary approach for its solid waste program. Soon after the initiation of the city-wide, single-stop collection program, the city began its own large-scale sanitary landfill operation: first in Griffith Park near downtown and, shortly thereafter, at two sites in the San Fernando Valley. Around the same period, the county also established several of its own sites. The landfill era in Los Angeles had arrived.[13]

TRASH WARS AND THE MOVE TO LANCER

With a sprawling land base, undeveloped canyons, and a growing waste stream, the city and county waste managers quickly moved to adopt the sanitary landfill as their preferred long-term waste disposal strategy. Both agencies decided that landfills were the cheapest and safest method of disposal and would later become even more attractive once the sites reached capacity and became "sparkling ex-dumps," as one local waste official put it, to eventually be converted into marketable real estate such as golf courses, botanical gardens, or even residential sites.[14]

By the mid-1970s, controversies surrounding municipal landfills began to surface in the Los Angeles area, even while the waste bureaucracies continued to seek new sites and increase capacity at operating facilities. Opposition emerged especially from community and environmental groups who had organized to protect the Santa Monica Mountains from development, one of the last and most valuable undeveloped areas within the city limits. This powerful, largely upper-middle-class movement, which was politically effective and well organized, had played an important role in both the election of Tom Bradley and in influencing the Los Angeles City Council on a range of environmental issues such as preservation of open space and coastal protection.[15]

The Santa Monica Mountains groups were especially successful in helping change the tenor of the landfill debate in Los Angeles. In effectively challenging the expansionist plans of both the county and city waste bureaucracies, these groups helped uncover and articulate new and growing concerns about groundwater contamination and noxious gases emanating from landfills. They pointed out that air emissions, which reached a peak level two to three years after a landfill opened, would continue at that same level throughout the life of the landfill and then for another 20 to 30 years after the landfill closed. The fear of methane gas explosions was also crucial in the debate, a fear later intensified when landfill gas began invading offsite properties, including a school in the San Fernando Valley. A small portion of these gases, furthermore, were known to be toxic and there were concerns that they could be contributing to the health-related air quality problems of nearby communities.[16]

By the late 1970s, the issue of hazardous materials had also entered the landfill debate. One of the attractions of the sanitary landfill was its compatibility with a simple, single-stop collection and disposal system that did not require information about the nature of the waste stream. Until the 1970s, there had been little attempt to monitor wastes for hazardous materials, which continued to be a significant—and growing—component of both household and industrial wastes. A wide range of household hazardous products, such as paints, pesticides, and household cleaning products, were deposited each year into trash bins and eventually trucked out to the city and county landfills. Furthermore, hazardous industrial wastes, which, after the passage of RCRA in 1976, were prohibited from disposal in municipal solid waste facilities, continued, nevertheless, to show up there from the continuing—and growing—practice of illegal dumping, mainly by small-quantity generators. One of the largest waste haulers in Los Angeles County, for example, was charged with "routinely" hiding hazardous wastes inside truckloads of trash bound for three landfills in the Los Angeles area.[17]

By the late 1970s, it had become clear that the ubiquitous presence of hazardous wastes in landfills posed a genuine, long-term threat to local groundwater resources. Of the first 800 Superfund sites, 100 turned out to be municipal landfills, with the 27,000 municipal landfills also accounting for almost 25 percent of the Superfund inventory of possibly contaminated sites. At the same time, an EPA study indicated that all landfills would eventually leak. These problems in turn were increasingly seen as having serious implications for public health. Initial concerns about contaminated groundwater at one particular location were transformed into fears about *all* landfills, fears that could no longer be allayed.[18]

Armed with increasing evidence of environmental problems, the anti-landfill activists were able to place the Bureau of Sanitation on the defensive. The bitter political debates at the time, dubbed the "Trash Wars" in the press, were considered potentially divisive by Mayor Bradley, who had come to power in part through his coalition of West Side environmentalists and inner-city residents. As the political heat increased, the Bureau of Sanitation eventually decided it needed to expand its options, and began a search for a new disposal strategy.[19]

It was, in fact, not much of a search. The BOS, like other waste bureaucracies, tended to be wary of recycling and reuse programs, considering them to be economically and technically unrealistic as well as too heavily dependent on active public participation. And, like other government officials at the state and federal level, BOS staff had also been skeptical of the developing waste-to-energy technology during the 1970s. But the more boxed in they became over the political unpopularity of landfills, the more they became willing to reconsider the option of burning wastes.

The impetus for this change was reinforced by the realization that time might be running out. BOS estimates during this period suggested that Toyon I, one of the city's two main landfills, could reach capacity as early as the mid-1980s, with the second, Lopez Canyon, following soon after. By 1979, the Bureau of Sanitation had decided to take a new look at waste-to-energy by reviewing existing and proposed plants in the United States, as well as in Western Europe and Japan. The BOS's senior sanitation engineer, Mike Miller, who would eventually head the LANCER effort, toured these plants and came back with a recommendation that the city more fully explore waste-to-energy, including both mass burn and RDF technology. The city's interest in waste-to-energy began to grow just at the moment this industry started to take off throughout the country.[20]

LANCER TAKES SHAPE

It quickly became apparent that the waste managers at the BOS were ready to ride the incineration bandwagon. The background information and sites selected for the city's review of waste-to-energy technology, its overall performance and feasibility, and the competing alternatives, were, for example, initially put together *solely* by representatives from the waste-to-energy industry itself. Miller would later comment that only the industry people would have had the background and expertise to frame the review.[21]

At the same time, the city successfully applied for a grant from the Environmental Protection Agency to study the "feasibility of locating a resource recovery facility in the central area of the City." This study was designed to review issues regarding the technology, the local energy market, and the costs associated with the construction and operation of such a plant. The EPA-financed study, completed two years later in 1981, set the stage for the specific choice of LANCER and its mass burn technology.[22]

In January 1982, Tom Bradley formed his Mayor's Citizens Advisory Committee to formally recommend a preferred technology and overall site for what was then called the Central City Resource Recovery Facility. The Mayor's Committee relied heavily on the BOS's industry-framed "feasibility study," which had concluded that a "mass burn system with [optional] preprocessing of waste [for aluminum removal] to be more viable than a Refuse Derived Fuel system." Though RDF was considered potentially more capable of reducing problems of air pollution and ash residue, given that metals, glass, and aluminum would be removed prior to the burn, overall the feasibility study still characterized the technology as having "excessive operational and maintenance costs, low reliability, and high safety risks."[23]

For BOS, the key factor was what industry advocates touted as mass burn's "proven track record" in Europe and Japan. Although such an operational history had not yet been established in the United States, the BOS selection of the technology was reinforced by the growing preference of a number of other communities for large, mass burn facilities. "The City needed a tried and true technology," Miller commented of the BOS decision, "something that worked and was not experimental. Something that could get rid of the trash."[24]

In February 1984, the Los Angeles City Council established a LANCER Steering Committee to coordinate varied aspects of the project, including permit requirements and public relations. The council also adopted its "Long Term Landfill and Alternative Disposal Policy," which called for a 1,600-ton-per-day mass burn incinerator for the central area of the city. BOS, meanwhile, also began to formulate plans to develop two other LANCER plants of equal size to serve the West Side and the Valley.[25]

By late 1984, the selection of the actual site began to preoccupy waste officials. That same year, the Cerrell Associates report was released, discussing differences in community response to siting issues. Aware of the difficulties of siting landfills, and the potential community opposition to incinerators, the BOS made a decision to reserve a potential site *prior* to any community input about the choice of the mass burn technology and site selection.

On December 1, 1984, shortly before the first public hearings on the *Draft Environmental Impact Report*, the city, through the BOS, entered into a "stand by property agreement" to reserve a 13.3-acre piece of land in the south-central area of the city. The property—known as the Alameda site, for the street on its eastern boundary—had once been part of a thriving industrial and commercial area more recently characterized by the deteriorating housing found throughout the south-central region. Though some industrial and commercial activity was still present, including a metal recycling operation, much of the area was single- and multiple-unit residences. Demographically, compared with the county as a whole, the neighboring community was young, poor, and heavily minority. The median age was 23.4 years, while that in the county was 30.8 years; 40 percent of the residents had incomes below the poverty level as compared with 13.4 percent in the county. Moreover, the community was 52 percent black and 44 percent Hispanic as compared to county averages of 12 percent black and 28 percent Hispanic. Also, the community suffered from a greater incidence of disease than the county at large (see table 6.1). Within LANCER's "impact zone"—defined by the city as the area encompassed by a half-mile radius surrounding the plant—were 3,359 homes, a number of schools and churches, and a recreation center.[26]

By the time the *Final Environmental Impact Report* (*FEIR*) for LANCER was released for public review in April 1985, the project had already become substantially defined. The first LANCER plant would serve the 1.4 million people residing in an area in the central part of the city, encompassing 130 square miles. It would burn 1,600 tons per day of

Table 6–1
COMPARATIVE HEALTH STATISTICS

	L. A. COUNTY	SOUTHEAST HEALTH DISTRICT
(rate per 100,000 population)		
Cancer	173	214
Heart disease	294	348
Strokes	58	66
Cirrhosis	18	39
Tuberculosis	18	41
(rate per 1,000 live births)		
Infant deaths	10	16
Neonatal deaths	7	10

SOURCE: County of Los Angeles, Department of Health Services, Health District Profiles, May 1986.

residential refuse generated within that area and produce 40 megawatts of electricity for sale to help offset project costs.[27]

As the venture moved forward through the environmental review process, the low-key, low-profile BOS strategy seemed to be paying off. In July 1985, LANCER cleared a major hurdle by receiving the unanimous approval of the City Council for certification of the project's *Final EIR*. The next month, the LANCER Steering Committee signed a memorandum of understanding with the municipally owned Los Angeles Department of Water and Power for the purchase of LANCER electricity at terms favorable to the project. In that same month, the L.A. City Council also revised its Municipal Solid Waste Plan, as required by state law, and approved in concept the burning of all city-collected waste in the three proposed LANCER facilities.[28]

In October and November of 1985, nearly a year after the Alameda site had been reserved and more than a year after the *Final EIR* had been certified, the LANCER Steering Committee organized a group of community workshops for residents in the immediate area to inform them about LANCER. The meeting was arranged in part by City Council area representative Gilbert Lindsay. Lindsay had already been pushing hard for the project and had become so heavily identified with it that his office had been transformed "into the City Hall command center for the lobbyists who pushed the project for the big national firms that build and run garbage incinerators."[29]

By 1986, however, both Lindsay and BOS officials had begun to worry about criticism of the project's ability to move quickly through the required procedures and obtain many of the necessary permits. This "fast track" approach was designed partly to beat deadlines likely to be established under the pending new tax law. The community workshops were finally held to "stimulate an opposition," as a BOS official later put it, in order to fend off any subsequent charges or court action about lack of community input, while at the same time managing and containing any dissent.[30] Toward that end, a $10 million "community betterment fund" was proposed (and eventually incorporated into the LANCER contract), to be controlled by a Lindsay-appointed Citizens Advisory Committee. After the fund was established, one Lindsay aide proclaimed that the money from it would be "used first to improve the Teresa Lindsay multipurpose community center," named after the councilman's late wife.[31] Once real community opposition did emerge, Lindsay, according to one of the neighborhood opponents of LANCER, was ready to apply pressure on individual opponents who might be vulnerable to such tactics.[32]

The first community workshops for the Alameda-site neighborhood were incongruous events, given the eventual emergence of an effective and widespread opposition to LANCER in this community. At the time, confusion and lack of information about the project prevailed, causing one community opponent to complain to the *Los Angeles Times* that "some of the elderly people around here think LANCER is a shopping center."[33] The workshops themselves were sparsely attended and included several Lindsay aides and backers. At the meetings, BOS officials provided a limited and environmentally optimistic view of the project. These officials even suggested to skeptical community residents that the health risks associated with LANCER were so minimal that a wedding could be held on the proposed grassy area at the front entrance of the proposed facility. One leading opponent who attended the first set of workshops came away from the meeting "bewildered and confused," as she put it, though she was also "determined to find out more about just what was going on."[34]

MONEY'S ROLE: LANCER'S PERIOD OF TURBULENCE

While the first signs of community opposition began to surface, the financing and ownership issues so crucial to the project also began to unfold. On December 30, 1985, in anticipation of proposed changes in the Tax Reform Act, the City Council rushed to approve a $235 million financing package for the project in order to be able to issue certificates of participation for industrial development bonds, the method of financing pushed by LANCER advocates. The $235 million in IDBs were then placed in escrow to qualify under the Tax Reform Act's "grandfathering" clause, which allowed the IDB's tax-free interest provisions to remain intact. This action, however, intensified the fast-track approach, since final LANCER approval needed to occur within two years or the IDBs would no longer be usable.[35]

With the financing package arranged, the council turned its attention to the selection of a vendor. The two finalists were Ogden and Signal/Wheelabrator, the two leaders in the industry. Both undertook aggressive and extraordinary lobbying campaigns, wining and dining and making hefty contributions to council members, especially the always available Gilbert Lindsay.[36]

This intensive lobbying scenario was also repeated during the selection of the project's bond lawyers and investment bankers, who also stood to gain huge commissions and fees if chosen. The council, in a controversial action, selected two high-powered brokerage firms to sell the bonds. Similar to the selection of the vendor, the process of awarding the bond contract had been rife with extensive lobbying, campaign contributions, and even allegations of conflict of interest. The bond houses recommended by the city staff, moreover, were passed over by the council in favor of the firms who had made the largest political contributions to council members and the mayor. These actions ultimately became symbolic of the critical role of money and power in the LANCER situation.[37]

When the dust had settled, Ogden was selected as vendor and both E. F. Hutton and Smith Barney were chosen to make the bond offering. In the first phase of the financing arrangement, Smith Barney, who had given $71,800 in political contributions to the mayor and council members, would receive $659,000 in commissions, $300,000 in fees, and expense reimbursements of $146,000, while E. F. Hutton, who had donated $29,650 to city officials, stood to reap commissions of $189,000, fees of $250,000, and reimbursements of $16,000. Moreover, in the second and final phase, which would involve the sale of about $200 million in long-term replacement bonds, the two, along with two minority firms with smaller shares, would divvy up a projected $3.8 million more in commissions (see table 6.2).[38] The vendor, Ogden, another big winner, with a contract estimated to be worth $250 million to $400 million, could look forward to annual fees of approximately $11 million, supplemented by a 5 percent profit on construction and 10 percent of the revenue from the sale of the electricity.

The Ogden arrangement, as spelled out in the LANCER service agreement covering the finance, construction, and operation of the project, stipulated a number of conditions for both the city and Ogden as the vendor. The city, for example, was obliged to deliver to the plant no less than 485,470 tons of nonhazardous wastes each year, while paying a service fee and assuming any risks brought about by changes in legislation (e.g., redesignating the ash as hazardous). Ogden, on the other hand,

Table 6–2

LANCER Bonds: Contributions and Payments

Company	*Political Contributions 1984 to mid-1986*		*LANCER Project Bond Sales, Earnings, & Expense Reimbursements*				*Recommended by Staff*
	COUNCILMEN	MAYOR BRADLEY	SOLD	COMMISSIONS	EXPENSES	FEES	
Smith Barney	$11,800	$60,000	$156 million	$659,175	$146,642	$300,000	No
E. F. Hutton	$11,300	$18,350	$45 million	$189,763	$16,850	$250,000	No
Grigsby Brandford	$8,000	$5,800	$28 million	$119,850	$8,330	$200,000	Yes
Daniels & Bell	$7,000	$11,900	$6 million	$29,963	$17,447	$250,000	No
Pryor, Goven	$3,000	$1,100	$0	$0	$0	$0	No
Goldman Sachs	$900	$5,100	$0	$0	$0	$0	No
Merrill Lynch	$700	$10,250	$0	$0	$0	$0	Yes
Kidder Peabody	$600	$16,000	$0	$0	$0	$0	No
Salomon Bros.	$0	$10,000	$0	$0	$0	$0	Yes

Firms donating the most to Mayor Bradley and the City Council were awarded the contracts to issue bonds for LANCER.

SOURCE: *Los Angeles Times*, August 11, 1986.

was obliged to build the facility to the city's specifications and within a given budget, and operate the facility at a guaranteed annual cost in accordance with specific, but yet to be defined, "performance guarantees." Ogden would invest $28.3 million out of the total project price tag of $133 million, with the $235 million in IDBs financing the construction, bond interest, and other expenses.[39]

Opposition to the project, meanwhile, began to coalesce. An organization of residents living near the Alameda site formed soon after the first set of community workshops took place. Calling itself the Concerned Citizens of South Central Los Angeles, the group consisted mostly of newcomers to environmental politics, while some of the more high-powered and experienced community groups from the south-central area, such as South Central Organizing Committee, steered clear of the issue, fearing the decision to proceed with LANCER could not be overcome. LANCER opponents also speculated that the proponents of all *three* LANCER projects, which were considered to be part of one single plan, anticipated far more problems siting LANCER 2 and LANCER 3 in the better-organized and wealthier West Side and San Fernando Valley communities. By rushing LANCER 1 through and successfully siting it in a poor, minority neighborhood without effective opposition, LANCER proponents could blunt opposition to LANCER 2 and 3 on grounds of "racism"; that is, you could not justify scrapping LANCER 2 and 3, particularly in terms of health and environmental considerations, if LANCER 1 was already operational.[40]

The Concerned Citizens group, however, achieved a major breakthrough when it began to develop alliances with incineration opponents outside the south-central area. These included a "slow growth" West Side coalition of environmental and homeowner groups that called itself Not Yet New York and an anti-incineration coalition, the California Alliance in Defense of Residential Environments (CADRE). CADRE members came primarily from white working-class and suburban middle-class neighborhoods such as the San Gabriel Valley to the east of the LANCER site, where several other incinerator plants were being proposed by the L.A. County Sanitation District. The development of the coalition and the extent of community opposition took the LANCER proponents by surprise, and their response tended to be clumsy and heavy-handed.

In March 1987, as part of their response to the growing opposition, LANCER supporters opened, under the aegis of Councilman Lindsay's office, a LANCER Information Center, located on a corner of the LANCER site. The center, with a $250,000 allocation from the city, included a display room, a LANCER video, pro-LANCER fliers, and a paid staff of eight community representatives chosen by Lindsay.[41] The opposition meanwhile attacked the use of public funds to support a public relations campaign for LANCER, and that same month, with help from an association of local south-central business leaders, held its most ambitious and widely attended event, a forum featuring Dr. Paul Connett, a leader of a national anti-incineration network then called the National Coalition Against Mass Burn Incineration. This group would also be drawn into the LANCER conflict, offering technical and tactical advice and sharing information regarding similar disputes between neighborhood groups and waste managers on the East Coast, where incineration plans had also become prominent.[42]

As the debate heated up, the growing alliance between the opposition groups became a critical factor in the politics swirling around LANCER. While the press and public officials had initially tended to view L.A.'s burgeoning "slow growth" movement as based on an elite, upper-middle-class politics, the connections between the south-central, West Side, Valley, and East Side groups were suggesting otherwise. This factor became particularly worrisome to LANCER supporters, and at LANCER hearings in the south-central area that were also attended by West Side allies of the Concerned Citizens group, leaflets were distributed by city officials with the warning "Don't let people who live outside your community tell you what to think."[43]

This approach ultimately backfired. By 1987, Concerned Citizens had developed an effective city-wide base of support. Their ties with West Side groups increased when the Center for Law in the Public Interest, a West Side–based environmental and consumer advocacy group, joined the growing coalition, which included Greenpeace, the international environmental group. By then, the coalition's attention had shifted to the project's environmental documents, such as the draft of the *Health Risk Assessment (HRA)* released in April of 1987.[44]

At a public hearing on the *HRA* and the *Supplemental Environmental Impact Report* held in the south-central community on June 6, technical presentations were made but no opportunities for questions or comments from the audience were provided. Heckling from the 200 people in attendance brought the proceedings to a halt, and the platform was "reluctantly turned over," as one newspaper account had it, to the congressman from the area, Augustus Hawkins. Hawkins, who had earlier broken with political allies Tom Bradley and Gilbert Lindsay to oppose LANCER, announced to the audience that his mail was running 95 percent against trash burning.[45]

That same day, in two City Council elections, strong LANCER opponents won stunning victories, including the upset defeat of City Council president Pat Russell, a longtime ally of the mayor and supporter of his growth policies. Russell, in the heat of the campaign, had switched her position to oppose LANCER, but to no avail. One reporter covering these events wondered whether LANCER would be able to survive this "rising wave of citywide opposition."[46]

ASSESSING LANCER: PROJECT ECONOMICS

With the LANCER issue injecting itself into the political discourse over the direction of city policies, one of the less visible, yet critical aspects of the incineration option concerned the project's economics. As with other large-scale mass burn facilities throughout the country, LANCER advocates argued that the facility, with its $235 million price tag, would eventually be cost-effective in comparison with landfills and other waste management strategies. Project opponents, on the other hand, raised the issue of financial risks and public subsidies and questioned its shifting economics.

The cost figures themselves were a matter of considerable controversy. The original LANCER feasibility study completed in 1983 had concluded that the project had only "marginal economic feasibility" and would only be cost-effective if "the real energy cost growth were in excess of 1 percent."[47] The price of energy, however, had gone down and not up since the early 1980s, declining as much as 225 percent from 1981 to 1987.[48] The energy calculation, nevertheless, tended to disappear in later financial evaluations, with project backers, including the vendor, Ogden, and the BOS, basically satisfied with the arrangement for the sale of energy established with the Los Angeles Department of Water and Power.

The energy contract with the DWP constituted both an implicit subsidy to the project and an explicit subsidy to Ogden, in terms of its share of the profits from the sale of the energy. At the urging of the mayor's office, the DWP, the largest municipally owned utility in the country, reluctantly agreed to make LANCER electricity available for publicly owned facilities at the same price the city would otherwise have to pay the DWP for the electricity. This ingenious arrangement was an unusual variation of the much coveted "avoided cost" procedure, since the LANCER project would in effect be "paid" a fee equivalent to the retail price charged a DWP customer.[49]

On paper, the DWP claimed to lose nothing, portraying the transaction as a simple "wheeling" of energy, which would be subtracted from the city's DWP account. Yet when the price tag for the LANCER energy deal was measured against the energy price available to DWP on the open market in 1987—about 2 cents/kwh—the difference was considerable. Based on those figures, DWP would in effect incur a $10,828,800 net loss in the first year alone of the project's operation. Even when measured against the DWP's most recent source of energy, a coal-burning facility in Utah, the net loss—or subsidy to the LANCER project—still amounted to over $1.3 million in the first year of the plant's operation.[50]

These figures were significant in that they represented two kinds of subsidies, while also disguising the true costs to the public. On the one hand, the rates paid by DWP customers, Los Angeles's residents and businesses, would be higher than they would have been had DWP not agreed to accept LANCER's power but had instead bought an equivalent amount of power at its wholesale market rate. Furthermore, since the vendor, Ogden, was guaranteed 10 percent of the total energy revenues, it would consequently receive 10 percent of the subsidy, which would be a little more than $1 million annually, based on the 1987 market price for energy.

The energy subsidy issue in fact paralleled other similar aspects of LANCER's economics where certain project costs and subsidies were not included as part of the projected total cost. LANCER project estimates, for example, were based on an assumption that 100 percent of the waste delivered to the plant could be processed. The California Waste Management Board, however, had generally indicated that about 15 percent of the waste would be unprocessable, including such large items as appliances, couches, and engine blocks.[51] LANCER cost estimates had neither included the cost of separating out nor disposing of such materials, although contractually it was responsible for both. BOS, furthermore, had not even developed adequate information to estimate these costs.

The costs of infrastructure improvements were also excluded from LANCER's cost estimates, although they could be directly attributable to the project. According to various estimates, the construction and operation of LANCER would require four major street widening and improvement projects whose costs were estimated at slightly under $2.5 million.[52] These costs did not include possible land condemnation or additional environmental studies, should they be required, a likely event given the far-reaching impact on the streets and adjoining properties and the degree of opposition and concern in the neighborhood. Community residents argued forcefully that the LANCER projections underestimated the amount of street widening that would ultimately be needed to accommodate the increased truck traffic resulting from the project. The traffic issue was even further compounded by the existence of railroad tracks lining the center median of two of the main thoroughfares near the project site. The estimates for LANCER in this instance made no provision for the costs of railroad grade separations or of a possible underpass at one of the busier intersections that might be necessary to avoid massive traffic tie-ups.[53]

Traffic and street widening were not the only significant infrastructure issue for LANCER. The incineration plant, for one, would be a heavy user of water in a water-short region, as well as a major discharger of wastewater into the city's already overburdened sewer system. The water use for LANCER 1 was first estimated at 821,000 gallons per day, a projection later increased to 1.1 million gpd, while the total daily wastewater discharge was estimated to be 213,000 gallons.[54] The city would have to absorb any costs necessary to handle improvements to the sewer system, costs that were also not included in the overall calculations.

Besides specific subsidies and costs not included or estimated, there was a range of issues, constituting a form of both acknowledged and unacknowledged financial risk for the city, that enormously increased the uncertainty of the project's economics. For example, any unanticipated costs that the vendor might incur could be passed on to the city through the service fee, the per ton tipping fee paid by the city. Furthermore, there could easily be changes in the regulatory environment affecting the plant, such as those involving air emissions. The city in this case would be responsible for the cost of bringing the plant into compliance with any new standards. In April 1987, in fact, a city-commissioned engineering report revealed that LANCER would indeed require additional pollution control measures not included in the original project costs or design, thus resulting in additional costs to the city of $4 million to $7 million *per year*.[55]

This significant cost addition, however, paled in comparison with the possible cost increases that could result from a reclassification of the ash. The LANCER cost estimates had been based on the assumption that all of the ash generated by the facility would be disposed in a proposed "ashfill" at the Lopez Canyon landfill site, though the costs of adequately constructing this additional facility had not been initially included. But if the fly ash component of the ash (about 10 percent of the total ash content) were to be classified as hazardous, LANCER's costs would rise considerably, by more than $2.50 per ton. And, then, should the combined fly and bottom ash be classified as hazardous, costs could rise by an astronomical figure of more than $25 per ton. Such a jump in costs would in effect make the project economically unfeasible by any measure.[56]

As the LANCER saga unfolded in 1986 and 1987, the controversy over what should be the safe and legally required method for ash disposal was raging nationally. The uncertainty of the situation threatened the economic feasibility of the entire industry. Though the focus of the debate would subsequently shift from the courts to Congress, there was little doubt that the cost of ash disposal would rise whatever the outcome. The only question was by how much.

Even without these additional calculations and the other hidden project costs, the LANCER price tag, by June 1987, was becoming prohibitive. In

March 1986, when the LANCER debate first started heating up, the LANCER Steering Committee, in a memo to the City Council, estimated the per ton cost of LANCER at $27.43 amortized over the 20-year period of the bonds. In January 1987, a new analysis provided to the council showed an increase to $34.38 per ton. Only two weeks later, the city administrative officer had to revise these figures once again to $45.54 per ton, after it was discovered that the previous estimates had not included the lower end costs of ash disposal. And then, in April 1987, the report concerning the need for new pollution control measures pushed the cost estimates even higher. An average of these new costs added $11.35 per ton to the original estimate, bringing it to the extraordinary figure of $56.89 per ton, no longer cost-effective even in comparison with the most costly landfill scenario at the time.[57]

These cost increases and economic uncertainties ultimately played a role in the final decisions about LANCER, though they tended to be played out away from the public debate and the media coverage. While the opposition groups raised questions about the project economics, they generally focused more on the social and environmental consequences of the project. The concern over traffic congestion, for example, was not primarily about hidden costs. They concentrated more on the enormous disruption from increased noise, air pollution, safety hazards, and congestion the LANCER plant would have on this residential neighborhood. The idea of hundreds of trucks rumbling through the neighborhood, carrying trash and ash residue to and from the facility, was a potent, negative image.

There were larger city-wide impacts as well. The consequences of significant increased water use and wastewater discharges were particularly poignant in 1987, while Los Angeles was in the midst of a dry year and an unresolved water supply situation, as well as a growing crisis over its wastewater/sewage treatment capacity. The sewage issue, in fact, would eventually lead to some of the most extensive growth controls the city had ever undertaken.[58]

Most importantly, however, health and environmental issues loomed over the LANCER debate. The problems with air emissions and ash disposal were not simply economic calculations, considerable though they were, but struck at the heart of the controversies around the project. Nineteen eighty-seven, for example, was the year that Los Angeles, along with 73 other areas then not in attainment with the provisions of the Clean Air Act, faced the possibility of severe sanctions, including a construction ban on new major sources of air pollution.[59] The ash issue, furthermore, had become, in its own right, another and perhaps more controversial *landfill* issue, raising the intractable problem of cross-media pollution once again. It was through a resolution of these issues and their effect on public health that the project, as L.A. mayor Tom Bradley indicated during 1986 and 1987, would ultimately rise or fall.

EVALUATING RISK: THE HEALTH AND ENVIRONMENTAL ISSUES

Aware of the volatility of the health and environmental issues, LANCER advocates relied on two key sets of documents—the *Environmental Impact Reports* (*EIR*s, *Final* and *Draft Supplemental*) and a *Health Risk Assessment* (*HRA*)—to bolster their argument that the LANCER plant was environmentally safe. But these documents dealt with the most troubling issues—what would be spewed out in the air and the residue that would remain behind—in a way that, in the end, intensified rather than reduced the controversies surrounding the project.

The *Health Risk Assessment* took on special significance as the debate over the health and environmental issues became intense. With community opposition growing, project proponents, such as the mayor and various members of the City Council, increasingly looked to the release of the *HRA* as a way to allay fears about health and environmental risks. Instead of providing scientific input to help frame a policy decision, the *HRA* was quickly becoming, prior to its actual release in April 1987, the central justification for the project's survival.[60]

Similarly, the various environmental impact documents were used to attempt to alleviate concerns rather than advance the discussion over the various environmental consequences of incineration and its alternatives. The *EIR*s, in fact, not only presented the consequences inadequately, but also increased the confusion and uncertainties regarding several key aspects of the project.

This was particularly true with the air emissions data. The air emissions from LANCER, both proponents and opponents agreed, would be substantial. In fact, based on the EPA definition of plants that generated more than 100 tons per year of any one pollutant, the emissions estimates clearly indicated that this plant would be a "major new source" of air pollution. In the LANCER case, it was predicted that *four* pollutants would exceed this threshold (see table 6.3). These predictions concerning the amount of emissions, moreover, had also changed over time but were unexplained in the various LANCER *EIR* documents. While predicted emission rates for carbon monoxide showed a decline in the estimates provided in the 1985 and revised 1987 *EIR* documents, the average daily emissions of the four other pollutants, including NOx and HC, which contributes directly to the smog problem, *increased significantly* (see figure 6.4).[61]

Table 6–3
ANNUAL COMBUSTION EMISSION RATES/CRITERIA POLLUTANTS

POLLUTANT	TONS/YEAR AT 100% AVAILABILITY	TONS/YEAR AT 91.5 AVAILABILITY
NOx	851.0	779.0
SO_2	305.9	279.9
CO	237.9	217.5
NMHC	20.2	18.4
TSP	138.0	126.2
Lead	1.3	1.2

SOURCE: LANCER, DSEIR (Draft Supplemental Environmental Impact Report), April 1987.

The question of how unsafe or safe these emissions would be ultimately became one of the prime factors in the contentious debate over LANCER. The environmental documents suggested that the incineration technology was changing rapidly and that much research was being undertaken "on the control of air emission from refuse combustion facilities," as the *Draft Supplemental EIR* for LANCER put it. Special attention was paid to dioxin by both sides, as LANCER proponents argued that significant advances had been made that reduced the amount of dioxin released into the environment and thus would almost eliminate the dioxin health risk. LANCER opponents, on the other hand, expressed strong concern over any dioxin emissions, given that dioxin had come to represent the most prominent example of the unknown and unpredictable consequences of incineration.

The issue was further heightened by the dioxin-induced 1985 Swedish moratorium on incinerator construction and the development a year later of that country's new and stricter emission limits for incinerators, or guidelines, as the U.S. industry liked to call them. The LANCER project, its state-of-the-art technology notwithstanding, would nevertheless *fail to meet three of the five Swedish guidelines*. This included the dioxin standard, which, it was estimated, could be exceeded by no less than 50 percent and possibly as much as 675 percent.[62]

The LANCER *Health Risk Assessment* jumped directly into this issue by using a set of assumptions about dioxin that resulted in a significantly lower cancer risk from the proposed LANCER plant than other available, more conservative assumptions would have provided (see table 6.5). According to the *HRA*, the probability of an individual contracting cancer as a result of exposure to LANCER, also known as excess lifetime individual cancer risk, came to 0.118 cases of cancer per million population, the lowest cancer risk estimate at that time of any health risk assessment for a mass burn incineration plant.[63]

Figure 6–4
LANCER AIR EMISSIONS ESTIMATES

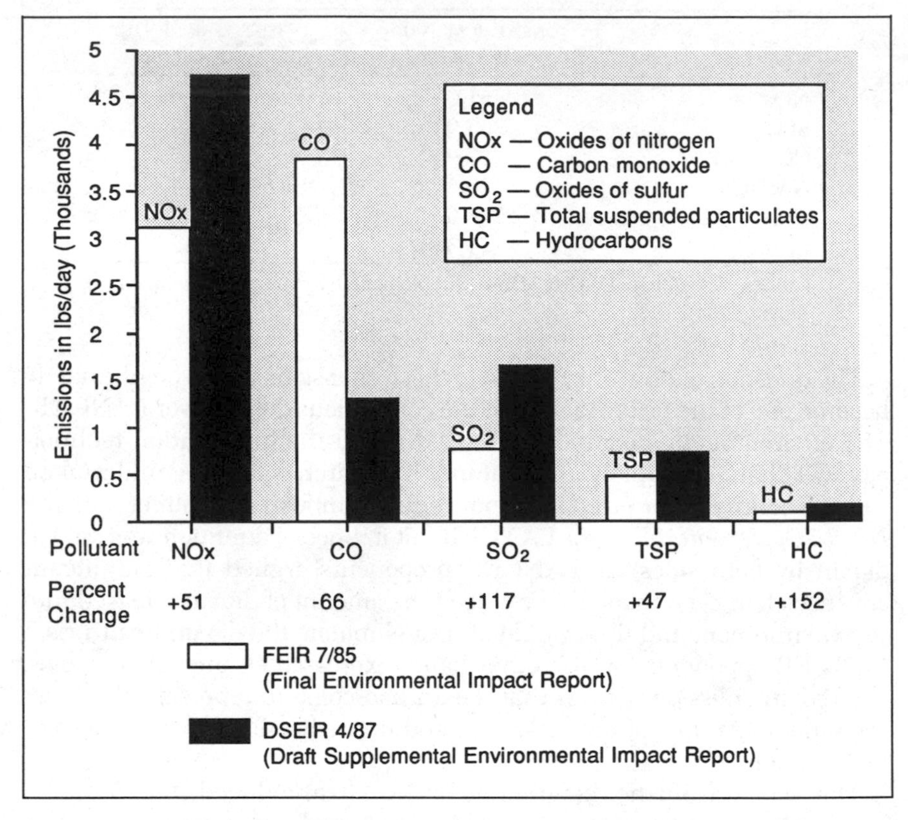

The LANCER *HRA* achieved this result for the most part by basing its emissions estimates on data taken from an operating incinerator in Wurzburg, West Germany, which was similar in design to LANCER, although about five times smaller. Furthermore, the *HRA* also selected a particular toxic "equivalency method" for dioxins used to establish the potential cancer potency of the emissions that was less conservative or protective than other methods, especially those used in the more stringent Swedish guidelines. The equivalency method is crucial, since the stack gas from incinerators will contain a mixture of several different varieties of dioxins and furans, and methods differ on how to compare the cancer potency of all the compounds present at any one time.[64]

The methods used in the LANCER *HRA* were quickly subject to criticism. The Wurzburg data, compiled by the vendor of the plant rather than

Table 6–5
COMPARISON OF SELECT ASSUMPTIONS

ASSUMPTION		LANCER HEALTH RISK ASSESSMENT	SOUTH COAST AIR QUALITY MANAGEMENT DISTRICT AND CALIFORNIA DEPARTMENT OF HEALTH SERVICES
Terminology:		Generally plausible expectations	Most and worst plausible case
Dioxin absorption by inhalation		39.3%	100%
Cancer potency for cadmium		2 ng/Nm³	12 ng/Nm³
Soil ingestion by children	by age:	2.5 years	1.5–3.5 years
	by amount:	250 mg/day	500 mg/day
Soil mixing depth		10 cm	15 cm
Half-life of dioxin in soil		10 years uncultivated 5 years cultivated	12 years
Cancer potency of inhaled dioxins		17μg/Nm³	38μg/Nm³
Emission rate of dioxin equivalents		0.307 ng/Nm³	2.115 ng/Nm³
		From Wurzburg plant using EPA 1985 method	From guarantee by Ogden in contract with the city using DOHS method

a German governmental agency, had previously been categorized by both the U.S. EPA and its West German equivalent as incomplete and inadequate due to the lack of information about the nature of the waste stream, the type of air pollution control equipment, and several of the testing procedures. The reliability of the data was further questioned since a number of key operational parameters were not kept constant during the tests.[65] One U.S. industry analyst who had developed a standardized data base for incinerator dioxin emissions had in fact excluded the Wurzburg data for many of the same reasons.[66]

In calculating the ultimate cancer potency of the LANCER plant's dioxin emissions, the *HRA* used a 1985 method developed by the Environmental Protection Agency. This approach differed considerably from other equivalency methods, including one developed by the California Department of Health Services (DOHS), which had been recommended, and would be required by the Air Quality Management District. Moreover, the California DOHS method had been used in drafting the original LANCER contract with the city of Los Angeles. The *HRA* failure to use the DOHS method was misleading, since the results would inevitably be recalculated for the final permitting process.

The difference between the EPA method and the California DOHS approach did indeed have a substantial bearing on the low *HRA* cancer risk figures. Based on emissions measured at Wurzburg, the EPA method yielded a far lower estimate for dioxin equivalents in the stack gas (0.307 ng/Nm^3) than the California DOHS number, 2.115 ng/Nm^3.

The dioxin emission issue had also been paramount when the LANCER contract was first being discussed with the vendor, Ogden. In an emotional interchange with a City Council member, an Ogden executive declared that his company would not be willing to guarantee meeting the Swedish guideline for dioxin, a figure of 0.12 ng/Nm^3. Instead, Ogden was only willing to guarantee 17ng/Nm^3 as part of its initial draft contract with the city, a figure nearly 170 times greater than the Swedish guidelines. Ogden, however, tried to reassure nervous council members that, despite its refusal to change the terms of the draft contract, the company felt, in its view, that the *Health Risk Assessment,* which had not yet been prepared, would base its analysis of dioxin risk on the higher figure in the contract. Such a procedure would then show, the official concluded, in justifying the company's unwillingness to meet the Swedish limit, that the "risk of having 17 nanograms in the community" would turn out to be "an insignificant risk."[67]

Less than three months later, the draft of the *HRA* was released, with its dioxin analysis playing a major role in the risk assessment. But, despite the Ogden reassurance, the 17-nanogram figure was never used in the calculations. Instead, the *HRA* used a value of 2.1 ng/Nm^3, more than eight times lower than the figure Ogden was willing to guarantee. Ogden's assurance had become another casualty of the narrowly conceived *HRA* analysis.[68]

The use of these assumptions was justified by the LANCER *HRA* in part by its conceptual approach in assessing health risks. The LANCER assessment relied on an approach that identified the most conservative or protective assumptions (for example, the use of the average dioxin emission rate of all plants operating at that time) as unrealistic, approximating a kind of worst-case scenario. The *HRA* also distinguished between what it characterized as realistic and unrealistic judgments about how risk should be quantified. Yet there are no set standards available to help determine such a "realistic" judgment about dioxin risks. The data regarding dioxin are highly variable and subject to great debate and thus create substantial uncertainties in any evaluation. Judgments about risk have therefore always involved certain choices: whether, for example, to assume a possible, "realistic" scenario that tended to minimize some of the risk variables, or, instead, in a more conservative fashion, to assume a scenario where those variables are included. In the LANCER *HRA*, the suppositions and methodology used ultimately minimized the degree of risk.[69]

This approach in the *HRA* was also compounded by important omissions as well as the assumptions used in the calculation of risk. As with most risk assessments, the focus of the LANCER *HRA* was limited primarily to *carcinogenic* effects caused by LANCER emissions. As a result, it dismissed possible effects of a number of noncarcinogenic air pollutants, such as acidic gases and NOx, both of which can cause irritation of the eyes and the respiratory tract as well as other health problems. Those most sensitive to such problems are children, older people, and individuals already suffering from respiratory diseases, while the LANCER *HRA* and other risk assessments have used healthy, adult males for their calculations.

The LANCER *HRA* evaluated possible plant emissions of acidic gases and then compared them to the levels demonstrated to cause adverse health effects in workers exposed over an eight-hour day. Since the LANCER plant would have significantly lower levels, the effects of these gases were not considered to be a factor in risks to the community. This method, moreover, implicitly assumed that these, and in fact all, emissions would be released into the equivalent of "clean air," that is, air that did not already contain high levels of air pollutants, clearly not the case in the Los Angeles air basin. The health risk assessment thus became a "best case scenario" that failed to take into account the specific circumstances of an already degraded air basin and high-risk population groups, many of whom in fact resided in the poor neighborhood surrounding the LANCER site.[70]

The LANCER *HRA* further limited its analysis of possible health effects solely to air emissions, thus excluding any comprehensive analysis of the potential health risks from the disposal of ash. This approach was similar to that used by health risk assessments for other incinerator facilities, none of which evaluated the impacts of ash, the single most important hazardous by-product of the incineration process.[71]

While the LANCER *HRA* avoided the ash issue, the discussion of ash in the environmental impact documents was dramatically deficient, undercutting the claim by project advocates that the LANCER ash residue would be both minimal and nonhazardous. There was, for example, some confusion in these documents about the weight and volume of the residual ash. In the *Final Environmental Impact* report, it is asserted that the volume of trash made available to LANCER would be reduced by 90 percent. These figures were based on comparing the *uncompacted* trash fed into the incinerator with the resulting ash, and failed to account for the large, unburnable items, ranging from 5 to 15 percent by weight of any load set aside (and eventually sent to landfills) before incineration. The comparison most often used in promoting incineration, however, involves the relative volume reduction that would be achieved if the wastes were landfilled. When trash is landfilled, it is compacted, first mechanically, then over time through biodegradation. In terms of compacted trash, the reduction by incineration at LANCER and most other plants would be 70 to 75 percent by volume and 50 to 70 percent by weight.[72]

A far more serious deficiency in the environmental impact documents had to do with the question of ash toxicity. While the *EIR* was being prepared, the Bureau of Sanitation arranged for a series of laboratory tests with a private consultant to determine both the moisture content and the toxicity of the ash residue. The timing was important since a nonhazardous classification of the ash by the California Department of Health Services prior to January 1, 1985, would be exceedingly difficult to change, according to legislation passed at the time.[73]

Though the documents are both murky and incomplete on how the conclusion was drawn, DOHS did indeed provide a nonhazardous classification based on the consultant's tests. The procedures used in the test, however, were quite extraordinary, given the importance of the issue.

The "samples" for the test arrived one day in several unsorted bags, containing about 30 pounds of trash. Neither the Bureau of Sanitation nor the chemist who performed the tests were able to produce any records indicating the percentage of the various waste stream components in the sample, nor what, from the sample, was actually tested. "These bags just arrived one day," the chemist recalled, without any notation of their contents. The letter describing the test that was reprinted in the *Final Environmental Impact Report* refers only to "aluminum cans, boxes, grass, paper, plastic, and glass in the sample." No mention is made, for example, of ferrous metals, which are generally found in about the same concentration as both plastic and glass. Neither the chemist nor BOS officials could later recall whether, in fact, ferrous metals were included, a serious omission given that an important purpose of the test burn is to determine the concentration of metals likely to leach out when the ash is placed in a typical municipal landfill. However, when arrangements were made to do a separate moisture content analysis, eight truckloads of refuse were delivered to the laboratory. The test priorities, at best, were skewed: eight truckloads of trash for moisture analysis but only a couple of bags for toxicity![74]

Taken collectively, the LANCER health and environmental documents produced more questions and doubts than information and answers. The slipshod and disorganized character of the various *EIR*s, with missing information, inaccurate or misleading estimates, shifting figures, unexplained discrepancies, and disjointed presentation, inhibited any attempt to assess accurately the health and environmental impacts of LANCER and to compare them with those of other solid waste management alternatives. The LANCER *HRA,* meanwhile, relied on an approach and methodology that rejected more conservative assumptions about possible health risks, while utilizing, on several occasions, questionable methods and procedures that produced an extraordinarily low estimate of health risks. The documents attempted to alleviate concerns in the community and neutralize the growing opposition, but their publication ultimately inflamed it.

THE DENOUEMENT

By June 1987, ten years and some $12 million after the Bureau of Sanitation was first prodded into exploring mass burn incineration, the decision regarding LANCER came to a head. The *Health Risk Assessment,* released in April, was immediately subject to heavy criticism, particularly for its extremely low numerical estimate of cancer risk. Community opposition groups, who chafed at their lack of access to question or challenge the report's author and BOS officials, had received a major boost when the Center for Law in the Public Interest and the city's advisory Environmental Quality Board intervened by requesting an extended comment period for the *HRA*. In early June, state senator Art Torres, the chair of the California Senate Toxics and Public Safety Management Committee, held a news conference announcing the findings of a report prepared within the UCLA Urban Planning Program on the LANCER project and the general waste management issue (from which this book is partially derived). Torres's statements, mirroring the report's conclusions, emphasized the environmental and economic uncertainties and the serious flaws in the LANCER decision-making process, and suggested that the project be canceled.[75]

For Tom Bradley, continued support for LANCER was quickly threatening to become a serious political liability. Besides Art Torres's strong opposition, a potential Bradley opponent in the next mayoral election, West Side City Council member Zev Yaroslavsky was slowly shifting his position away from his earlier support for the project. Yaroslavsky, who was also trying to identify himself with the emerging "slow growth" constituencies, had begun to realize that opposition to LANCER would have the benefit of linking environmental with minority community issues, just the kind of alliance that had brought Tom Bradley to power in 1973.[76]

Bradley, on the other hand, was constrained both by his own proclivity for high-tech, construction-oriented solutions and the continuing insistence by his Bureau of Sanitation that a solid waste crisis was at hand. There were no serious alternatives to LANCER, the BOS argued; both the *EIR* and *HRA* documents had little if anything to say about such alternatives. In public appearances at the time, BOS officials and LANCER advocates Delwin Biaggi and Mike Miller would dismiss recycling as a serious option, arguing that even under the most ideal circumstances, including fully developed markets for recycled products, recycling could account for at most only 25 percent of the city's waste stream. In their planning documents, BOS presumed, in fact, that recycling would provide a far smaller contribution. One operating LANCER plant, in contrast, Biaggi argued, would more than match recycling's best efforts. And three operating LANCER plants, it was assumed, would basically take care of business.[77]

Bradley's deliberations were further compounded by the multitude of lobbyists and power brokers who had by now become associated with the LANCER project. These included politically connected law firms, the bond houses, and a variety of other consultants and subcontractors, all of whom stood to lose hundreds of thousands of dollars in fees if the final contract for the project was not signed. The political adage that money could dictate policy turned out to be only partially true in the LANCER case, as the early successes of the lobbyists were undermined by the intensity and breadth of the opposition. Boxed into a political corner, with his desire for an unprecedented fourth term as mayor potentially at stake, Bradley, hoping to rebuild bridges with the West Side environmentalists, decided he had to switch his position.

With a crestfallen Gilbert Lindsay at his side, the mayor, at a June 16 news conference, announced he was withdrawing his support from LANCER. Using some of the arguments of the opposition, Bradley called the project's health and environmental impacts too uncertain and said he would not subject the south-central community, where LANCER was to be located, to further concern. Lindsay echoed the mayor's comments, suggesting that community opposition was a product of "fear." "They [community residents] are just frightened to death," Lindsay complained. "I can't have my constituents unusually unhappy." When asked about alternatives, Bradley was vague, but suggested that an expanded recycling program could be the basis of a new start.[78]

In the weeks and months following the June 16 press conference, BOS officials, along with Ogden executives and other LANCER lobbyists, maneuvered to see if some sort of project could be salvaged, possibly in another location or at a later date. The City Council, however, which now included its two new strong LANCER opponents, finally decided, in August 1987, to put the project to rest. Mass burn incineration in Los Angeles was no longer a viable public option, at least for the short term.[79]

The demise of the LANCER project had been a major setback, not only for Los Angeles's waste bureaucracy and the plant's industry and financial backers, but for those forces nationally as well. LANCER had been touted as having the most sophisticated and advanced technology of the day. The plant, according to one of its consultants, would be "safer than any operating trash burner" to date.[80] It would be large, cost-effective, and ultimately eliminate the city's growing and intractable waste problem. Unlike the Vicon situation, the project involved some of the biggest players in the industry, and it had become, in the process, a model to emulate. Its defeat signaled, as much as any other plant failure, the eclipse of the incineration option.[81]

With the end of LANCER, a kind of cat-and-mouse game began to develop around the politics of solid waste management in Los Angeles and throughout the country. A few months after his June press conference, L.A. mayor Bradley announced, with great fanfare, that the city might be ready to pursue some form of mandatory recycling, though he offered no specific proposals. Under Bradley's urging, the city also banned the purchase of Styrofoam by city departments and talked of plans to expand a West Side pilot recycling program. The BOS, meanwhile, also suggested publicly that it was ready to jump on the recycling bandwagon. Mike Miller, once LANCER's most visible spokesman and defender, told the *Wall Street Journal*, "We made a mistake. Reducing the waste stream first makes a lot more sense."[82]

L.A.'s new romance with recycling, however, was still fraught with difficulties. According to City Hall insiders, despite the public pronouncements, the BOS continued to drag its heels on a series of feasibility studies for recycling. At the same time, BOS officials, along with other waste planners at the county and state level, quietly began to raise another set of scenarios, including talk of long-distance hauling of wastes, other possible new landfill sites, and a scaled-down incineration program, only, of course, after a recycling effort had also been initiated.[83]

A year after LANCER had been killed, both industry and the waste bureaucracy had decided to bide their time. With the new emphasis on recycling in vogue among elected officials and the media, a shift in position began to occur among the incineration advocates. Recycling, they argued, was *not* incompatible with a burn or bury strategy: instead, all three constituted the elements of an integrated waste management strategy. Yet, underlying their position was another set of assumptions and biases that continued to divide the waste bureaucracy and industry from the community opposition. The lessons of LANCER were still to be drawn.

NOTES

1. Author's notes, discussion with Bob Morales, administrative assistant to state senator Art Torres, June 20, 1987; "Bradley Abruptly Kills Trash-Burning Plant," Kevin Roderick, *Los Angeles Times*, June 18, 1987.
2. "LANCER Slated to Be Los Angeles' 21st Century Solid Waste Management Solution," Bureau of Sanitation, City of Los Angeles, 1986, Press Release.
3. "Report No. 2," Department of Public Works, Bureau of Sanitation, City of Los Angeles, January 23, 1987; *Solid Waste Program in the City of Los Angeles: Issues and Recommendations, Overview,* Delwin A. Biagi, Bureau of Sanitation Director, City of Los Angeles, January 1987; *Refuse Collection Tonnage by District by Month for Sections A-E Only S, For Calendar Years 1980–1986,* City of Los Angeles Bureau of Sanitation, computer printout, April 29, 1987; Interview with Robert Alpern, Los Angeles Bureau of Sanitation 1987.
4. "Impact Assessment of Waste-to-Energy Projects and Alternatives in the South Coast Air Basin," Brian Farris et al., Planning Division, South Coast Air Quality Management District, March 1987; "Waste to Energy in California," Joan Melcher and Paul Relis, Community Environmental Council, Santa Barbara, 1986; "Waste-to-Energy Update," California Waste Management Board, April 1987.
5. "Stakes Are High in Battle of L.A.'s Lancer Contract," Kevin Roderick, *Los Angeles Times,* March 9, 1986; "City Politicians Reap Harvest in Bond Sale Field," Victor Merina, *Los Angeles Times,* August 11, 1986.
6. *Thinking Big: The Story of the Los Angeles Times, Its Publishers and Their Influence on Southern California,* Robert Gottlieb and Irene Wolt, New York, 1977.
7. Letter from S. Smith Griswold to the Los Angeles County Board of Supervisors, September 27, 1955, regarding the "Establishment of Uniform Ordinances on Burning of Combustible Rubbish in the Los Angeles Basin"; also "Smog Circus in L.A.," Frank Stuart, *Frontier,* December 1954; *Los Angeles Times,* September 28, 1957.
8. *Rubbish Disposal Facilities in the Los Angeles Metropolitan Area,* Office of the Los Angeles County Engineer, 1957; "Mock Funeral Held as Incinerators Burn Out," *Los Angeles Times,* October 1, 1957.
9. "Future Trends Which Will Influence Waste Disposal," Abel Wolman, *Environmental Health Perspectives,* December 1978.
10. "Feeding Garbage to Hogs Spreads New Disease," *Public Works,* November 1952.
11. *Los Angeles Times,* May 8, 1961; May 5, 1961; also *Yorty: Politics of a Constant Candidate,* John C. Bollens and Grant B. Grant B. Geyer, Palisades Publishers, Los Angeles, 1973.
12. Author's interviews with Norris Poulson, 1975; Sam Yorty, 1976; also *Yorty: Politics of a Constant Candidate,* John C. Bollens and Grant B. Geyer.
13. "Dorn Assures City on New Dump Sites," *Los Angeles Times,* January 31, 1959; "Sanitary Landfill Operations in Los Angeles," *Public Works,* vol. 96, no. 11, November 1965.
14. "Landfills and People: A Volatile Mix," Greg Braxton, *Los Angeles Times,* May 22, 1983.
15. *Backroom Politics: How Your Local Politicians Work, Why Your Government Doesn't, and What You Can Do About It,* Bill and Nancy Boyarsky, Los Angeles, 1974; also interviews with Sue Nelson, 1981; Rubell Helgeson, 1987.

16. "Air Quality Impacts of Waste-to-Energy Facilities in the South Coast Air Quality Management District," Sanford Weiss, testimony before the California State Assembly on Environmental Safety and Toxic Materials, October 16, 1985; "Control of Gaseous Emissions from Active Landfills," Staff Report to Rule 1150.1, South Coast Air Quality Management District, March 28, 1985; Staff Report Rule 1150.2, SCAQMD; see also "Waste Hazards Raise Doubts Over Landfills," *Los Angeles Times,* November 18, 1984; "Study Finds Toxic Gases in Ordinary Trash Dump," Paul Jacobs, *Los Angeles Times,* November 26, 1986.
17. "Hazardous Waste Found in Household Garbage," *Chemical and Engineering News,* October 12, 1987; "Consideration of Cleanup and Abatement Orders for Solid Waste Disposal Sites," Regional Water Quality Control Board, Los Angeles, January 26, 1987, item 5, p. 6.
18. *New York Times,* February 12, 1987; *The Prevalence of Subsurface Migration of Hazardous Chemical Substances at Selected Industrial Waste Land Disposal Sites,* Geraghty and Miller, EPA/530-SW-634, 1977.
19. "Showdown Imminent on Mission Canyon Landfill in Santa Monica Mountains," *Los Angeles Times,* June 14, 1978; *Los Angeles Times,* May 22, 1983.
20. Presentation by Mike Miller, UCLA, February 4, 1987; *Phase II Feasibility Study,* Central City Resource Recovery Facility, 2 vols., Los Angeles, March 1983.
21. Presentation by Mike Miller and Drew Sones, UCLA, February 4, 1987.
22. *Phase II Feasibility Study,* vol. 1, March 1984, pp. 1–2.
23. *Phase II Feasibility Study,* vol. 2, March 1983, p. 92.
24. Presentation by Mike Miller, UCLA, February 4, 1987.
25. "Search for 2 More Trash Plants Sites OKd," Rich Connell, *Los Angeles Times,* June 14, 1986; *LANCER DATE FACTS,* Bureau of Sanitation, City of Los Angeles, n.d.
26. 1980 United States Census; *Phase II Feasibility Study,* volume 3, March 1983; *Final Environmental Impact Report,* Los Angeles City Energy Recovery (LANCER) Project, Cooper Engineers, Department of Public Works, Bureau of Sanitation, April 1985.
27. *Final EIR,* City of Los Angeles, April 1985; *LANCER Quick Facts,* Bureau of Sanitation, City of Los Angeles, n.d.; City of Los Angeles, *Request for Proposals from Qualified Vendors for Waste-to-Energy Disposal Service,* LANCER, n.d.
28. City of Los Angeles Solid Waste Management Plan (CoSWMP); Letter from Norman E. Nichols, Los Angeles Department of Water and Power, to Keith Comrie, Chairman, LANCER Steering Committee, August 29, 1985; interview with Dennis Whitney, 1987; interview with Daryl Mills, Los Angeles Department of Water and Power, 1988.
29. *Los Angeles Times,* June 18, 1987.
30. Mike Miller/Drew Sones, UCLA Presentation, February 4, 1987.
31. *Los Angeles Times,* March 9, 1986; *Request for Proposals from Qualified Vendors for Waste-to-Energy Disposal Service,* LANCER, City of Los Angeles.
32. Sheila Cannon UCLA Presentation, March 4, 1987; "Lindsay Sees No Wrong in City-Paid Publicity Drive for LANCER Project," Victor Merina, *Los Angeles Times,* October 25, 1986.
33. *Los Angeles Times,* March 9, 1986.
34. Interviews with Cynthia Hamilton, 1987; Sheila Cannon, 1987.
35. Drew Sones, UCLA Presentation, February 4, 1987; see also on IDBs, "The New Tax Law: How It Will Affect Financing and Ownership of W-T-E Projects," Charles Samuels, *Solid Waste and Power,* February 1987.
36. "L.A. Armed for Award of Rich Pact on Trash Plant," Kevin Roderick, *Los Angeles Times,* March 18, 1986; "New Jersey Firm Chosen for Trash-Burning Plant," John Chandler, *Los Angeles Herald-Examiner,* March 27, 1986.
37. *Los Angeles Times,* August 11, 1986; March 9, 1986.

38. *Los Angeles Times,* August 11, 1986.
39. *LANCER Service Agreement Summary,* LANCER Information Packet, Department of Public Works, Bureau of Sanitation, City of Los Angeles, 1986.
40. Presentation by Ken Thorbourne, South Central Organizing Committee, Liberty Hill Community Funding Board, May 17, 1987; interviews with Laura Lake, 1987; Cynthia Hamilton, 1987.
41. "Lindsey Sees No Wrong in City-Paid Publicity Drive for LANCER Project," Victor Merina, *Los Angeles Times,* October 25, 1986.
42. Author's notes, presentation by Paul Connett, National Coalition Against Mass Burn Incineration, March 21, 1987, Office of Los Angeles City Council Member Joel Wachs.
43. "Trash Burning LANCER Plan Draws Heat," Kevin Roderick, *Los Angeles Times,* June 7, 1987; "Cross-County Unity Served Up at Triumphant Lancer Foes' Party," Lynn O'Shaughnessy, *Los Angeles Times,* August 2, 1987; interview with Laura Lake, 1987.
44. Discussion with Joel Reynolds, Center for Law in the Public Interest, 1987; "Comments on the Draft Outline of Methods for the HRA Analysis for the Proposed Los Angeles City Energy Recovery Facility (LANCER)," prepared by Michael A. Dowell, National Health Law Program on Behalf of Concerned Citizens of South Central L.A., August 6, 1986; "Foes of Trash-Burning Project Get Support of Legal Advocates," Kevin Roderick, *Los Angeles Times,* April 30, 1987.
45. *Los Angeles Times,* June 7, 1987.
46. *Los Angeles Times,* June 7, 1987; June 9, 1987.
47. Memo from Keith Comrie, City Administrative Officer, to Los Angeles City Council, June 28, 1983, File no. 81-3395.
48. "Market Prospects for Refuse-to-Energy Development in California," Paul R. Peterson et al., presented at "Waste Tech '87," October 26–27, 1987, pp. 5–6; *Economics of Refuse-to-Energy in California,* California Energy Commission, Sacramento, 1988.
49. Interviews with Dennis Whitney, Los Angeles Department of Water and Power, 1987, 1988.
50. "Los Angeles City Energy Recovery Project (LANCER)," letter from Norman E. Nichols, Assistant General Manager—Power, Department of Water and Power, to Keith Comrie, Chairman, LANCER Steering Committee, August 29, 1985; interview with Daryl Mills, 1988; *Request for Proposals from Qualified Vendors for Waste-to-Energy Disposal Service,* City of Los Angeles.
51. *Waste-to-Energy,* California Waste Management Board, Technical Information Series, June 1983.
52. *Supplemental Environmental Impact Report,* Los Angeles City Energy Recovery (LANCER) Project, Cooper Engineers, Bureau of Sanitation, City of Los Angeles; *Final EIR,* April 1985; interview with Carl Haase, Assistant LANCER Project Manager, 1987.
53. *Final EIR,* April 1985.
54. *Final EIR,* April 1985; *Supplemental EIR.*
55. "Price Tag Might Trash LANCER Project," John Schwada, *Los Angeles Herald-Examiner,* April 23, 1987; interview with Drew Sones, 1987; "Lancer Project Air Quality Issues," letter from Carl Haas to Michael Miller, November 9, 1984.
56. Interview with Drew Sones, 1987.
57. Memo from the LANCER Steering Committee to the Los Angeles City Council, March 17, 1986; also memo from the Chief Administrative Officer to the Finance and Revenue Committee, Los Angeles City Council, January 26, 1987.
58. "Advice on Alternative Building Permit Allocation Strategies," Draft Interim Report, Katherine Stone, Project Director, submitted to City of Los Angeles, October 27, 1988.
59. Clean Air Act §110 (a) (2) (I); & § 173 (4) as amended 1977; also 42 U.S.C. §7401-7642;

40 CFR Part 81, Subpart C (1987); FR 50686 and 50695-6, November 2, 1983 ("The 1983 Sanctions Policy").

60. "Pact Removes Obstacle to Use of Trash Incinerators," Kevin Roderick, *Los Angeles Times*, February 3, 1987; author's notes, presentation by Dr. Robert Valdez, LANCER Peer Review Committee, UCLA, May 1987.
61. *Statement of Findings and Overriding Considerations, Final Environmental Impact Report*, July, 1985; also *Draft Supplemental EIR*, vol. 2, appendix A; "Application for a Permit to Construct and Operate a New Source of Air Pollution," LANCER Project, Ogden Martin Systems of Los Angeles, submitted to the South Coast Air Quality Management District, October 2, 1986; *Los Angeles Herald-Examiner*, April 23, 1987.
62. DSEIR, vol. 2, appendix D; these figures are based on the assumption that the emissions of dioxin from LANCER would not have exceeded those of the Wurzburg, West Germany, plant, or of an equivalent Ogden plant at Marion, Oregon.
63. *Health Risk Assessment of the LANCER Project*, Dr. Alan H. Smith, Health Risk Associates, Draft, April 17, 1987; "Critique of the LANCER Health Risk Assessment Report," LANCER Peer Review Committee, Los Angeles, July 1, 1987; "Transcript of Public Works Committee Meeting," July 1, 1987, Item No. 12 (LANCER); "A Suggested Approach to Overcome California's Inability to Permit Urban Resource Recovery Facilities," Steven A. Broiles, *Risk Analysis*, vol. 8, no. 3, 1988, pp. 357–66.
64. For a discussion of dioxin equivalency methods see Staff Report, California Air Resources Board, *Public Hearing to Consider the Adoption of a Regulatory Amendment Identifying Chlorinated Dioxins and Dibenzofurans as Toxic Air Contaminants*, June 6, 1986, Technical Support Document, Part B: *Health Effects of Chlorinated Dioxins and Dibenzofurans;* see also *Journal of Occupational Health*, J. H. Milby et al., vol. 27, 1985, pp. 351–55.
65. *Draft Incinerator Emissions Data Base*, Environmental Protection Agency, at 3.12; interview with Mr. Bert Johnke, German Environmental Agency, April, 1987.
66. "A Data Base for Dioxin and Furan Emissions from Municipal Refuse Incinerators," Milton Beychok, *Atmospheric Environment*, vol. 1, no. 1, 1987.
67. Transcript of Special Meeting of the Finance Committee of the Los Angeles City Council, January 26, 1987.
68. *HRA*, vol. 2, table 2.
69. LANCER *HRA* for a discussion of risk analysis; also "A Suggested Approach to Overcome California's Inability to Permit Urban Resource Recovery Facilities," Steven A. Broiles, *Risk Analysis*, vol. 8, no. 3, 1988.
70. "Reasonable Further Progress Report for 1984 on the South Coast Air Quality Management Plan," South Coast Air Quality Management District and the Southern California Association of Governments, June 1986, pp. VII–27, VII–36; see also "Comments on the Draft Outline of Methods for the HRA Analysis for the Proposed Los Angeles City Energy Recovery Facility (LANCER)," prepared by Michael A. Dowell, National Health Law Program on Behalf of Concerned Citizens of South Central L.A., August 6, 1986.
71. "Fundamental Objectives of Municipal Solid Waste Incinerator Ash Management," Richard A. Denison, Environmental Defense Fund, prepared for publication at the 81st Annual Meeting of the Air Pollution Control Association, Dallas, Texas, June 20–24, 1988.
72. FEIR, vol. 4; see also presentation by Don Avila, Los Angeles County Sanitation Department, City of Commerce plant, April 10, 1987; also *To Burn or Not to Burn*, Environmental Defense Fund, New York, 1985.
73. SB 2292 (the Campbell bill), codified as California Health and Safety Code 25143.5 Section 25143 (d).

74. Interview with Ellen Tiedemann, 1987; interview with Carl Haase, 1987; FEIR, appendix G, also letter from David J. Leu, Chief, Alternative Technology and Policy Development Section, California Department of Health Services, to Delwin Biaggi, L.A. Bureau of Sanitation, April 18, 1985; interview with Norman Riley, Department of Health Services, 1987.
75. Author's notes: State Senator Art Torres press conference held June 9, 1987; "UCLA Group Urges City to Drop Lancer Trash Plan," Kevin Roderick, *Los Angeles Times,* June 10, 1987.
76. Interview with Laura Lake, 1987; also *Los Angeles Times,* February 3, 1987.
77. Author's notes, presentation by Delwin A. Biaggi at Not Yet New York meeting, January 1987; Mike Miller presentation at UCLA, February, 1987.
78. "Bradley Abruptly Kills Trash-Burning Project," *Los Angeles Times,* June 18, 1987.
79. Transcript, Public Works Committee of the Los Angeles City Council, July 1, 1987.
80. *Los Angeles Times,* June 7, 1987; April 23, 1987.
81. "California Recycling for the '90s: Garbage Burning Dead," *Sierra Club Yodeler,* Oakland, California, July 1988, p. 4; "Energy from Garbage Loses Some of Promise as Wave of the Future," Bill Richards, *Wall Street Journal,* June 16, 1988.
82. *Los Angeles Times,* September 20, 1987; June 29, 1988; *Wall Street Journal,* June 16, 1988; *Los Angeles Herald-Examiner,* October 27, 1987; June 16, 1988; *Los Angeles Daily News,* August 19, 1988.
83. *Los Angeles Times,* January 3, 1989; December 22, 1988; April 6, 1988; February 28, 1988; also, "Burying Garbage Is No Longer Enough," Winston Porter (EPA), in *Rapid City Journal,* South Dakota, July 17, 1988; also Letters to the Editor, J. Winston Porter, *Wall Street Journal,* July 15, 1988.

PART IV

THE ALTERNATIVES: CONFRONTING AN ENTRENCHED ORDER

7

RECYCLING'S UNREALIZED PROMISE: "A NICE THING TO DO"

RECYCLING AMERICA

The executives at the office of the Waste Management subsidiary in the small town of Carsonville, located in the "thumb" of Michigan about 60 miles from Detroit, wanted to improve the company's public image. This Waste Management operation, which had recently bought out the Tri-City Landfill in nearby Bridgehampton Township, had been taking a beating, similar to many of Waste Management's 771 other subsidiaries throughout the country. Waste Management, Inc. (WMI), the world's largest landfill and waste disposal company, had, in fact, been finding itself increasingly challenged and confronted on the local level, as in Carsonville. There, a group of residents, the Citizens Against Rural Exploitation (CARE), had been fighting WMI's proposal to expand its landfill, located on the Black River, by another 46 acres.[1]

The CARE activists were particularly irked by how the Tri-City Landfill fit into Waste Management's future plans. Most of the trash, upwards of 70 percent, was brought to the landfill from communities outside the county, including Detroit, which was experiencing its own landfill capacity problems while completing construction of one of ten incinerators planned for the state. At a capacity of 4,000 tons per day, the Detroit facility would be the largest mass burn plant ever built or planned in the country.[2] Meanwhile, however, Detroit and other surrounding communities were hurriedly seeking new disposal options, and the expansion of the Tri-City Landfill would provide a crucial additional repository for their wastes. The possibility of expansion, which only exacerbated the out-of-county issue for the Carsonville area residents, ran directly counter to CARE's own approach, which called for an expanded local recycling program and limits on imported garbage.[3]

Waste Management officials knew they had to respond to the criticism, particularly since the state Department of Natural Resources would soon rule on the expansion plans. One day, a Waste Management secretary, who later would quit the company and join the CARE group, overheard these officials complaining about their problems. The company had decided to develop its own "recycling program," which consisted of distributing hundreds of plastic bags to be used for recyclables. On the bags, the name of Waste Management would be displayed, along with a picture of Iron Eyes Cody, the Los Angeles–based Indian who had been widely seen in advertisements promoting antilitter campaigns. The company executives were hopeful that this effort would be sufficient, as one official remarked, to "keep these yah yahs [recycling advocates] off our backs."[4]

It wasn't long after that episode that the Waste Management officials showed up at a county meeting to discuss the five-year update of the County Solid Waste Management Plan, required by Michigan solid waste laws. "They came in with their three piece suits, their attorneys, and all their flash and glitter," recalled CARE activist Renee O'Connell. "All through the meeting they talked about how they recycled. When we asked them why they didn't have anything more to show for all their talk about a big recycling push, they exclaimed, 'Oh, don't worry, we're just beginning to set it all up.' In fact, it seemed clear to us they just wanted to distract attention from their expansion plans."[5]

Despite this approach, the state Department of Natural Resources would ultimately deny WMI's request for the 46-acre expansion of the Tri-City Landfill, citing possible contamination problems and proximity to the Black River. Several months later, WMI changed the name of its landfill to the Tri-City Recycling and Disposal Facility to account for what it called its "Recycling Drop Off Center," where the plastic bags with the recyclables were now supposed to go. The name change occurred within weeks of WMI's application for another state construction permit, this time for a 15-acre expansion of Tri-City, but at a slightly different location. The landfill-cum-recycling connection had been drawn once again.[6]

The approach in Carsonville is, in fact, representative of new directions taken by both landfill operators and the incineration industry. This strategy, in response to the rise of community movements, involves the promotion of a public relations–oriented policy declaring recycling to be compatible with landfills and/or incineration. Such an approach has been designed to create an image of environmental responsibility, while seeking to diminish opposition to the burn or bury options. For the waste industry, this combined approach has become the basis of the newly conceived "integrated waste management" strategy.

By the late 1980s, all the major waste companies had begun to advocate this concept of the integrated approach. The chairman of Wheelabrator, Paul Montrone, for example, sought to define his company as the "true environmentalist—not in an obstructionist sense but as an enterprise that leads the way to environmental solutions." In that light, Montrone talked of the company's support of community recycling programs, particularly encouraging citizens to remove such items as batteries, certain plastic items, and noncombustible products that would increase the Btu value of the wastes and reduce the level of contaminants in the ash.[7]

Wheelabrator's main competitor, Ogden Martin, also began to promote its own "recycling policy," which it introduced with great fanfare in September 1987, just a few weeks after its crucial LANCER project was canceled. Ogden's approach emphasized a "materials recovery facility," or MRF, a small- to medium-sized industrial plant that primarily uses magnetic separation and other sorting techniques at an incinerator site or separate location rather than at the curbside. This recycling effort emphasizes the removal of various commingled recyclable materials such as glass and metal cans, which in turn also increases the efficiency of incinerators and potentially reduces the amount of chlorine-bearing materials to be burned.[8]

But it was Waste Management, the most visible of all the waste companies, that moved most aggressively in fashioning a recycling strategy. Waste Management had been a special target of the anti-landfill/anti-incineration coalitions. A number of its subsidiaries had been convicted of price fixing and violating various environmental regulations. And some of its companies had developed a reputation for a rather heavy-handed manner of conducting business, exemplified by a 1988 incident in New Orleans where Waste Management representatives were reported to have threatened two city officials over an investigation of alleged overcharging, suggesting that the officials might end up with "lead cement boots" and would "meet their maker."[9]

At the same time, however, Waste Management also became adept at countering this image by publicizing its recycling activities. These had originated with the company's acquisition of the recycling program in San Jose, California, and subsequent acquisitions of other municipal and privately run programs. As a result, Waste Management created yet another subsidiary, which it called "Recycle America," and moved quickly to capitalize on its new recycling presence.[10]

"1988 will be the year America turns the corner on recycling, and Waste Management is providing the leadership," proclaimed one Waste Management ad in a New Jersey recycling publication.[11] What the company in fact developed was an adept *takeover* policy, especially targeting curbside programs already initiated by municipalities. These acquisitions complemented WMI's own collection business while eliminating any potential recycling competitor. By combining landfills with recycling and a possible inroad in the incineration market with its newly acquired holdings in Wheelabrator, Waste Management moved a step closer toward its ultimate goal of control of *all* waste services in many communities.

As part of this search for monopoly status, Waste Management sought to convince skeptical municipalities and residents that it was indeed committed to recycling. Around the same time as the Carsonville episode, for example, the Waste Management subsidiary in Lima, Ohio, made a successful appeal to take over a mandatory recycling program being discussed by city officials under pressure from local citizens. At the meeting, the local Waste Management executive, Jack de Witt, made a big pitch for Recycle America, based on the company's claim that Waste Management's recycling efforts were a "fully operational, viable enterprise." Yet, in an interview a few days after his presentation, de Witt admitted that he in fact was "not familiar with Recycle America," nor did he have any idea how to operate a curbside recycling program. Indeed, de Witt wasn't even clear just what Recycle America did in the first place.[12]

De Witt's confusion was not just a function of his lack of familiarity with this Waste Management subsidiary. The company was, as it turned out, not much more than "just a trademark," as one Recycle America official described it; a device to link various collection subsidiaries that had incorporated curbside recycling as part of their operations. Instead of actively seeking to develop markets for recyclables or play a meaningful role in developing the overall recycling industry, the Recycle America operations have largely been designed to sell the recyclables they collect to another party, who would then market the products. In effect, Recycle America had become a minor component of Waste Management's waste hauling and landfilling business.[13]

Most importantly, Recycle America is becoming for Waste Management a core part of a larger public relations effort to provide an environmental cachet to its operations. This includes recent attempts to woo more "responsible" environmental organizations such as local Sierra Club chapters and the National Wildlife Federation, whose board Waste Management chairman Dean Buntrock joined in 1987. Its recycling pursuits have become essential to that strategy, since the decline of landfills and the eclipse of incineration has now elevated recycling to a central place in everyone's waste management strategy, whether public officials, the waste industry, or various community or environmental groups. But, as the Recycle America effort has also suggested, recycling can become a kind of catchall phrase, lumping together different programs and different concepts.

For the waste industry, recycling has come to represent a salvage operation for their own core programs, landfills and incineration. For a number of community groups, on the other hand, recycling, particularly the concept of "intensive recycling," represents the most effective means of reducing the waste stream and offers a comprehensive alternative to incineration and landfills. Far from presenting a coherent and identifiable strategy, recycling has evolved, by the end of the 1980s, into yet another waste management battlefield.

FROM SCOW TRIMMING TO COMPACTION TRUCKS

Recycling in this country has always had a diverse set of constituencies. With the advent of the Progressive Era and its emphasis on professionalization and increased productivity, the utilitarian concept of production for use became directly linked to the solid waste issue. The growth of various "conservationist" movements, from the public sanitarians who emphasized problems of pollution to the newly developing field of sanitary engineering with its interest in resource efficiency, brought increased attention to both the problems and opportunities of solid waste collection and disposal methods. As the Progressive reformers came to shape the discourse around resource management, the waste issue especially became an important part of their concerns.

The early reform efforts, such as those inspired by New York City's first commissioner of street cleaning, Colonel George E. Waring, sought two key goals: a more systematic public role in the collection and disposal of wastes, and the development of a variety of programs that could segregate various types of wastes and deliver them in some marketable or disposable form. Until the World War I period, collection and disposal was accomplished by a wide variety of techniques and activities. The concept of recycling or resource recovery, for example, tended to reflect such varied efforts as source separation, reduction, sorting, swine feeding, and scavenging. These efforts, furthermore, involved both municipalities, private contractors, and independent labor, such as scow trimmers and ragpickers, also known as chiffoniers, who helped establish an important role for scavenging in the informal waste recycling activities of that era.[14]

These diverse types of recycling activities all relied on source separation—the separation of wastes at the household level—as their foundation. By the World War I period, as many as 70 percent of cities in the United States operated some kind of source separation program. Most of the key disposal technologies such as reduction or the "cooking" of the garbage and even various types of land disposal were based on separating different components of the waste stream. Source separation also came to be seen as a civic responsibility by middle-class reform movements that emphasized the need for better sanitation, greater resource efficiency, and more effective utilization of waste as a resource. Such an approach had an important economic dimension as well, since waste by-products could be sold. For these movements, improper waste management, at either the household or community level, hampered efforts at reforming the urban and industrial environment, which underlay the Progressive approach.[15]

The strategy of the Progressives, however, conflicted at times with the informal, secondary labor market associated with solid wastes that had emerged by the turn of the century. Most communities relied upon a wide variety of scavenging activities, including animals who prowled the city streets and a large number of individuals who were at first encouraged but eventually squeezed out of the scavenging business. Several cities, prior to the turn of the century, actually developed programs and legislation to protect animal scavengers and encourage human scavenging, including scow trimming. In this practice, workers employed on garbage boats "skimmed" through the load, collecting salvageables before dumping the remaining wastes into the ocean. But as separation programs became more popular and municipal sanitation bureaus were organized to manage wastes more systematically as a resource, municipal governments such as the city of New York reversed policy and began to charge workers for the right to keep the materials they recovered from the scow.[16]

The urge to find profits in garbage extended to a range of other separation and disposal activities. Wet garbage or "swill," the organic component of municipal wastes, became the raw material—with a varying economic value—for the swine-feeding business. Other garbage and sewage was collected and redistributed as soil fertilizer for agricultural use, an important activity that dated back to pre–Civil War days. Papers and rags, which were also important recycled products, were in fact the target of the very first trash-sorting programs that sprang up in the late nineteenth century.[17]

As Progressive reformers promoted public control and professionalization of solid waste management, many of these early recycling activities were either disbanded or reshaped by public agencies. Much of the motivation was economic, as waste by-products were increasingly seen as a source of revenues—the garbage-as-gold concept—to help pay for waste management programs at the municipal level. Scow trimming and other, less formal scavenging and separating activities were discouraged by public agencies. Scavengers were considered "dirty" and "unsanitary" by middle-class reformers, an attitude extended in part to racial and class biases. European immigrants in the North and rural blacks in the South, for example, were thought incapable of complying with rules requiring source separation, while assumptions were made (incorrectly) that European immigrants generated more waste than their middle-class "American" counterparts. Class more than ethnicity, as some studies at the time suggested, influenced the characteristics (rather than the volume) of wastes.[18]

With cities increasing their control and the Progressives setting the agenda, consideration turned to the variety of disposal methods, many of which were selected because of their ability to recover resources. Reduction, of course, was popular in this period for its ability to recover and then market both grease and "tankage" as salable raw materials for the production of soaps, perfumes, and as the base for fertilizer. Some early incinerators were designed as part of a recovery system linked to the development of sorting technology. New York City's Waring, for example, established the country's first mechanized sorting plant driven by the first U.S. power-generating incinerator system. A distinction developed at the time between "crematories" or "destructors" and power-generating incinerators connected to sorting plants that were primarily resource recovery operations.[19]

By World War I, despite the prevalence of the Progressive approach, a set of often contradictory trends emerged. On the one hand, the war interrupted the trend toward overproduction and overconsumption that would dramatically reassert itself in the 1940s and 1950s. During the war years, there was an overall reduction of about 10 percent of the waste collected, while various materials and resource recovery programs were actively promoted. But changes in the selection of collection and disposal options were influenced more by economic than social factors. The cost of land, the economics of incineration, and the external costs increasingly associated with ocean and surface water dumping became primary in determining new choices in disposal technology. These rapidly changing options also reflected changes in capital and labor costs, as well as in price factors governing the sale of waste by-products and materials separated at the source.[20]

During the Depression, the availability of cheap labor further impacted the waste issue, effectively undermining source separation and recycling by driving waste management toward one-stop, comprehensive solutions involving unprocessed and unsorted wastes. The low cost of land and labor especially favored open dumps. Cheap labor was also a factor in encouraging the development of sorting operations, many tied to incineration facilities built during the 1930s. Furthermore, scavenging began to decline, as municipalities sold the salvage rights at dumps to expand their revenue base. Meanwhile, household source separation, due partly to the decline in reduction and the growing use of landfills, was increasingly becoming an outmoded form of waste sorting for purposes of recovery.[21]

The growing tendency to base decisions on cost also meant that once labor costs increased, waste management activities would focus on labor displacing technologies for both the collection and disposal of wastes. This trend quickly emerged during and after World War II, with the shift from recovery and recycling toward mixed collection and landfilling. The rise of motorized collection vehicles and compactor trucks reduced labor costs and provided more efficient collection systems for both land disposal and incineration.[22] Changes in the quantity and nature of the waste stream, which increased the volume of nonrecyclables, also influenced this shift away from separation and recovery. And, finally, marketing and technological changes stimulated the development of convenience packaging and a variety of single-use, throwaway disposable items. This change, in turn, played an enormous role in the ideological shift and economic restructuring of solid waste away from a *product* or resource, to a discarded material for disposal.

These shifts were interrupted by a brief resurgence of recycling during World War II. The heavy emphasis on materials recovery, especially, played a crucial role in elevating recycling into a kind of patriotic duty. Paper collection drives and appeals to save scrap metal and aluminum cans were nearly daily occurrences. As much as 35 percent of all wood fiber products, such as paper and cardboard, were recycled by the end of the war.[23] This flurry of interest in recovery, however, was driven by materials shortages. And some of the technological breakthroughs of the war years, including development of new, petrochemical-based products, ultimately undermined this temporary preoccupation with scarcity.

By the late 1940s and through the 1950s, the structural changes of the war and postwar period had been set, as illustrated by the experiences of the San Francisco–based Sunset Scavenger Company.

Sunset Scavenger had collected refuse in the San Francisco area since 1920. Through the Depression, the collection business had been highly competitive, seeking every possible source of revenue to keep rates low. Companies like Sunset Scavenger engaged in salvage operations, relying on a manual sorting process while the collection truck made its rounds. This was performed by one of the four crew members inside the truck, squeezing and shaking out the bags to sort and ultimately salvage rags, bottles, aluminum, newspaper, cardboard, and other recyclables. Salvaged material was then graded, classified, and resold. Bottles, a significant source of income, would be washed, packaged, and sent back to the distributor for reuse. By the early 1950s, the company's salvage operations netted as much as half a million dollars.[24]

By the mid-1950s, however, there developed a "decided decline in the sales of salvageable materials." It was, a Sunset executive lamented, "the beginning of the 'no deposit, no return' era of packaging materials." At the same time, changes in the composition of the waste stream, such as the growing use of plastics, began to force a change in the technology of collection. Also, the wet or "garbage" component of household waste had begun to decline, in part due to the widespread use of garbage grinders, which had become especially popular during the 1950s. Changes in the other components, including paper, cardboard, glass, metal, and especially plastics, also affected the nature of the waste stream, dramatically increasing the volume of wastes. As a consequence, collection companies turned to the compaction truck, designed specifically to handle increased volume while keeping labor costs down. These trucks, which have a hydraulic mechanism to squeeze the trash inside an enclosed compartment, basically eliminated the salvage business based on manual separation at the pickup point. Most collection companies subsequently eliminated their salvage operations. Those that continued, such as Sunset, tended to limit it to newspapers and corrugated paper, among the few still-viable recyclable products. Newspapers, moreover, required special pickup points and trucks, since, once inside a compaction truck, newspapers absorb moisture from other wastes and thus become "contaminated," or no longer resalable. These constraints, in turn, contributed to increased costs while reducing the supply of this recyclable.[25]

For public agencies, these changes during the 1950s and 1960s had yet to make an observable impact. The new collection technologies, which also reinforced a reliance on landfills, still kept waste management costs at an acceptable level. But the growing awareness of the problems related to landfills, combined with problems facing other disposal technologies, especially incineration, had begun to preoccupy not just local communities, but the federal government as well. With pollution issues forcing their way onto congressional agendas, and new social movements beginning to question the changing patterns of production and consumption, recycling, another discard of the postwar era, was ready to make a comeback.

THE RESOURCE RECOVERY CONUNDRUM

The passage of the Resource Recovery Act in 1970 marked a transition point in the approach toward recycling and larger waste management issues. Already, by the start of the decade, a range of new recycling programs, nearly all of them run by volunteers, had sprung up in a number of cities across the country. One estimate suggested that as many as 3,000 volunteer-based recycling centers were organized in the months immediately preceding and following Earth Day in 1970.[26]

The focus of these efforts was initially on newspapers, bottles, and aluminum cans. Organizations were established less as small businesses than as an extension of social and cultural movements. These small community recycling efforts were motivated largely by concerns about a conservation "ethic" and the proliferation of roadside litter. Recycling, in fact, had split into two very different camps: the small-scale, mostly voluntary, efforts, and the holdover scrap businesses, primarily metals reprocessing, which were still able to maintain their limited markets.[27]

The rapid rise of environmental politics, meanwhile, had caught public officials by surprise. Most government recycling efforts at the municipal level, which tended to be small-scale drop-off collection centers or pilot curbside programs, were developed as a result of pressure from activists or as extensions of existing volunteer programs. In 1968, only two municipalities had any kind of newspaper curbside collection program, for example, but by 1974, as a result of such pressure, 134 cities had developed some kind of effort.[28] Local public agencies, however, still tended to dismiss recycling as more symbolic than substantive, and governmental resources were almost exclusively directed at increasing landfill capacity or exploring other disposal technologies.

At the federal level, the consolidation of environmental organizations into a lobbying force, as well as the rise of the volunteer-based movements, played a direct role in the promotion of recycling as part of a larger waste management focus that came to be called "resource recovery." This concept, which, like recycling, meant many things to many people, came to signify a range of nonlandfill alternatives, including materials recycling and energy conversion. The definition that EPA ultimately provided in the mid-1970s—"a wide variety of technical approaches for retrieving or creating economic values from waste streams"—reinforced this broad interpretation.[29]

During the early and mid-1970s, the EPA Office of Solid Waste, empowered under the Resource Recovery Act to explore waste management options, began to develop a variety of resource recovery programs, the first significant federal intervention in this area. Studies were initiated on a range of possible recycling programs for newspaper, paper products, aluminum, glass, tires, and even plastics. (Other studies focused on preparing waste as a fuel for energy production, such as resource-derived fuel.) Yet during the 1970s, recycling rates changed little, only hovering between 5 and 7 percent of the waste stream. Of that amount, 90 percent was newspaper, wastepaper, paperboard, and aluminum cans.[30] Aluminum recycling was driven as much by considerations of energy conservation as waste management concerns. By 1975, one of four aluminum cans was recycled, with the recycled product utilizing only 5 percent of the energy needed to produce a can from raw ore. Pressure from citizen groups—aluminum can collection centers grew from just one in 1967 to 1,300 less than ten years later—helped develop the supply side of this recyclable product. Consequently, the market for recycled aluminum continued to remain strong through the 1980s.[31]

By the late 1970s, at the height of EPA's activity, the agency argued that the current 5 to 7 percent recycling rate could be increased to 25 percent, matching the recycling rates for wastepaper, metals, and rubber established during World War II.[32] This could be accomplished, according to EPA officials, by a range of new government programs, most of them based on economic incentives as well as modest government intervention, such as purchasing guidelines, restructuring rail freight rate procedures, or a national beverage container deposit tax. These proposals were among the ten proposed policy initiatives undertaken for review by the Resource Conservation Committee (RCC), the single most important entity of the 1970s, to help define the federal government's approach toward recycling, reuse, and reduction strategies.[33]

The Resource Conservation Committee directly evolved out of the debates over recycling and other waste management strategies prior to the passage of RCRA. Though Congress failed to enact as part of RCRA more concrete conservation-oriented measures such as a 5-cent beverage container deposit, both EPA and the environmental lobby were pleased with the provision in RCRA establishing this new committee. The committee was to consist of the secretaries of the Interior, Treasury, Labor, and Commerce, and the heads of the Office of Management and Budget, EPA, and the Council on Environmental Quality. The Carter administration added the secretary of the new Department of Energy as well as the chair of its Council of Economic Advisors, contributing to the substantial economic orientation of the committee.[34]

In its first year in office, the Carter administration sought to elevate "conservation" into a kind of ideological homily, the "moral equivalent of war," which would, among other objectives, strengthen Carter's political standing in this environmental era. In his first environmental message, in May 1977, the president raised the waste issue, emphasizing the newly constituted RCC and urging it to speed up its timetable and offer recommendations within the year.[35]

The Resource Conservation Committee's task, a 1977 EPA report to Congress suggested, could become a "prime forum for debating and formulating a national materials policy." "Committee recommendations," the report asserted, would "essentially represent an Administration consensus, and, since Congress has a long-standing interest in this area, the likelihood of important national materials policies being enacted on the basis of the recommendations would be promising."[36] The RCC, in fact, came to symbolize the possibilities—and limits—of federal involvement in the recycling and reuse component of resource recovery.

Once constituted, the committee defined its task in both economic and environmental terms, seeking to address the problems of industrial pollution and environmental degradation as well as "the material welfare of future generations." Though a wide range of programs could have been considered, the committee limited its deliberations to ten specific policy areas for review, to represent "a full spectrum of policy alternatives."[37]

These ten specific programs involved various regulatory, tax, and subsidy measures to encourage recycling, reuse, and reduction efforts. "Each of the policies studied," the committee declared, "is focused on a particular stage in the materials life cycle." If the policies were effective, however, "each would cause a 'ripple effect' throughout the cycle, affecting people's choices to consume or conserve at the extraction stage, through processing and use, to disposal or recovery."[38]

The committee spent a relatively large amount of time on two areas in particular—beverage container deposits and a national solid waste disposal charge. The other eight policy areas included a national litter tax, local user fees, deposits and bounties on durable and hazardous goods, product regulations, the elimination of railroad freight rate discrimination favoring virgin materials, subsidies for resource recovery programs, extraction taxes on virgin materials, and eliminating existing federal tax subsidies for virgin materials.

Each policy review was framed by four general criteria established to guide the committee's research and findings: (a) consistency with "free market principles," that is, would the regulations or taxes impinge on the workings of the market system; (b) the "polluter-pays" principle, that is, would those responsible for the pollution bear the costs; (c) "economic efficiency," an important buzzword in the Carter administration; and (d) did the policy result in any undue administrative cost or feasibility problem. These four guidelines, some potentially contradictory to others, became the basis for caution and at times opposition to specific policies for many of the committee members, confounding the search for consensus. With research by committee staff and senior advisers from several of the departments and agencies involved, especially EPA, the committee sought to reach a consensus in each of the ten areas in order to develop possible recommendations.[39]

Despite the expectations of EPA and others, the RCC failed to reach agreement on how to pursue direct action for *any* of the ten measures. At most, the committee suggested that further research was needed, or it endorsed a concept without specific recommendations or means of implementation. The committee, nevertheless, was unanimous in *rejecting* reduction-oriented concepts such as product regulations or a national solid waste disposal charge. On other programs, the committee split. This was most pronounced over beverage container deposits, clearly the most controversial and divisive of the policy areas, which paralleled acrimonious debate in Congress and in several states over the same issue. Ultimately, the lack of agreement and inability to develop recommendations served to discourage even the modest initiatives resulting from other conservation-oriented resource recovery studies.

When the Resource Conservation Committee's report, entitled *Choices for Conservation,* was published in July 1979, the attention of Congress, the Carter administration, the public, and the media had already turned toward hazardous waste. The approach concerning resource recovery, furthermore, was changing. Recycling and reuse elements were increasingly being shed while a more exclusive focus on waste-to-energy began to be adopted. This shift, orchestrated by the waste management industry, soon came to preoccupy public agencies and officials as well. Though EPA still provided only limited support for mass burn incineration as a desirable resource recovery activity as late as 1980, the incineration industry was already moving aggressively to appropriate the term "resource recovery" as synonymous with incineration-based energy recovery. By the new decade, EPA's growing bias in favor of the private sector playing a more dominant role in waste management, a position strongly reinforced by the Reagan administration's promotion of privatization, ultimately led to a shift in the agency's own position around mass burn. As a consequence, while incineration moved up the hierarchy, federal intervention as well as municipal preference for recycling had all but disappeared.

THE RECYCLING REVIVAL

By the early 1980s, recycling still managed less than 10 percent of the waste stream, while predictions for "resource recovery," now just a thinly disguised term for waste-to-energy, assumed a much greater role in future waste disposal (see table 7.1). The Institute for Resource Recovery, an industry trade group set up exclusively to promote waste-to-energy, became the foremost advocate of mass burn technology. The rise of incineration, which came to preoccupy local, regional, state, and federal agencies, was tacitly understood to be incompatible with more intensive recycling. Incineration technology, for one, benefited from the presence of some key—and growing—components of the waste stream such as plastics and computer paper because of their high Btu value. Moreover, an incinerator had to have a guaranteed supply of waste to be financially viable. Consequently, public officials became convinced that the large, industry-touted trash burners would assume the role of the sanitary landfill, namely, one single disposal method handling all the unsorted waste. Such a singular approach obviated the more complicated and politically sensitive requirements of a multiple reduction, reuse, and recycling-based strategy involv-

Table 7–1
RECOVERY OF MATERIALS FROM MUNICIPAL SOLID WASTE, 1984

MATERIAL	DISCARDS[a]	RECOVERY[b]	RECOVERY RATE (%)
Aluminum	2.1	0.6	28.6
Paper and paperboard	62.3	12.9	20.7
Glass	13.9	1.0	7.2
Rubber and leather	3.4	0.1	3.0
Iron and steel	11.3	0.3	2.7
Plastics	9.6	0.1	1.0
Total	100.6	14.4	14.3

SOURCE: Franklin Associates, 1986.

[a] 1984 MSW figures, millions of tons.
[b] Postconsumer, millions of tons.

ing a substantial waste stream analysis, source separation, and other programs engaging the public, and/or direct government intervention in the pricing of products, packaging, and disposal alternatives. Mass burn incineration, instead, became the high-tech, high-hopes, but high-cost (as it turned out) answer to the waste problem, further marginalizing recycling's more complex requirements.

As a political factor, however, recycling was still presumed to have important public relations value by waste officials. It became a general truism in the industry that the public would be more favorably inclined toward waste-to-energy if recycling appeared to be integrated into such an approach. "We got our permits because the public liked the idea [that] we were recycling," one official told a recyclers' conference in 1986, characterizing such activity as "a nice thing [for the public] to do."[40]

The kinds of recycling programs favored by the waste industry and public officials tended to be educational in intent or experimental in scale and didn't figure significantly in the waste management plans being developed on the state and local level in response to RCRA. Those waste management plans that did call for specific recovery rates from recycling often had small targets, around 10 percent, just slightly higher than existing rates. Recycling programs, moreover, were never provided a proportionate share of funding to achieve their goals; programs for 10 percent recycling rates never received 10 percent of a city's waste management budget. The reemergence of recycling as a major option in the mid- and late 1980s can be traced less to such low-budget public programs than to the confrontations over incineration and landfill projects that were then taking place.[41]

In numerous communities throughout the country, community groups opposed to landfills or incinerators became the primary advocates of more intensive recycling, or even, in some instances, initiated their own programs. Such was the case in Sauk County, Wisconsin, the site of a controversial landfill that state agencies had been forced to close. Several local activists, led by Millie Zantow, a middle-aged housewife, decided to develop a recycling program as an alternative approach, while the landfill decision was still being debated. As a first step, Zantow and her group decided that it was necessary to know what was being unloaded into the dump. For two days, Zantow sat on a hill by the dump and watched, taking notes. She was particularly struck by the large volume of plastic products and containers being disposed of and wondered why such materials were not recycled.[42]

Zantow decided to find the answer herself. She quickly became a self-taught expert on the variety of plastic products and their potential for recycling. She began collecting containers in her pickup truck, removed the labels, hand-washed and -dried them, then separated them into different categories. She created Wisconsin Intercounty Nonprofit Recycling, or WINR, and started making arrangements with different plastic companies to accept her recyclables.

Once the plastics recycling got off the ground, Zantow expanded her effort, taking on a number of new recyclables, including cardboard, office paper, waste oil, newspaper, batteries, cooking oil, chipboard, aluminum cans, and three colors of glass. She created a staff of volunteers, and, through her group, handled more than 1,000 tons of recyclables a year. This saved the financially strapped county an estimated $20,000 a year in landfill fees. Zantow became convinced that upwards of 75 to 80 percent of residential waste (excluding food and yard wastes) could be recycled, arguing that markets could be established if the recyclables were of sufficient quality. For Zantow and her supporters, their recycling operation became more than just a business or an ethic of conservation, but an extension of antilandfill activism.

The experience in Sauk County was not unique. In several other communities, even in such places as Zip City, Alabama, and Natcitoches, Louisiana, activist groups have been at the forefront of a new recycling movement.[43] Extending the earlier efforts of the volunteer-based groups of the 1970s, these community activists have explicitly questioned the overall solid waste strategies of the public agencies and challenged the role of the waste industries. Their presence has forced recycling back onto the solid waste agenda and expanded the terms of the debate that had been shaped by the rise of incineration in the early 1980s.

By the late 1980s, the different wings of the recycling movement and recycling business—and the various programs that they had created—had become even further differentiated. The remnants of the 1970s volunteer-based efforts, some of them now small businesses tied to the

National Recycling Congress, still survived. Many had lost their ideological fervor while seeking to improve the operational aspect of their recycling activities. The National Recycling Congress, in fact, failed to develop any visible lobbying presence and even refrained from taking any position on the incineration controversy, the very issue that was spurring the new community movements, now the leading advocates of a new, intensive recycling approach.[44]

The other main force of the 1970s, the old-line scrap businesses that coalesced into the National Association of Recycling Industries, still functioned during the 1980s as modest-sized businesses. Some of these recycling businesses, ironically, worried about new pressure from expanded recycling efforts, particularly that such supply-side efforts might overtax existing markets for recyclables and cause declining prices for their products.[45]

By the late 1980s, the supply side of recycling had indeed revived, as a number of municipalities and states quickly developed highly visible programs. This expansion of recycling was significant as much for its timing (much of it defined in crisis management terms) as its scale. Some communities, like Los Angeles, appeared to undergo a 180-degree reversal regarding recycling's contribution to waste management objectives. Targeted goals of 25 percent and even as much as 50 percent, in just a short period of time, were proposed by elected officials pressed by the community opposition. This heightened expectation contrasted with the ongoing presumptions of many in the waste bureaucracies that a 10 to 15 percent recycling rate was the most realistic goal possible.[46]

Similar to the public agencies, and for somewhat parallel reasons, waste industry companies, particularly incinerator and landfill operators, also began to explore recycling strategies and programs. Waste Management's Recycle America was, of course, the most prominent of such efforts, which were extensions of their garbage collection programs. Others were organized as waste processing programs, similar to the older sorting procedures of a previous generation. These emphasized a "one-stop source separation" effort, whereby all recyclables would go into the same container and sorting would be done (mostly through mechanical means) at an incinerator, or materials processing plant. These materials recovery facilities could eliminate those recyclables such as glass and cans and thereby improve the efficiency of an incinerator. Though waste processing plants were often designed independently of any incineration strategy, they were quickly embraced by some waste companies as a "convenience" approach to separation rather than curbside programs. "Householders," one Ogden official argued in favor of this approach, would be "relieved that recycling can be made so easy for them."[47]

In each case, moreover, the waste industry companies were quick to incorporate such programs as part of their newly developing public relations approach, one shared by the public agencies. This involved the adoption during the late 1980s by both the waste industry and the public waste bureaucracies of the concept of a *horizontally integrated* solid waste management strategy. According to this approach, there would be a place for recycling *and* mass burn, reuse and landfills, but without any real hierarchy or prioritization of programs.

The original concept of an integrated strategy had been raised earlier by activists hoping to slow down the incineration bandwagon. A vertical hierarchy was essential to this strategy since reduction, reuse, and recycling, in that order, were generally accepted as more environmentally benign options. But with the shift in political fortunes of waste-to-energy, incineration advocates moved quickly to appropriate the integrated waste management concept while dropping the notion of hierarchy. They argued that the waste crisis left little time for experimentation with new recycling strategies. The waste industry and the public agencies, while acknowledging recycling's new respectability, did not always translate respectability into concrete programs, other than the stated goals of this new, "integrated" approach. Nor was there any meaningful attempt to tackle in conceptual or ideological terms how wastes were created by challenging, for example, the short use life of a particular product or developing an extensive household separation program.

The integrated approach underlined as well the limited response at the federal level toward recycling during the Reagan years. With the passage of RCRA in 1976, Congress mandated that EPA establish federal procurement guidelines for products made from recycled materials. In 1980 and again in 1984, Congress further established deadlines for EPA action. But by the end of Reagan's second term, EPA had issued guidelines for only two categories: paper products (promulgated in 1987 in response to a lawsuit by environmental groups) and concrete containing coal fly ash (which, according to critics, remained continually underemphasized even after promulgation). The Commerce Department, furthermore, which had been explicitly mandated by RCRA to stimulate the secondary materials market essential for recycling, had, in the same time-span, "done almost nothing to fulfill this statutory mandate," as one recycling study noted.[48]

At the local and state level, however, the combination of pressure from community and environmental groups and the increasing problems facing incineration and landfills had caused public agencies, by the late 1980s, to jump into the recycling arena. But this effort to get "serious about recycling" was taking place, as one *New York Times* article noted, "for lack of options." In New York's case, for example, both New York City and the state of New York initiated recycling programs that were designed to eventually include a mandatory system of separation of certain recyclables at the household level. The passage of these programs was accompanied by a major publicity campaign suggesting that recycling had become a major enterprise deserving public support. The investment in these efforts, however, remained far less substantial than New York's push for incineration. New York City, for example, adopted its recycling program only after approving the 3,000-tpd Brooklyn Navy Yard incinerator, and then allocated $10.2 million, less than 2 percent of its $725 million Sanitation Department budget, for recycling.[49]

Mandatory recycling, the ultimate objective of the New York program, had, by the end of the 1980s, become something of a buzzword. More and more management plans were put forth with mandatory recycling as their centerpiece, though few specified in detail the design and implementation of such an effort. In this period, dozens of cities and counties, as well as several states, including Rhode Island, Connecticut, New Jersey, and Maryland, had adopted mandatory recycling statutes. Though the mandatory programs tended to obtain higher participation rates and often increased overall recycling rates by a substantial margin, they were structured with few if any means to insure compliance. At most, fines were issued or warnings given that failure to comply could lead to a refusal to pick up unsorted curbside trash. Despite the minimal enforcement, advocates of mandatory recycling argued that such programs essentially had an important educational value, demonstrating the seriousness of the recycling effort.[50]

Another initiative encouraging recycling that emerged around the same time was the disposal ban, which precluded the landfilling of certain wastes—yard trimmings, for example. Such programs were begun in Portland, Oregon, which banned the landfilling of yard waste by 1989; Broome County, New York, which, by May of 1986 had banned leaves from landfills; Delaware County, Pennsylvania, which banned newspaper and other recyclables from its transfer stations; and Madison, Wisconsin, which set a one-year deadline in 1987 for its landfill ban of yard trimmings and other compostable waste. The Dukakis administration in Massachusetts also proposed a timetable to ban all leaves from solid waste facilities after communities in that state had established composting programs.[51]

In Madison, the disposal ban also generated a process of reexamination of the community's overall waste disposal strategy. Waste Management had been the major hauler for a number of Wisconsin communities, including Fitchburg, located in the center of the Madison metropolitan area. The Fitchburg leaders, unhappy with Waste Management's service, decided to try another hauler in conjunction with a new recycling effort. This program had been developed in response to the Madison disposal ban as well as to the growing problems facing local landfills. The new hauler first proposed a centralized collection point along with the distribution of clear plastic bags to hold the recyclables, similar to Waste Management's plan in Carsonville, Ohio. But in Fitchburg, residents balked at the idea, another set of council members were elected, and the new council set about to develop instead a participatory source separation program.[52]

"We were careful to move deliberatively, while informing people as best we could of the options," then Fitchburg mayor Jeane Seiling recalled. Fitchburg decided to use the approach that had proven so successful in San Jose, California, where bright-colored collection crates (a different color for each recyclable) were distributed to everyone participating in the program. At first, the program was voluntary and the participation rate was relatively low. But once the program became mandatory, community interest became far more extensive and participation rates tripled, reaching 66 percent in the first several months of operation. "It might seem a little old fashioned," Seiling told a group of University of Wisconsin students in 1988, "but this project has become a source of pride for myself and many others in the community. People often feel they can't do very much about all the enormous problems we face in our society, but here they could demonstrate that something *could be done,* even if our contribution to dealing with the overall solid waste problem was still small." The Fitchburg program, initiated by local elected officials and backed by constituents, many of whom had initially been skeptical of the idea, quickly became a model for other communities in Wisconsin.[53]

The Fitchburg experience, in fact, was being duplicated in a number of communities throughout the country, particularly in some smaller towns where the need for citizen action was most compelling. In East Hampton, New York, for example, landfill capacity problems created an interest among community activists as well as sympathetic local officials to explore a substantial recycling effort. Facing numerous proposals for incinerators in the county, the town turned for guidance to the Queens College Center for the Biology of Natural Systems (CBNS), headed by the well-known writer and scientist Barry Commoner. Commoner's group had been in the forefront of the incineration issue and was also designing a more intensive approach to recycling based on an exhaustive waste stream analysis. By developing such data, various distinct categories of recyclables could be directed to a variety of possible uses, as in the Japanese approach to recycling. Thus, recycling goals could be set at far higher figures than previous targets by striving to achieve what was theoretically possible to recycle. In the East Hampton case, CBNS calculated that the maximum percentage of the waste stream potentially recyclable was 84 percent (by weight). Then, the target goal was determined by multiplying this figure by the percentage of people in the community who would separate their wastes. In East Hampton's case, of the 100 families that volunteered for the pilot program, participation rates averaged an exemplary 95 percent, resulting in an overall record-setting recycling rate of 73 percent.[54] These are phenomenal numbers and the results have led to the characterization of East Hampton's recycling effort as "the nation's most ambitious curbside recycling program."[55] A CBNS survey in the larger and more diverse city of Buffalo supported these findings, where nearly 80 percent of the randomly sampled residents indicated a willingness to separate their trash. Moreover, the response was similar in all three neighborhoods sampled, despite their quite distinct class and ethnic backgrounds.[56]

The major difficulty facing the new recycling movement has been the continuing skepticism by the waste industry, public officials, and even the media that intensive recycling could provide a major response to the waste dilemma. Such a method, as the Commoner group insisted, could be as substantial, in terms of volume reduction, as incineration, and, ultimately, when combined with a waste reduction program, the *primary* method of handling the entire municipal solid waste stream.[57] In fact, the growing popularity of recycling as the preferred solution had already forced incineration and landfill advocates to embrace the concept of an integrated waste management strategy as a means of keeping their options viable.

As early as 1986 and 1987, industry figures and public officials had begun to make a distinction between a modestly conceived recycling program with its place in a nonhierarchical integrated strategy and intensive recycling as potentially *the* major component of such a strategy. A Spokane, Washington, official, for example, wrote in a waste-to-energy trade publication that "the choice is not 'recycle or burn.' " Rather, he said, "the only option is to strike a balance between all four legs of the solid-waste management chair: waste reduction, recycling, solid waste combustion and landfill disposal. . . . Every program will be unique." The Spokane official concluded, "Every 'chair' has to serve a unique location. But each 'leg' of every chair must be the proper length to reach the ground."[58] It was unrealistic, incineration advocates argued, to set recycling goals too high, as in the East Hampton case: such a position, it was feared, might alter the industry concept of partners in an integrated strategy ("partners" who in fact had never been equal in terms of subsidies, funding support, and other forms of government intervention) toward that of a waste management *hierarchy*, with the various strategies ranked in priority order for funding and support.[59]

The rapid interest and contending views regarding recycling that emerged in the late 1980s had in fact tended to obscure the distinction between the "horizontal" integrated approach, which had become the waste industry's position, and the "vertical" waste hierarchy policy, which the community groups put forth. A waste hierarchy necessarily implied different priorities and new approaches, including a possible indefinite moratorium on new incinerators. Such an approach reversed most existing policies, which were based, in effect, on *inverted* hierarchies. Reduction, reuse, and recycling, the three top rungs in that order in the hierarchy advocated by community groups (and accepted in the abstract by public officials and the waste industry) had over the years been reversed, with landfills and incineration receiving far greater support and commitment from local and federal agencies alike. What remained largely absent from the public debate over recycling, particularly in terms of the larger public forums such as City Council meetings, congressional hearings, and press coverage, were precisely those features of recycling that would be most compelling—and most controversial—in providing the kind of support that incineration had once had and continued in part to still enjoy. These included, among other possible forms of support, the development of detailed, standardized, and community specific waste stream analyses, new pricing mechanisms, and direct government intervention in the establishment of markets. Without these, recycling would likely remain an unfulfilled promise.

SOURCES OF SUPPORT

The essential starting point for any intensive recycling program is a systematic waste stream analysis. Such a detailed analysis, for example, has long been recognized by the Japanese as central in achieving their impressive 50 percent recycling rate.[60] Waste stream analysis in the United States, however, has tended to involve only approximate estimates, primarily based on a materials flow methodology. This approach, first developed during the 1970s by EPA analysts, essentially involves an estimate of what is likely to be discarded based on what is manufactured rather than actual measurements made at disposal sites.[61] These studies were originally designed to provide background information for national programs in materials recovery (such as paper and glass recovery programs) and energy recovery (that is, waste-to-energy projects), rather than detailed information for local recycling programs designed specifically as part of a community's waste management strategy. These studies, furthermore, relied exclusively on weight parameters, such as tons per day, tons per year, and so on, and not on volume factors. The most recent EPA-sponsored waste stream analysis, the 1986 Franklin Associates study, suggested, in fact, that the "relationship between volume and weight of the components of [municipal solid waste] have not been well established, so far as is known."[62] But it is the volume of wastes that has become an increasingly critical factor for landfills in recent years, since their capacity is primarily a function of volume, dependent on the remaining available space. Incinerators, on the other hand, have defined their capacity in weight terms, and have been better served by materials flow analysis, which has also produced information on a discard's Btu content (heat value) as well.

The focus on weight in current waste stream analyses has served to underestimate the importance of changes in the volume of wastes. For example, EPA analyses have shown that, by volume, the paper and plastic components of the waste stream—both highly conducive to mass burn technologies—have been increasing more rapidly than any of the other categories. At the same time, these studies also found that containers and packaging showed a slightly declining percentage of the total by weight, not because of less generation of such wastes, but from the use of *lighter-weight materials;* for example, the plastic soft drink container.[63] In some respects, substitution can be considered an advantage in the operation of incinerators (less weight, the plastics offer a higher Btu, less metals, and so on), while presenting distinct disadvantages for landfills (greater volume, nondegradable products, more toxic leachate).

This situation is compounded by other problems in methodology for waste stream analysis. EPA's own standardized survey form available to local communities for estimating, characterizing, and sampling municipal refuse, for example, has been criticized for its inaccuracy and unreliability. Meanwhile, at the local level, where waste stream analysis is at a minimum or can represent a delaying tactic for establishing new programs, those communities that have undertaken their own studies have utilized a hodgepodge of methods. This in turn makes any comparison of estimates difficult.[64]

A detailed and consistent waste stream analysis, however, lies at the heart of the intensive recycling strategy. As the East Hampton study demonstrated, the various categories developed in a waste stream analysis are essential in the design of a successful recycling program. (See table 7.2 for selected waste composition studies.) The city of Seattle, for example, developed an extensive schemata to characterize their waste in order to facilitate a new recycling initiative. This included five separate categories for paper, plastics separated by recyclable and nonrecyclable products, the diversion of household toxics, and two categories of compostable wastes, as well as a three-way scheme for metals (ferrous, aluminum, and others), glass containers (though not segregated by color as some other recycling programs require), tires, other organics (in addition to the compostables), and other inorganics (separated into construction materials and all others). The state of Washington's Department of Ecology, along the same lines, produced a detailed waste stream analysis for a recycling study that had 27 different categories and, importantly, distinguished

Table 7–2
CLASSIFICATION SCHEMES IN SELECTED WASTE COMPOSITION STUDIES

NYC	CT	HUDSON FALLS, NY	SEATTLE
Paper	Paper	Paper	Paper
Glass and ceramics	Newspaper	Water	High grades
Metals	Books and magazines	Glass	Newspaper
Ferrous	Office	Metal	Corrugated box
Aluminum	Corrugated	Yard waste	No. 1 mixed
Other	Glass containers	Wood	Nonrecyclable
Textiles	Metal	Food waste	Metals
Rubber and leather	Ferrous appliances	Misc.	Ferrous
Plastics	Ferrous cans	Plastics	Aluminum
Wood	Ferrous other	Leather and rubber	Other nonferrous
Food and yard	Nonferrous		Glass containers
Misc., including bricks, rock, and dirt	Aluminum	Textiles	Tires
	All else		Plastics
			Recyclable
			Nonrecyclable
			Compostable waste
			Yard waste
			Other compostable
			Other organics
			Other inorganics
			Construction materials
			Other
			Hazardous materials

SOURCE: *Coming Full Circle*, Environmental Defense Fund, 1988.

between commercial and residential sectors. And the city of Portland, which developed an extensive recycling proposal with a 52 percent reduction goal, based it on a detailed study of the waste stream undertaken by the city's environmental agency. Moreover, groups like Commoner's CBNS and the Garbage Project at the University of Arizona have developed the capacity to detail a community's waste stream compatible with an intensive recycling approach.[65]

While a sophisticated waste stream analysis helps lay the framework for a successful "supply side" operation, the "demand side" of recycling can be most directly influenced by various economic measures, including pricing mechanisms. Such mechanisms, in fact, have been employed over time to *discourage* markets for secondary (recycled) materials and to favor the use of virgin materials, thus undercutting recycling. Rail freight had long been priced to favor such key virgin materials as glass, wood, and ferrous metal. Furthermore, subsidies for energy production and tax policies, such as depreciation and extraction allowances, have encouraged use of raw materials and inhibited recycling.[66]

During the 1970s, several of these antirecycling programs promoting the use of virgin materials came under attack primarily because of their costs and energy requirements. The Resource Conservation Committee study, for example, discussed possible economic disincentives such as a national solid waste disposal tax in order to reduce raw materials use, partly for the energy savings involved. Such proposals, however, tended to disappear from the legislative and public debate over solid waste programs, particularly by the early 1980s, when talk of an energy crisis began to fade.[67]

The key issue underlying any discussion of pricing incentives and disincentives has remained the question of the real—and hidden—costs of disposal. Only recently have both landfills and incineration begun to assume a few of the key external costs that are passed on to the public. These include costs of cleaning up and preventing pollution (though often not the health costs borne by exposed populations). Moreover, the price of disposable products, some of the fastest-growing items in the waste stream, continue to ignore the true costs of disposal. Such costs include the decline and replacement of recyclable and reusable products. By the late 1980s, some communities tentatively responded to this issue of external costs by modestly adjusting their disposal fees, partly to encourage underfunded recycling programs. Examples included a Los Angeles County surcharge at landfills and incinerators for recycling programs, and a Lane County, Oregon, direct subsidy to waste haulers in the form of reduced landfill fees if they also offered a recycling program. The increased landfill and incinerator fees were generally modest, from New Jersey's $1.50 per ton surcharge to a $6 a ton charge in Vermont.[68]

These surcharges, however, still fail to reflect the full extent of the resources communities need to allocate to handle the range of solid waste disposal problems. One method for calculating such costs is the use of a marginal cost pricing system to determine the tipping fee schedule for landfills or incinerators. Marginal cost pricing in the waste arena has been defined by some industry analysts as the "unit cost of the extra capacity to handle the last annual ton of waste."[69] Such a method would include the opportunity cost of land used for landfilling, for example, as well as costs posed by environmental risks and higher cost replacement facilities.

The reverse side of this approach is the concept of "avoided costs" attributed to any effort to reduce the size of the waste stream.[70] Such avoided costs represent a potential economic benefit to any government agency involved in waste collection and disposal and should correspond to what an agency would be willing to spend to reduce the amount of wastes otherwise earmarked for disposal. The avoided cost formula generally corresponds to the savings generated from reduced landfill use, the ultimate disposal repository.

An accurate measure of avoided cost takes into account a number of different variables. These can include the costs of collecting, transporting, and disposing of the wastes, otherwise known as the "base" costs, and the "value of diversion" costs related to extending the landfill's life and therefore postponing the day when present capacity runs out and much higher costs have to be expended for a replacement. A more developed avoided cost approach, however, should also include "pollution costs" related to costs of closure, postclosure operations, and maintenance, remedial action, and compliance with new environmental regulations; the "resource depletion cost" associated with the value of the material recovered or recycled, the value of the material lost if landfilled or burned, and the cost to develop new supplies of raw materials and the energy saved; the "opportunity" cost related to the loss of the landfill property to future use; and finally the "social costs" in terms of the number and kinds of jobs created.

The implementation of a complete avoided cost approach would play a significant role in the development of any intensive recycling program. It would more accurately measure the economic value of recycling and provide a cost-effective rationale for the significant financial and policy support such a program needs in order to reach the goals it is capable of achieving. A few communities have accepted in principle the concept of avoided costs, but have still defined it in minimal terms, thus undercutting the full value of recycling in their overall approach.

Los Angeles, for example, credited only a portion of the base costs (transportation and disposal) and resource depletion costs (value of material recovered) against the gross costs of its recycling program. As pointed out in the UCLA LANCER study, however, the city did not recognize any savings on collection costs, suggesting that all such costs are fixed. But other communities with curbside collection programs have indicated that they have achieved between a two-thirds and one-to-one reduction in collection costs. In other words, for every ton that is recycled, collection costs are reduced between two-thirds and 100 percent. By applying a two-thirds reduction factor against the city's collection costs during 1984–85, for example, an additional $20.46 for each ton recycled would have been saved. Furthermore, an additional cost ranging from $37.44 to $56.89 per ton could have been saved if L.A.'s avoided cost formula had also included those figures then available for calculating the replacement "pollution" and "opportunity" costs, categories only partially included in L.A.'s calculations. This estimate, moreover, does not include any of the unknown environmental costs, social costs, or the resource depletion costs. By including only the known variables, the city's calculation of avoided costs came to $33.39 per ton while the UCLA study estimated a minimum figure of between $72.70 and $88.90 per ton. This differential alone suggests that use of an avoided cost formula by itself could easily support a modest recycling effort.[71]

The major contention of critics of intensive recycling is that the demand-side markets for recycled materials and products, more than the supply-side of recycling, is either undeveloped or potentially incapable of handling any large traffic in recyclables. The absence of sufficient buyers of recycled materials as well as a sufficient number of production facilities, these critics argue, will mean that the increased supply developed through mandatory programs and related efforts will only undermine the markets that currently exist. The problems experienced during the mid- and late 1970s by recycling industries, including wastepaper, scrap metal, and aluminum, are seen as examples when the demand side of the recycling business bottomed out. A demand-side equivalent of a mandatory program would mean, as the executive director of the Institute of Scrap Recycling Industries, a major intensive recycling critic, put it, that "the government would have to dictate which materials manufacturers would have to use in their products. It would have to specify which market alternatives are available and which are no longer available. It would have to dictate to consumers that they must purchase certain products and not others. . . . These demand-based 'mandatory recycling' needs," the scrap industries recycling director concluded, "are simply not possible, if they are even desirable."[72]

By the 1980s, however, most large-scale recycling programs that did exist, with the possible exception of newsprint, were not hampered by inadequate markets.[73] Furthermore, *assumptions* about "inadequate markets," as one recycling study pointed out, "may be a more significant barrier to increased recycling than markets themselves."[74] The decline of markets in the mid- and late 1970s, for example, was primarily a result of the major economic recession from 1974 to 1975, which brought about, among other consequences, a crash of the paper market, including recycled paper products. The recession and postrecession period, as EPA studies pointed out, witnessed a parallel decline in recycling activities on both the supply and demand side, as well as the overall per capita generation of wastes themselves.[75] Many of the recycling-oriented measures from RCRA, moreover, such as procurement guidelines, failed to develop over time to help offset the postrecession market fears that had undercut the momentum for recycling. When these programs revived in the 1980s, such key recycling industries as paper and aluminum were able to resume their growth—by the late 1980s, for example, more than 50 percent of all aluminum cans were made from recycled aluminum[76]—despite the assumption that market growth was still only minimal.

Markets, whether for virgin or recycled products, do not exist in a vacuum in terms of government intervention. The federal government, for example, accounts for about 2 percent of all the paper consumed in the United States.[77] Decisions about what to buy and what to support through government policies have always been basic decisions affecting the waste stream as well as the overall economy.

Government procurement policies could include "set aside" programs, such as one developed in the state of Maryland, where at least 40 percent of all paper purchased by the state has to contain at least 80 percent secondary fibers, a figure often used to define a recycled product. They could also include a "price preference" approach, such as the one developed in New York that requires the purchase of recycled paper if the bid is within 10 percent of that of the virgin paper manufacturer.[78]

By the late 1980s, government procurement programs were still relatively small, most having been developed in the previous years or even months, despite the fact that the huge number of people employed by federal, state, and local agencies—more than 16 million as of 1985—spent the equivalent of 35.4 percent of the gross national product. But as of 1988, EPA had issued only its two sets of procurement guidelines, while 18 states, three cities, and only one county had, in this same period, developed their own guidelines. Most of these programs applied to paper, though nine states had devised some procurement laws for additional products, such as tires and motor oil.[79]

Government intervention on the demand side has also been justified as a form of job creation and economic development, reasons often used for economic support and intervention in industries utilizing virgin materials. Most recycling programs tend to be more labor- than capital-intensive, with the exception of waste processing or materials recovery facilities that rely heavily on mechanical sorting technologies. One study estimated that for every 600 tons per year of waste material to be disposed of or recycled, one recycling job is created, while the equivalent figure for landfilling is .04 jobs, with .05 jobs for incineration.[80]

The support mechanisms for intensive recycling, as the programs themselves, are underutilized or even undercut at times. Recycling, like other waste collection and disposal activities, represents a form of government intervention. It requires both more complete implementation of existing policies as well as the initiation of new approaches, such as pricing mechanisms. What these forms of intervention fail to address—

significant as they are in developing a much vaster recycling effort—are the production and consumption values associated with the broader question of wastes in general. To recycle then becomes only one part of the issue; the definition of waste itself sets the framework.

FROM WASTE TO DISCARD: A QUESTION OF TERMS—AND IDEOLOGY

The variety of recycling programs reflects key values and assumptions about the waste issue. The difference between source separation at the household level and materials recovery facilities where separation is done at one centralized facility can, for example, suggest contrasting attitudes about public engagement. Government agencies and waste companies that have belatedly, often reluctantly, become involved in recycling continue to argue that a public backlash is possible, pointing to the decline of household separation in the post–World War II years. The community movements that have raised recycling as an extension of their anti-landfill or anti-incineration activism insist, on the other hand, that increased public involvement is a *precondition* for the success of recycling, and that such programs must be designed with democratic input and popular support. "We need community projects as opposed to large scale projects that try to recycle too mechanically," Wisconsin's Millie Zantow argues. "You just have to break down the system to make it simple, to make it a little more personal."[81]

The distinction between an *integrated* waste management program and the waste management *hierarchy* has also produced different conceptions of the role of recycling. Within the last few years, the waste industry and government agencies have revised their waste management scenarios to accord a larger but still modest role for recycling while arguing that incineration and landfills remain essential.[82] Waste reduction, also included in such integrated strategies, is given lip service at best, "an option to consider," as one waste industry publication put it, but "one over which local governments have limited or no control."[83]

Adherence to the waste management hierarchy addresses precisely which programs to support first and with what share of available resources. These efforts, however, are hampered when there is an unwillingness to implement the hierarchy. For example, though several states such as Oregon, New York, Washington, Vermont, Massachusetts, and Michigan have all endorsed the hierarchy concept, they often utilize a broad concept of "resource recovery" in defining their priorities.[84] The use of the term suggests multiple approaches, some of which are potentially incompatible, and can often preclude the adoption of more extensive support mechanisms such as the use of comprehensive avoided cost calculations.

Vague waste hierarchies also often fail to account for different recycling modes and the critical distinction between a product that is *reused* in its existing form and one that undergoes a product transformation, albeit from material that is *recycled*. The former, defined as the reuse of a waste, involves far less energy and often requires no additional raw materials to produce the final "recycled" product. The latter, on the other hand, often requires both additional energy and materials and may result in some environmental pollution in the production of the "new" product.

Most importantly, all current waste management planning has been designed to accept, by definition, the current structure within which wastes are generated and then disposed. "As long as you call it garbage," one recycling activist declared, "people will think of it as garbage."[85] The term "garbage," and the more neutral sounding "solid waste," emerged historically from a system based on a tendency toward overproduction, with its heavy use of cheap and subsidized virgin materials, erosion of product durability, and rapid growth of convenience items or one-use disposable products. This system of production was accompanied by the rise of collection and disposal technologies, mainly landfilling and incineration, which allowed the exponential growth of wastes to be easily placed, as it has been classically stated, "out of sight and out of mind." Incineration, especially, offered hope that the growth of the waste stream, considered by some an essential measure of economic growth, could continue unabated. The notion that a "postconsumer product," the more technical term for a waste that has completed the consumption cycle, could be and indeed ought to be a product potentially available for reuse in its own or another form conflicted with basic assumptions about growth and productivity and the technological means to sustain them.

The key to a successful intensive recycling strategy involves the redefinition of this postconsumer product and the assumptions about productivity that relate to it. If postconsumer products entail a bag of unsorted garbage, as recycling theorist Mary Lou Van Deventer argues, recycling becomes just about impossible. "Recycling mixed garbage," she proclaims, "is like unscrambling eggs."[86] "Waste," "trash," and "garbage" have all become dirty, taboo words in our culture. Sorting from a garbage bin is left to contemporary scavengers, the homeless or underemployed.

In the Progressive Era, postconsumer products were seen as having economic value. But after World War II and the triumph of new production and consumption values and waste management technologies, the derivative value of wastes were once again minimized, subsumed by other economic imperatives. By the 1970s and 1980s, the concept of "resource recovery," a variant of the old Progressive theme of production for use, became essentially an argument about energy use and production. The economic value was in terms of cost avoidance (lengthening the life of a landfill), with no intrinsic value to the waste itself.

The new recycling movements, bred of community activism in the 1970s and 1980s, began to develop the rudiments of a new conception of waste. Their focus has been primarily on community engagement, with source separation appropriately viewed as the essential, initial activity. In this setting, the postconsumer product, becomes, as Van Deventer puts it, a "discard," a product that retains some of its own intrinsic value.[87] "Discard" implies availability for reuse, and it also assigns value to products that have a longer use life prior to ultimate disposal. The criticism of the production and use of polystyrene "Styrofoam" containers, particularly the clamshell package used in fast food outlets, for example, is derived not only from Styrofoam's negative environmental consequences (impact on landfills and incinerators, harm to the ozone layer, and so forth), but also from their extraordinarily limited use life (perhaps as little as 60 seconds) and inability to be successfully recycled or reused. For the community activists, the clamshell has been produced as a waste and not a discard.

The conflict over appropriate definitions, then, reflects the conflict over the conception of waste and strategies for its management. The varying and differentiated forms of recycling, moreover, provide an important illustration of this conflict. And while recycling in the late 1980s might still be considered "a good thing to do," the question of what should be done remains both compelling and unanswered.

NOTES

1. For background to the Carsonville episode, see *Sanilac County News*, April 6, 1988; April 13, 1988; December 9, 1987; May 4, 1988; March 16, 1988; January 18, 1989; September 28, 1988; November 23, 1988; see also interviews with Sally Teets, 1988; Rene O'Connell, 1988; Peter Montague, 1988; also "List of 772 Subsidiary Corporations Owned by Waste Management Inc. as of 12/31/1987," Environmental Research Foundation, Princeton, New Jersey, July 2, 1988.
2. "Planned and Projected Municipal Waste Combustors Profile Update," Draft, Radian Corporation, May 18, 1988, EPA Contract No. 68-02-4378, Research Triangle Park, North Carolina; also *New York Times*, June 3, 1987; "News Break," *Waste Age*, April 1987.
3. "Report: Meeting with Blue Water Cardboard-Paper Recycling," Citizens Against Rural Exploitation, Recycling Committee, Lowell Rainge, Spokesperson, October 26, 1988, Carsonville, Michigan; "Five Year Solid Waste Plan for Sanilac County," memo from Rene O'Connell and Mary O'Simon, Co-Chairs, Citizens Against Rural Exploitation, to Sanilac County Solid Waste Committee and Sanilac County Board of Commissioners, November 28, 1988.

4. Interview with Rene O'Connell; see also "Community Report," Tri-City Landfill ads in the *Sanilac County News,* April 13, 1988; April 20, 1988; April 27, 1988.
5. Interview with Rene O'Connell, 1988.
6. *Sanilac County News,* January 25, 1989; January 18, 1989; November 23, 1988; June 18, 1988.
7. "A True Environmentalist," Paul Montrone, in *Annual Report* for 1987, Wheelabrator Technologies, Danvers, MA, 1988.
8. "Recycling Integrated with Waste to Energy," Lewis Ott Ward and Garrett A. Smith, Ogden Martin, 1988; interview with Garrett A. Smith, 1988; "Recycling Policy," Ogden Martin Systems Inc., September 1987; see also "Materials Recovery Facilities," Richard J. Kattar, *Resource Recycling,* January/February 1988.
9. "Officials Say Garbage Collector Threatened Lives," Frank Donze, *Times-Picayune* (New Orleans), October 7, 1988; "Waste Management Accused of Gangster-Style Death Threats Against New Orleans Officials," *Hazardous Waste News,* October 17, 1988; no. 28, June 8, 1987; see also *The Greenpeace Report,* Greenpeace, Washington, D.C., 1987.
10. Interview with Stuart Clark, Western Regional Director, Empire Waste Management (Recycle America), 1988.
11. "Recycle America Project Profiles," advertisement appearing in *Recyclenet Gazette,* Quarterly Newsletter of the Association of New Jersey Recyclers, August 1988, Absecon, New Jersey.
12. Interview with Jack de Witt, 1988; "Recycling Plan's Success Rests in User Participation," Lori Nims, *The Lima News,* January 15, 1989, p. 1.
13. Interview with Stuart Clark.
14. *Recovering the Past: A Handbook of Community Recycling Programs, 1890–1945,* Monograph, Suellen Hoy and Michael Robinson, 1979; "Currents in the Waste Stream: A History of Refuse Management and Resource Recovery in America," Daniel Thoreau Sicular, 1984, master's thesis, University of California at Berkeley.
15. *Pollution and Reform in American Cities, 1870–1930,* Martin Melosi, ed., Austin, 1980; also "Some Statistics of Garbage Disposal for the Larger American Cities in 1902," *Public Health Papers and Reports,* American Public Health Association, vol. 24, p. 904; "Refuse Collection and Disposal," *Municipal Journal,* vol. 39, no. 20, November 1915; *Municipal Housekeeping,* William Parr Capes and Jeanne D. Carpenter, New York, 1918.
16. *Street Cleaning and Its Effects,* George E. Waring, Jr., New York, 1898; *Garbage in the Cities,* Martin Melosi, 1981.
17. *Collection and Disposal of Municipal Refuse,* Rudolph Hering and Samuel A. Greeley, New York, 1921; "Currents in the Waste Stream," Daniel Thoreau Sicular, 1984.
18. "Refuse Disposal in Southern Cities," *American Journal of Public Health,* vol. 5, no. 9, September 1915; *Collection and Disposal of Municipal Refuse,* Rudolph Hering and Samuel A. Greeley. A more recent bias has emerged concerning the correlation of income and ethnicity with recycling behavior. This has also proven to be illusory in terms of actual studies of refuse data and recycling participation by neighborhood. See, for example, "Recycling: Great Expectations and Garbage Outcomes," Randall H. McGuire, *American Behavioral Scientist,* September/October 1984, volume 28, no. 1; see also "Intensive Recycling: Preliminary Results from East Hampton and Buffalo," Barry Commoner et al., presented at the Fourth Annual Conference on Solid Waste Management and Materials Policy, January 27–30, 1988.
19. "Disposal of Garbage by the Reduction Method," Irwin S. Osborn, *American Journal of Public Health,* vol. 2, no. 12, December 1912; "History of Solid Waste Management," David Gordon, in *The Solid Waste Handbook,* ed., Michael Robinson, 1986; Capes and Carpenter, *Municipal Journal,* November 1915, p. 190.

20. "Currents in the Waste Stream," Daniel Thoreau Sicular, 1984; *Collection and Disposal of Municipal Refuse,* Rudolf Hering and Samuel Greeley, 1921.
21. "Committee on Refuse Collection and Disposal," *Refuse Collection Practice,* American Public Works Association, Chicago, 1941; "Experimental Sorting of City Refuse," Rolan D. Stoelting, *The American City,* vol. 53, no. 8, August 1938, pp. 42–43.
22. "Refuse Collection and Disposal Developments in 1945," Carl Schneider, *Public Works Engineers' Yearbook,* American Public Works Association, Chicago.
23. "Technical Problems Associated with Wastes from Paper and Paperboard Products," G. Keith Provo, in *Proceedings: First National Congress on Packaging Wastes,* SW-9rg, EPA, 1971.
24. Remarks on Panel, "Natural Biases Toward Packages and Packaging Wastes Problems," Leonard Stefanelli, in *Proceedings: First National Congress on Packaging Wastes,* 1971; "Salvage Is Big Business," H. P. Godwin, *Purchasing: The National Magazine of Industrial Purchasing,* September 1944, pp. 92–94.
25. Remarks, Leonard Stefanelli, in *Proceedings,* 1971; also, for a discussion of the problem of contamination of recyclables in computer trucks, see *Resource Recycling: What Recycling Can Do,* Daniel Knapp, Berkeley, 1982.
26. Cited in "The Rise and Fall of Recycling," Neil Seldman, *Environmental Action,* January/February 1987.
27. "The United States Recycling Movement, 1968 to 1986: A Review," Neil Seldman, Institute for Local Self-Reliance, Washington, D.C., October 1986.
28. *Resource Recovery and Waste Reduction,* Third Report to Congress, 530/SW/161, EPA Office of Solid Waste, 1975.
29. Third Report to Congress, 1975, ibid., p. 1.
30. *Resource Recovery and Waste Reduction,* 4th Report to Congress, 530/SW-600, EPA Office of Solid Waste, 1975.
31. *Materials Recycling: The Virtue of Necessity,* William U. Chandler, Worldwatch Paper No. 56, October 1983, Worldwatch Institute; also *Resource Recovery and Conservation Activities: A Nationwide Survey,* 530/SW-142, EPA, 1977.
32. Fourth Report to Congress, EPA, 1977; "Recycling in the U.S.: The Vision and the Reality," Oscar W. Albrecht (EPA) and others, in *Proceedings of the 7th Annual Research Symposium,* EPA, March 1971.
33. *EPA Activities under Resource Conservation and Recovery Act, Fiscal Year 1977,* 530/SW-663, EPA, 1977.
34. *Hazardous Waste in America,* Samuel S. Epstein and others, San Francisco, 1982; also *Choices for Conservation,* Final Report to Congress and the President, Resource Conservation Committee, 530/SW-779, EPA, July 1979; interview with Frank Smith, 1988.
35. *New York Times,* May 24, 1977; *Los Angeles Times,* May 24, 1977.
36. Cited in *EPA Activities under RCRA, FY 1977,* EPA, 1977.
37. *Choices for Conservation,* Resource Conservation Committee, EPA, July 1979.
38. *Choices for Conservation,* Resource Conservation Committee, EPA, July 1979.
39. Interview with Frank Smith, 1988; see, for example, "Product Regulations as a Resource Conservation Strategy," Staff Background Paper No. 18, March 18, 1979, EPA, prepared for Resource Conservation Committee.
40. "Waste to Energy in California," Joan Melcher and Paul Relis, Report of Conference Held February 21, 1986, Community Environmental Council, Santa Barbara, 1986.
41. *Recycling . . . The Answer to Our Garbage Problem,* Stephen Lister and others (contributors), Citizen's Clearinghouse for Hazardous Wastes, Arlington, Virginia, May 1987; "Solid Waste Industry Enters Recycling Arena," BioCycle, volume 29, no. 4, April 1988.
42. Interview with Millie Zantow, 1988; see also "Recycling: In Wisconsin, a Real WINR,"

Everyone's Backyard, vol. 5, no. 4, Winter 1987; "Sauk County's Rural Waste Reduction Center," Wisconsin Intercounty Non-profit Recycling, Inc., 1987.

43. Interview with Will Collette, 1988; *Recycling . . . The Answer to our Garbage Problem,* Citizen's Clearinghouse for Hazardous Waste, Arlington, Virginia, May 1987; "Hazardous Waste Facility Siting: From Not in My Backyard to Not in Anybody's Backyard," Michael K. Heiman, presented at the 30th Annual Conference of the Association of Collegiate Schools of Planning, October 27–30, 1988, Buffalo, New York.
44. Interview with Mary Lou Van Deventer, 1988; "The Rise and Fall of Recycling," Neil Seldman, *Environmental Action,* January/February 1987.
45. "Is Mandated Recycling Possible?" Herschel Cutler, Executive Director, Institute of Scrap Recycling Industries, *Solid Waste and Power,* vol. 2, no. 4, August 1988; interview with Tania Lipshutz, California Department of Conservation, 1988.
46. *Solid Waste Management Status and Disposal Options in Los Angeles County;* "Bradley Willing to Study Mandatory Recycling Program," Jim Tranquada, *Los Angeles Daily News,* August 19, 1988; "EPA Sets Strategy to End 'Staggering Garbage Crisis,' " Philip Shabecoff, *New York Times,* September 23, 1988; "EPA Urges States to Reduce Use of Landfills," Lori Silver, *Los Angeles Times,* September 23, 1988.
47. "Recycling Integrated with Waste to Energy," Lewis Ott Ward and Garrett A. Smith, 1988; also for early advocacy of MRF approach, see "The Place of Incineration in Resource Recovery of Solid Waste," J. H. Fernandes and R. C. Shenk (Combustion Engineering), *Combustion,* October 1974; see also "Materials Recovery Facilities," Richard J. Kattar, *Resource Recycling,* January/February 1988.
48. *Coming Full Circle,* Environmental Defense Fund, Berkeley, 1988; also *EDF Letter,* September 1988.
49. "For Lack of Options, New York Gets Serious About Recycling," David C. Anderson, *New York Times,* May 15, 1988; *The Burning Question: Garbage Incineration vs. Total Recycling in New York City,* a Study by the Public Information Research Center and the Toxics Project of the New York Public Information Research Group, New York, 1986; also *The Rush to Burn: America's Garbage Gamble, Newsday,* 1987; the State Department of Environmental Conservation subsequently refused to grant a permit for the incinerator, *New York Times,* January 29, 1989.
50. "Recycling at Curbside," Cynthia Pollock, *Worldwatch,* January/February 1988; "Recycling Comes by Necessity to East Bay Counties," Elizabeth Fahey, *San Francisco Chronicle,* October 10, 1988; *Coming Full Circle,* Environmental Defense Fund, 1988.
51. "The Commonwealth's Solid Waste Master Plan: Toward a System of Integrated Solid Waste Management," Division of Solid Waste Management, Commonwealth of Massachusetts, December 20, 1988; *Coming Full Circle,* Environmental Defense Fund, 1988.
52. "Fitchburg Prepares to Sort Through Recycling Effort," *Wisconsin State Journal,* January 20, 1988; "Fitchburg Recycling Program a Success," *Capital Times,* January 20, 1988; Mandatory Recycling Ordinance No. 86-41, City of Fitchburg; interview with Jeane Sieling, 1988.
53. Author's notes: presentation by Jeane Sieling, Department of Urban and Regional Planning, University of Wisconsin at Madison, March 1988.
54. "Development and Pilot Test of an Intensive Municipal Solid Waste Recycling System in the Town of East Hampton," Barry Commoner et al., Center for Biology and Natural Systems, Flushing, New York, 1988; "Intensive Recycling: Preliminary Results from East Hampton and Buffalo," Barry Commoner and others, presented at the Fourth Annual Conference on Solid Waste Management and Materials Policy, January 27–30, 1988, New York City; also interview with Tom Webster, 1988.
55. Cited in *Coming Full Circle,* Environmental Defense Fund, 1988, p. 24.

56. "Development and Pilot Test of an Intensive MSW Recycling System," Barry Commoner et al., CBNS, 1988; "Intensive Recycling," Barry Commoner and others, 1988.
57. Interview with Tom Webster, 1988; "Waste Not, Want Not," Jim Schwab, *Planning*, October 1988; "Don't Let City Garbage Go Up in Smoke," Barry Commoner, *New York Times*, January 29, 1989; *Hazardous Waste News*, no. 108, December 19, 1988.
58. "Choosing All of the Above," David W. Birks, Executive Director, Spokane Regional Solid Waste Disposal Authority, *Solid Waste and Power*, vol. 2, no. 3, June 1988.
59. See, for example, "WTE Development, A Zigzag Course," Harvey Gershman and Nancy Peterson, *Solid Waste and Power*, August 1988; "Sierra Club Solid Waste Management Policy," November 1986.
60. *Garbage Management in Japan*, Allen Hershkowitz and Eugene Salerni, INFORM, 1987, especially pp. 39–66.
61. *Comparative Estimates of Post-Consumer Solid Waste*, Frank A. Smith, 530/SW-148, EPA, 1975; *The Private Sector in Solid Waste Management: A Profile of Its Resources and Contribution to Collection and Disposal*, vol. 2 ("Analysis of Data"), Applied Management Sciences, SW-51d.1, EPA, 1973.
62. *Characterization of Municipal Solid Waste in the United States: 1960–2000*, Franklin Associates, EPA, 1986.
63. Franklin Associates, 1986, especially pp. 1–16, S-2.
64. "How Garbage Analysis Can Reduce Collection Costs," Brian W. Evans, *BioCycle*, vol. 26, no. 2, March 1985; *Coming Full Circle*, Environmental Defense Fund, 1988, see pp. 39–44; for examples, see *Air Pollution Control at Resource Recovery Facilities*, California Air Resources Board, May 24, 1984, especially pp. 42–43, 45.
65. *Washington State Recycling Survey*, Department of Ecology, State of Washington, April 1987; *Integrated Solid Waste Management*, Interim Report, Joint Select Committee on Preferred Waste Management, Washington State Legislature, January 1988; "The Garbage Decade," William L. Rathje, in *American Behavioral Scientist*, vol. 28, no. 1, September/October 1984; *Coming Full Circle*, Environmental Defense Fund, 1988.
66. *Mining Urban Wastes: The Potential for Recycling*, Cynthia Pollock, Worldwatch Paper No. 76, April 1987, Worldwatch Institute.
67. *Choices for Conservation*, Resource Conservation Committee, EPA, July 1979.
68. "Opportunity to Recycle Knocks in Oregon," Ann Mattheis, *Waste Age*, July 1987; *Coming Full Circle*, Environmental Defense Fund, 1988.
69. "Sanitary Landfills Are Too Cheap," Frederick C. Dunbar and Mark P. Berkman, *Waste Age*, May 1987.
70. "Integrated Waste Management and the Next Economy: A Presentation to the Local Government Commission of California," Paul Relis, Community Environmental Council, January 30, 1987; "Environmental Allowance: Redefining Traditional Cost-Benefit Analysis Applied to Evaluation of Waste Reduction and Recycling Program," Robin Kordik, Project Manager, Seattle Solid Waste Utility, 1987; *To Burn or Not to Burn*, Environmental Defense Fund, New York, 1986, see p. F-1.
71. "Status of City Sponsored Recycling Projects—Waste Reduction Projects," Report No. 3 CF Nos. 81-4086, 4426, and 84-0638, Transmittal No. 1 and 2, Bureau of Sanitation, City of Los Angeles, March 14, 1986; "The Dilemma of Municipal Solid Waste Management," Terry Bills and others, a Comprehensive Project Report, Graduate School of Architecture and Urban Planning, UCLA, 1987, see especially pp. 354–59 and appendix IV.
72. "Is Mandated Recycling Possible," Herschel Cutler, *Solid Waste and Power*, August 1988.
73. "Surfeit of Used Newsprint Erases Profit in Recycling," Andrea Ford, *Los Angeles Times*, February 19, 1989; "Plan to Ship Trash to China—It Looks Good on Paper," Kevin

Roderick, *Los Angeles Times,* January 26, 1989; *Mining Urban Wastes,* Cynthia Pollock, Worldwatch Paper No. 76, 1987; *Materials Recycling,* William Chandler, Worldwatch Paper No. 56, 1983.

74. *Coming Full Circle,* Environmental Defense Fund, 1988, p. 89.
75. *Resource Recovery and Waste Reduction,* 4th Report to Congress, 530/SW-600, EPA, 1977.
76. *Materials Recycling,* William Chandler, 1983.
77. "On the Capital Paper Trail," Renee Cravens, *Environmental Action,* July–August 1988.
78. *Mining Urban Wastes,* Cynthia Pollock, 1987, pp. 39–41.
79. "Recycled Materials Procurement," Nancy Vandenberg, *Resource Recycling,* part 1, September/October 1986; part 2, November/December 1986; see also "Legislation and Involved Agencies," William L. Kovacs, in *The Solid Waste Handbook,* ed. William D. Robinson, 1986; *Coming Full Circle,* Environmental Defense Fund, 1988, pp. 97–104, table II.
80. "The Rise and Fall of Recycling," Neil Seldman, *Environmental Action,* January/February 1987; *Coming Full Circle,* Environmental Defense Fund, pp. 113.
81. Interview with Millie Zantow, 1988.
82. "The Solid Waste Dilemma: An Agenda for Action," 530/SW-88-05, EPA Office of Solid Waste, 1988; see also "Recycling May Be the Equal of Incineration in Waste Disposal," Environmental Defense Fund, *New York Times,* Letter to the Editor, December 6, 1988.
83. "WTE Development: A Zigzag Course," Harvey Gershman and Nancy M. Peterson, *Solid Waste and Power,* August 1988.
84. *Coming Full Circle,* Environmental Defense Fund, 1988, p. 59.
85. Interview with Mary Lou Van Deventer, 1988.
86. Interview with Mary Lou Van Deventer, "A Recycler's Lexicon," Materials World Publishing, Albany, California, March 1988.
87. Interview with Mary Lou Van Deventer, 1988.

8

THE SQUEEZE ON REUSE STRATEGIES: RECIRCULATING THE STREAM

"YEAR OF THE SMEAR"

As ad campaigns go, the mobilization by the glass container and beverage industry to defeat the California bottle bill initiative, placed on the November 1982 ballot as Proposition 11, was striking. The bottle bill opponents raised $5.6 million, about ten times more than the $600,000 raised by its supporters. Most of the opposition funds came from industry trade groups like the Glass Packaging Institute, which contributed more than $1.1 million, and the Can Manufacturers Institute, which added another $653,000, or from individual bottlers such as Anheuser-Busch, which provided a hefty $325,000, Coca-Cola, a $143,000 contributor, and Pepsi-Cola, which added $191,000.[1]

The initiative, sponsored by several consumer and environmental organizations, was designed to establish a 5-cent deposit on all soft drink and beer containers, in cans or bottles. It was largely similar to measures in several other states, most of which had been defeated in previous elections. By 1982, only a handful of bottle bills had been adopted, most notably in Oregon and Vermont through their legislatures, and in Michigan and Maine through the initiative process.[2]

When the initiative was first proposed, polls showed that large majorities in the state favored the idea. Much of that support, however, was "soft," based on general support for the environmental goal of reducing the waste stream. The bottle bill supporters had anticipated an expensive opposition campaign and a variety of tactics, including possible misleading or deceptive messages, a trademark of the bottle industry's counterattack. California, once again, had become the battleground, the culmination of more than a decade of organized efforts and successful opposition to this one-time standard procedure for product reuse.[3]

As anticipated, the opposition relied upon and even exceeded the kinds of egregious tactics used in previous campaigns. They created an election environment that the *San Jose Mercury-News* characterized as the "year of the smear." Phony EPA reports were cited (understating the volume of beverage containers in the waste stream), proindustry figures posed as consumer advocates, and endorsements were attributed to political figures, such as Los Angeles mayor Tom Bradley, when no such endorsements had been made. Opposition ads predicted price increases, lost jobs, and health problems, while existing government studies indicated that in states where similar legislation had already been enacted, there had been no price increases, no associated health problems, and *an increase* rather than decrease in the overall number of jobs. Perhaps most misleading was the statement by the bottle bill opponents that "recyclers [would] suffer a crippling blow." Proposition 11, as messages insisted, would "put almost all volunteer recycling programs out of business," while robbing "charitable groups, service clubs and religious organizations of a valuable source of revenue."[4] In other states, in fact, bottle bills had provided an enormous stimulus to recycling programs, with volunteer groups, such as the Scouts, actually increasing the scope of recycling activities.

The most striking and effective set of ads, released just a week prior to the election, was a strong attack on the Oregon bill, designed to convince Californians that passage of their own initiative would duplicate extensive alleged problems experienced in Oregon. These claims were made through a series of ads based on "interviews" inside a shopping mall with five presumably randomly selected consumers from Oregon. These five argued that the Oregon bottle bill was a disaster, creating long grocery lines, higher beverage prices, and enormous problems with the sorting process. One individual declared that he, like a lot of other Oregonians, bought his beer across the border in Washington.

In an extensive television blitz, the industry ads created what turned out to be insurmountable damage. It was later revealed that four of the five Oregonians featured in the ads in fact worked for a major beverage distributor, and that the one who talked of crossing the border to buy beer was the operational director of Columbia Distributing, a longtime opponent of the Oregon law. The fifth "random shopper" was a Safeway employee, another major bottle bill opponent. The five "consumers" were each paid $150 for their effort.[5]

The Oregon ad, reminiscent of Mark Twain's famous line that a lie can travel halfway round the earth while the truth is putting on its boots, was not rebutted sufficiently prior to the election to contain the damage. In contrast to the explicit message that the Oregon legislation was unpopular, surveys conducted by the Oregon Department of Environmental Quality showed that upwards of *90 percent* of that state's residents supported their bottle law. "Ask any Oregonian and you'll learn that the Oregon bottle bill is the most popular piece of legislation ever enacted in the state," commented one state legislator. "Young and old, indoor and outdoor types, liberals and conservatives—all know about and would like to talk about the bottle bill. So much citizen awareness about a piece of legislation is truly unusual."[6]

The failure of Proposition 11 in California led to a new round of lobbying and maneuvers in the state legislature during the next few years over the question of beverage container deposits. It also brought up the broader issue of whether it would be possible to change the disposal patterns for bottles and cans established by production choices made several decades earlier outside the public arena. The new bottle bills that were debated and the legislation that was finally enacted five years after the defeat of Proposition 11 were far weaker than that beleaguered initiative and ultimately failed to address the question of product reuse, central to the disposal issue. The forceful intervention of the industry once again had set the terms of debate and resolution. The broader intent of the bottle bills to reverse the trend toward nonrefillable containers remained unacceptable to the industry.

By the late 1980s, the battles over beverage container deposits, among the earliest attempts to address the problem of "disposability," had come to symbolize the political character of the reuse debate. Forty years after the single-use throwaway bottle had entered a market then dominated by reusable products, the refillable bottle proved to be an elusive goal.

THE RISE OF THE ONE-WAY DISPOSABLE

The current interest in recycling, and even in the waste management hierarchy, has failed to translate into effective programs supporting reusable products. In almost every version of the hierarchy, reuse occupies a higher rung or priority than recycling. A reused product is quite different from a recycled product, which requires process transformation. A reused product such as a refillable bottle has a second (and often multiple) life as a reusable commodity. A laundered diaper has dozens, even hundreds, of use-lives. Far less energy and often no additional raw materials are used to produce the recirculating product. Recycling, however, often involves grinding or crushing the material and then a new production cycle that can require both additional energy and materials, as well as the recycled item, to produce essentially an entirely new product. Bottles that are cleaned, washed, and then refilled are reuse products; those that are crushed, ground, and remolded are new products using (partly) recycled, or secondary material.

Industry resistance to reuse is substantial, reflected not only in opposition to bottle bills but also in hostility to any design change in the production system. The wide variety of reusable products common through the 1940s and 1950s were made from many different materials, including glass, rubber, paper, cloth, oil, and metals. Changes in production technology, the development of entire new product lines such as synthetics and other plastic-related materials, and the shift in distribution and marketing all contributed to the decline of reuse. Once these changes evolved, the resistance of industry groups to reorient their modes of production and distribution to account for the increased problems of disposal became most pronounced. Trade groups claimed they were only responding to consumer- or market-driven prerogatives, at the same time manipulating the political and legislative process to undermine even the most tentative initiatives in the direction of reuse. A kind of inverse hierarchy for industry was established: the higher the rung in the typical waste hierarchy, the greater their resistance.

The most visible and expansive shift in the post–World War II years involved the dramatic changes brought about by the petrochemical revolution. These changes took place at nearly every level of production, from the substitution of synthetics for natural materials in a wide array of products to the development of new consumer product lines and extensive changes in packaging. And although the percentage of plastics in the waste stream by weight remained, through the 1960s and 1970s, only in single digits, it was the fastest-increasing waste category and the single most significant factor in the decline of reusable products.[7]

Petroleum-based products fundamentally restructured the consumer products industry, displacing a wide variety of reusable/recyclable materials. Classic reuse products, such as the wood-corked champagne bottle or the cloth rag, were outmoded by minor changes in product design, such as the plastic cork. Industrial interests with a stake in eliminating or undermining reuse practices were also aided by government activities framed on behalf of those industries. Such was the case with waste or re-refined oil, another reusable product, that was either recycled (usually in some other form, such as a dust suppressant on rural roads), or blended with other oils for heating. With the decline of re-refined oil and a substantial increase in the amount of waste oil needing to be managed, government regulators have discovered that they must now contend with a waste product representing significant potential health and environmental hazards.[8]

Up through the 1950s, re-refined oil maintained a modest share of the oil products market, with over 300 million gallons of waste oil re-refined annually. By the 1980s, however, that figure had plunged to less than 20 million gallons a year, due in part to industry lobbying and government action. In 1964, the Federal Trade Commission ruled that all recycled waste oil had to carry the label "Made from Previously Used Oil," suggesting an inferior product, according to critics of the action. More than 15 years later, a House of Representatives report asserted that the FTC's "discriminatory labelling requirements" ultimately had "an adverse impact on consumer acceptance of re-refined oil."[9] Congress had also instructed EPA in 1976 to establish federal procurement guidelines for waste oil, but 12 years later, they had not been issued.[10]

Waste oil is a particularly troublesome waste. It contains high concentrations of heavy metals such as lead, cadmium, and chromium, and other toxic pollutants including benzene, PCBs, and polycyclic aromatic hydrocarbons. Used as a dust suppressant, waste oil has transformed several rural locations into hazardous waste sites, including the infamous Times Beach, Missouri. In a landfill, oil absorbs large quantities of some of the most hazardous materials found there, such as pesticides and heavy metals.[11]

In 1986, EPA decided not to list waste oil designated for recycling as a hazardous waste. The year before, it had ruled that a broad category of used automotive and industrial oils warranted such designation. EPA changed its position on the basis that gas stations and refineries, the primary repositories of waste oil, would no longer agree to collect the oil if such a designation were issued, an approach disputed by the U.S. Court of Appeals in a 1988 decision upholding used oil's designation as a hazardous waste.[12] Beyond the problem of designation, government programs to restimulate a re-refined oil program remained minimal. The economic and environmental costs of disposal, meanwhile, remained with the public.

By the 1970s and 1980s, nearly every kind of product actively reused by an earlier generation had become a casualty of industry activity, government biases, and restructured production processes. The importance of reuse, considered the most direct and environmentally benign form of product recirculation and thus waste management, continued to be undervalued through the 1980s, its place in the hierarchy presumably secure but unimplemented.

THE PLIGHT OF COMPOSTING

Composting, the biological decomposition of the organic constituents in the waste stream and the subsequent return of nutrients to the soil, is one of the oldest reuse strategies. Composting of organic materials provided the pretechnological/prechemical fertilization of land in most cultures, particularly rural-based societies where manure, food, grass, and other organic wastes were spread on fields. With the development of urban and industrial societies, this process of recirculation tended to disappear, though many gardeners continued to compost and a few composting plants with mechanical equipment were established during the 1930s, 1940s, and 1950s in the United States and Europe.[13]

The more technologically complex mechanical composting has often been designed as an enclosed system in which a digester supplies air and mixes the materials and stimulates fermentation by various microorganisms while destroying the pathogenic ones.[14] Such plants were most successful in Europe, with its rapidly declining landfill space. They soon spread to several countries in Asia and South America, and then made a brief reappearance in this country in the 1960s. The few experimental municipal composting plants of that time, however, quickly became casualties of inexpensive landfill space, weak demand for products made from compost such as soil amendments, and poorly constructed plants aimed at establishing a quick return on investment. These plants, "which tarnished the reputation of the [composting] industry as a whole," were closed down in the planning stage or soon after they had begun operations. But with the decline in landfills and the eclipse of incineration, interest in composting in this country recently has been renewed.[15]

The potential for composting is obvious and enormous. At an average $20 to $60 per ton, composting costs less than incineration and even landfilling in some locations, particularly rural communities. Organic materials typically make up as much as one-half to two-thirds of a community's residential waste stream; if paper is excluded, the organic component might still be as much as 35 to 40 percent. Yard and leaf wastes alone can account for 18 to 33 percent. In a place like Los Angeles, where yards are maintained year-round, compostable wastes can comprise as much as 81 percent of the city's wastes, and 59 percent if paper is eliminated, according to a 1981 waste stream analysis. Included in the category of compostable wastes are food and other putrescible wastes, wood, grass clippings, leaves, garden waste and bushes from households and parks, as well as paper and paperboard, and some textiles. Those elements of the organic waste stream that are not readily compostable, such as materials containing synthetic fibers, might constitute 5 to 10 percent of this total, still leaving upwards of 30 percent (and more than 60 percent in the case of Los Angeles) of the waste stream potentially suitable for some form of composting. While the compostable portion of the waste stream varies from community to community and region to region, composting affords an enormous opportunity for recirculating wastes (see table 8.1).[16]

Compost has a number of existing and possible applications. It is best

Table 8–1
PERCENTAGE OF MUNICIPAL SOLID WASTE STREAM WHICH IS COMPOSTABLE

MATERIAL DISCARDED INTO MUNICIPAL WASTE STREAM	
Compostable Component	*Percentage of Total*
Paper and paperboard	37
Yard waste	18
Food wastes	8
Wood	4
Subtotal	67
Noncompostable Component	
Glass	10
Metals	10
Plastics	7
Rubber and leather	3
Textiles	2
Miscellaneous	1
Subtotal	33
Total	100

SOURCE: Franklin Associates, 1986.

known for its use as topsoil, mulch, and soil nutrient. Compost can also be useful as landfill cover, extending landfill life by replacing about 17 percent of the capacity by volume taken up by ordinary landfill cover.[17] Parks, golf courses, public buildings, highways, community gardens, athletic fields, and other recreational areas are potential markets for compost material used as soil conditioner for landscaping purposes. Co-compost material, the mixture of sewage and household organics, can also be used for highway sound-wall barriers, levy repair, or erosion control. Even composted manure from zoos, called "zoodoo," is potentially marketable, as is compost from horse manure, both of which are free of heavy metals and chlorinated by-products.[18]

All of the various composting technologies, which vary in both cost and processing time, involve basically the same three steps. Some preprocessing is undertaken, usually grinding or shredding to reduce the particle size, followed by some separation to remove metals, plastics, and other noncompostables. After preprocessing, the remaining organic component is composted, the principal by-products of the fermentation process being carbon dioxide and water vapor. In mechanical composting, which usually occurs in covered areas or buildings, especially in urban areas, a variety of machines, such as drums, silos, and tunnel reactors, pulverize, turn, and aerate the compost. Additional shredding or grinding often takes place during this stage. After the material has decomposed, various post-processing activities take place, including curing, stabilizing, screening, and more grinding and separating, depending on the intended use of the final product.[19]

The mechanical composting procedure takes the least time—between five and ten days—although the actual "maturing" or "curing" of the compost, necessary to all composting systems, takes an additional several weeks and sometimes several months. Mechanical composting also produces the least amount of odor and is aesthetically the least offensive. The capital and operating costs of mechanical composting have been relatively high compared with other composting methods, but still compare favorably with mass burn.[20]

Two other composting processes, windrow composting and co-composting, have had only limited application in this country. Windrow composting, where waste materials are set in piles with lanes between them, is normally performed outdoors and is typically applied on a much smaller scale. Air is distributed through the piles either by blowers and piping, known as the high rate method, or by mechanically mixing the piles, the low rate method. The high rate method takes between two to four weeks, while the low rate method can take as long as five to ten weeks, and both have encountered operational difficulties.[21]

Co-composting, mixing of household waste with sewage sludge, is a process that is particularly attractive to communities facing a sewage capacity problem. These two wastes are mixed in set proportions, and then remixed and shifted over time while natural microbial aerobic fermentation changes the material into natural additives or "bugs." The sewage sludge is used as a moisturizing agent, and substitutes such as fish and chicken waste, or horse manure, as well as water, can also be used.[22]

The most serious problem confronting composting involves the presence of a wide range of possible unwanted inorganic materials integrated into the organic wastes. These include plastic, metal, glass, large stones, and wires, all of which reduce the usefulness and nutritional value of the compost, and could harm grazing animals once the compost is applied as a soil amendment. Unless metals are removed before processing (through source separation), they will also leach into the compost. Trace amounts of various kinds of toxic gases might also be released as a by-product, particularly due to the presence of various chlorinated compounds among the inorganics. If sewage is used, the co-composting process runs the increased risk of potential contamination by heavy metals and toxics that industries discharge into the wastewater treatment systems.[23]

These environmental concerns are compounded by the significant amount of land needed for composting, depending on the technology. A windrow compost system typically requires about 118 acres of land for every 10,000 cubic yards of compost processed each year. A fully enclosed mechanical plant, on the other hand, can process as much as 1,000 tpd on about 20 to 25 acres. Though the availability of land is an increasingly difficult problem in major metropolitan areas, the requirements for composting are still significantly less than those for landfills. And while incineration plants typically need less land, their siting is compounded by far more significant emission problems.

Similar to recyclables, compost products are reputed to have only limited markets. Their use as a soil amendment, for example, can be undermined on the one hand by the potential presence of trace amounts of chlorinated organic compounds (unless presorting is adequate), particularly for composting's most obvious constituency, organic or sustainable farming. On the other hand, more conventional farmers are reluctant to substitute compost for chemical fertilizers. Public agencies, meanwhile, have shown only limited interest in composting.[24] By the mid-1980s, no major metropolitan area had undertaken a systematic survey of its own potential land available for compost application. The city of Los Angeles, for example, maintains 15,000 acres of parkland, including 13 golf courses and about 200 ballfields. Of these, 55 park and recreation properties are larger than ten acres, three are several hundred acres, and three more are thousands of acres in size. All are potential markets for compost, though the city, despite its new position on recycling, has been reluctant to move aggressively on this issue.[25]

Composting efforts in this country have been small and mostly experimental. By the mid-1980s, prior to the latest recycling boom, only five commercial-sized composting facilities were operating. The largest of these is the Delaware Reclamation Plant, located at Pigeon Point in New Castle County, Delaware, which started operation in 1984. The facility extracts plastics and paper for resource-derived fuel pellets as well as ferrous metals, and sends the remaining wastes through further processing to recover glass, nonferrous metals, and eventually to produce cocompost. Since 50 percent of the plant's compost has been sewage, the plant has had difficulty marketing the compost product as soil conditioner due to moratoria in neighboring states on all application of sewage. As a result, the compost has been marketed for landfill cover.[26]

A number of other communities also began to develop more modest composting programs, including Portland, Oregon, and Berkeley and Palo Alto, California. An experimental program in Seattle, Washington, based on the windrow composting method has received some national attention, as have programs in Hennepin and Ramsey counties in Minnesota. With the new interest in recycling, composting has received some new attention as well, and by 1989, 42 cities and counties had made plans to develop composting plants, especially in Minnesota and Massachusetts. According to a report by the city of Portland's environmental agency, approximately 300 programs have gone into the planning stage, of which only 12 are for mixed solid waste.[27]

The new interest in composting has also been stimulated by landfill disposal bans on leaves and yard wastes initiated in several communities, as well as by broader recycling goals. Most of the bans, such as those proposed in Madison, Wisconsin, Portland, Oregon, Broome County, New York, and the state of New Jersey, were quickly established in response to landfill capacity problems as a form of crisis management, and were not designed as part of any systematic composting strategy or broader, integrated waste management plan.

The Madison situation is an interesting case illustrating the potential—and current plight—of composting.[28] For a number of years, both the city and the greater Dane County area had relied heavily on local landfills for disposal of both residential and commercial wastes. In the mid- and late 1980s, volatile organic compounds were discovered in wells close to several of these sites, forcing the state Department of Natural Resources to shut down the wells and set a timetable for closing the landfills. Public agencies at the city and county levels now faced a landfill capacity problem, and rapid action seemed necessary. The city of Madison, aware of the community's strong environmental sentiment, in 1987 banned the landfill disposal of yard trimmings, effective by the end of 1988. The city planned two composting facilities to handle the banned yard waste, and adopted a user fee to help finance them. Each household would be charged for the disposal of specially bagged yard trimmings, collected separately.

The new Madison policy, developed without public input, immediately ran into homeowner opposition. The main objection was the creation of the user fee to pay for a service (a form of solid waste "collection") that was presumed to be covered by the property tax levy. One Madison councilman, bombarded by irate calls from his constituents, declared, "People are in support in terms of the environmental concerns, but they felt very strongly this was a basic service that should be provided to the taxpayers."[29] Absent was any discussion of avoided cost mechanisms or surcharges on landfill fees, as well as a broader debate over how to best develop and structure a complete waste management plan.

The lack of any strategic agenda included the failure to more systematically link composting to the community's overall approach to waste management, and also failed to generate discussion about hierarchy and priorities in which composting would likely play a central role. The foundation of such an agenda would be the high priority given to intensive source separation programs, which had only been organized on a more limited basis in the adjacent community of Fitchburg. The more plastic, glass, ferrous metals, and other materials segregated from the organic wastes, the greater is the reduction of potential contaminants in the compost. A comprehensive source separation before shredding can remove the chlorinated organics and heavy metals, a likely precondition for its acceptance by farmers and others for soil application. In this sense, composting makes an ideal companion to intensive recycling.

The Japanese, for example, in rural areas, separate their trash into compostable and noncompostable bags. These bags are often examined by crew members on collection day. Those bags marked as compostable but containing too many noncompostables are simply not picked up, a tactic that can be used here to enforce mandatory recycling. Furthermore, the Japanese Ministry of Agriculture, Forestry, and Fisheries, like the U.S. EPA, has issued national compost standards dealing with such potential contaminants as lead and cadmium, the bane of incinerator ash.[30]

By the mid-1980s, another potentially crippling problem emerged to confront composting. Several leading solid waste incinerator firms began to pursue plans to construct incinerators for agricultural waste and wood chip waste. General Electric, for example, developed projects in Stanislaus County, California, Bangor, Maine, Schenectady, New York, and Telogia, Florida, to burn compostable waste from industrial waste streams and also accept waste from residential sources. This potentially polluting, single-source, high-tech option, competitive with composting, functions as a disincentive for a more environmentally benign waste management practice.[31]

These and other problems facing composting are symptomatic of larger reuse concerns. For example, the hazards generated by trace amounts of contaminants can be dealt with by recycling, both in advance and as part of the process, and by an effort toward reducing the level of contaminants at their source. The market issue also raises a red flag: neither supply nor demand has been sufficiently developed, nor is there much indication of greater government intervention except indirectly through disposal bans or more directly (and unpopularly) through user fees. In Europe, long-term contracts that guarantee minimum prices for recyclables (including compost) assure a market and take some of the wild fluctuations out of the demand curve.[32] Most composting programs have focused on the technology involved rather than on the policy consideration of how an overall program could fit in a community's waste management planning.

The development of the incineration option for compostables also demonstrates the central need for a hierarchy. Waste managers continue to insist that markets have to exist before the product does, forgetting that the primary value of compost is getting it out of the waste stream. In this way, the hierarchy is applicable: create and subsidize markets rather than waiting for them to happen. As yet, composting, perhaps the most practical and extensive form of product recirculation available to a local community, still receives far less attention—and support—than landfills and incineration, despite their presumed place several rungs below composting at the bottom of the hierarchy.

THE SAGA OF BOTTLES

There was indeed a time, not so long past, when the soft drink or beer container had a long and successful life outside the waste stream. Recirculating beverage containers were a fact of life throughout the country through the 1940s and 1950s, and the refillable bottle, with its return deposit, was not only widely accepted but also integral to the market structure of both the beer and soft drink industries. A bottle bought was a bottle meant to be returned, and by some accounts, a typical beer bottle would make as many as 30 to 40 trips from consumer back to producer to return once again onto the market as a new, refilled bottle of beer.[33]

The decline of the refillable bottle largely took place outside the solid waste arena. At its root were production choices and directed consumer options, changing distribution and marketing arrangements and new product lines, and a revolution both in packaging and disposal. In 1945, with the end of the war and the take-off of new consumer markets, the refillable bottle still dominated. In less than 20 years, that position began to change radically. The introduction of the flip-top, nonrefillable can by national breweries accounted for one major shift. Rather than differences in taste or price, this change was more a reflection of competition through product differentiation accomplished by packaging innovations and related advertising campaigns. Such changes, moreover, paralleled the decline of the local and regional breweries, often the last bastion of the refillable beer bottle.[34]

The decline of the refillable in the soft drink market took longer to accomplish. This was due in part to Coca-Cola's market position, its belated interest in packaging innovations, and the highly segmented and localized soft drink franchised bottlers. These operations were economically disadvantaged by the need to switch production from the refillable to the one-way, "disposable" bottle. Packaging changes, however, accelerated during the 1970s, with greater concentration among franchise bottlers and the introduction of the plastic, polyethylene terephthyalate ("PET") soft drink bottle in 1978. The overall market position of beer and soft drink refillables significantly declined, as the percentage of one-way glass bottles surpassed those of refillables, while "throwaway" metal cans accounted for nearly half of the rest of the beverage container market. A high percentage of the growth in packaging wastes was represented by the displacement of the refillable by the one-way container. By the early

1980s, the decline became precipitous, as the market share of refillables in the soft drink market shrank to just 16 percent, its lowest point ever. As a result of the displacement of the refillable, the increased percentage of glass placed in the trash ultimately exceeded the rapid increase in the national consumption of the beer and soft drinks themselves.[35]

The production of nonglass containers, such as the aluminum can and PET bottle, played a significant role in further carving up the market segments. Since 1978, the PET bottle has captured nearly 100 percent of the two-liter bottle market while making inroads in markets for both the one-liter and three-liter size. By the late 1980s, the PET bottle accounted for over 22 percent of the entire beverage container market.[36] Changes in distribution in the food business, including the development of regional distribution centers, the decline of the mom-and-pop store and the rise of the supermarkets, as well as the overall concentration of the grocery industry, all contributed to the decline of markets for refillables, driven in part by the increased transportation costs created by consolidation. The supermarkets, furthermore, became a leading opponent of refillables (and consequently of bottle bills) primarily because of the premium on storage space. And changes in the glass container industry itself played a role, as companies decided that their new growth had become tied to disposables. One-way bottles now provided the margin for improved sales and profitability.[37]

The impact of these changes on solid waste was slow to be recognized. In the early 1950s, as the disposable container began to capture market share in the beer industry, the first "ban-the-can" and "bottle bill" movements appeared. Early concerns, however, were primarily over increased litter, and the strongest actions came in those states most sensitive to the tourism implications of littered lands and highways. Vermont passed the first deposit law in 1953, actually banning the sale of beer and ale in nonreturnable bottles. That law expired in 1957 and was not reinstituted for another 15 years, as the container industry successfully argued that education and private initiative were sufficient.[38]

During the early 1970s, bottle bills began to reappear, this time more as a direct response to growing fears about landfills and dirty incinerators. Most legislation and referenda proposed at the time provided for a 2- to 5-cent deposit on beer and soft drink containers, but seldom required collection centers for refillables. Oregon joined Vermont as the first two states to pass legislation, and since both had strong environmental constituencies and relatively small markets, they were able to resist the growing counterattacks of the industry. By 1974, the industry lobbies had thwarted efforts to pass laws in 15 states, almost all initiated by volunteer and grass roots groups rather than by public officials. Industry ally Senator Jennings Randolph (D—W. Va) even introduced legislation seeking to deny federal solid waste grants for any state with an Oregon-type law.[39]

Then, in the late 1970s, a second wave of bottle bill measures were adopted in states such as Maine, Michigan, Iowa, Connecticut, and Delaware. Attempts to pass a national beverage container deposit law, however, were successfully opposed on several occasions. By the time of California's "Year of the Smear" campaign in 1982, bottle bills had reached a kind of impasse, with the substantial popularity of such measures often offset by the huge expenditures and increasingly manipulative campaigns waged against them.[40]

The most successful tactics utilized by industry groups at the height of these political battles of the 1970s centered around economic arguments, including job loss and potential price increases. A series of studies, including an extensive model performed by EPA as part of the congressional reports required under the 1970 Resource Recovery Act, sharply disputed industry contentions. Basing their study on the experience of the Oregon and Vermont laws, EPA found that, after a five-year implementation of a national deposit law, there would be a 60 to 70 percent reduction of roadside bottle litter and a 20 to 40 percent reduction of *all* litter; that more than 7 million tons of municipal solid waste would be reduced, which would account for 5 percent of the total waste stream and 8.5 percent of all manufactured goods discarded; that there would be a reduction in energy use equivalent to 245 trillion Btu or 40 percent of the total energy required to supply all beer and soft drinks; that an annual savings of virgin raw materials would include 5.2 million tons of glass, 1.5 million tons of steel, and 500,000 tons of aluminum; that there would be a net national economic savings for supplying beverages of upwards of $2.5 billion; that it would cost about 2.5 cents *less* in consumer costs for each 12-ounce serving; and that there would be a net employment *gain* of more than 80,000 jobs, consisting of an increase in beverage industry employment of 165,000 jobs and a manufacturing loss of 80,000 jobs.[41]

These extraordinary figures, presented to the Resource Conservation Committee as well as to members of Congress in the late 1970s, were not persuasive enough to create consensus at the federal level. Industry advocates successfully assembled a coalition of business groups, labor organizations, and elected officials as part of an attempt to portray bottle bills as economically marginal and socially elitist. The industry was also adept at pushing the revolving door, hiring personnel such as the staff director of the "Environmental Program" of the Glass Container Manufacturing Institute away from the Bureau of Solid Waste Disposal, EPA's predecessor for solid waste policy. Since advocates of the bottle bill focused their arguments primarily on recycling and litter issues rather than the implications for employment and industrial restructuring, they were often placed on the defensive, even though a refillable/reuse strategy had compelling labor and consumer advantages. In several states where bottle bill legislation had been enacted, in fact, labor groups reversed their position, once it became apparent that more rather than fewer jobs had been created.[42]

With the decline in landfills and the retreat of incineration in the late 1980s, interest in bottle bills emerged once again, though industry resistance continued to remain as formidable as ever. A few other states—New York, Iowa, and Massachusetts—enacted measures that established a deposit and required beverage container manufacturers and distributors to facilitate recycling of cans and bottles. Other states, such as California, witnessed contentious and protracted battles, with industry lobbying efforts intense and often successful, at least in modifying the structure and intent of the legislation. A 1986 law passed in California is a case in point.

Legislation in California had been introduced in several previous sessions and contained a number of standard—and quite modest—means for implementation, including the 5-cent return deposit, which many considered the minimal amount necessary to attract sufficient consumer response. This legislation was eventually scaled down enormously and riddled with exemptions and loopholes. The 5-cent return deposit was replaced by a penny deposit, with a provision to increase the deposit gradually if the initial specified return rate of 65 percent for each type of container was not met. Furthermore, a number of exemptions such as for liquor and wine cooler containers were incorporated into the bill, undercutting the potential supply of bottles to be returned.[43]

Within a year of implementation, it had become clear that the program was achieving poor return rates, and was strongly resisted by both bottlers and retailers. At the same time, no formal mechanism had been provided to encourage refillables. Attempts to close loopholes, such as the exemption of wine cooler bottles, were in turn reamended and reshaped. In the wine cooler case, a new loophole was established through extraordinary jockeying by industry interests. One striking provision established a new exemption for sparkling apple juice containers—the "fruit loophole."[44]

The maneuvering that characterized the California legislation had become more the rule than the exception. Echoing EPA predictions, where bottle bills were implemented, an average of 5 percent of the total waste stream was in fact diverted, and nearly all the systems had proven effective at controlling litter.[45] But few states had any effective provision to encourage the *reuse* of containers. Those that did, such as Michigan and Oregon, relied on pricing mechanisms, such as a lower deposit fee for refillables, hoping that this would make the price of refillables more attractive and thus increase demand and stimulate markets.[46] None of the legislation, however, designed the actual collection programs to directly favor refillables. Most, furthermore, provided for only limited separation, which made it difficult to segregate the returned bottles. The old era of reuse remained, even in the era of bottle bills, a curiosity rather than a guide.

TIRES TO BURN

If the distinction in bottle bills is between reuse and recycling approaches, the debate over what to do with the vast piles of tires thrown out each year—about 1.5 percent of the waste stream—has developed into an argument between reuse and recycling perspectives and the advocates of incineration.

Every American today throws away the equivalent of one 20-pound tire a year—more than 240 million altogether. In addition, 2 billion tires are illegally discarded around the country.[47] The average disposal cost of this product is 10 cents per pound. It would cost a city the size of Newark, New Jersey, or Toledo, Ohio, about $700,000 a year just to get rid of that year's annual supply of junked tires. Only about 10 percent of that amount are currently retreaded and even fewer are burned. Most, 75 to 90 percent, are sent to landfills or left in unlicensed tire piles, while the small percentage remaining are recycled into a variety of rubber products.[48]

Prior to World War II, when tires were manufactured primarily from natural rubber, worn tires were reprocessed for other uses or recirculated as retreads. This practice peaked during World War II, but with the development of synthetic rubber and the steel-belt radial and the subsequent decline in the percentage of natural rubber used, tire reprocessing and retreading decreased significantly. The lack of government intervention or regulation invariably caused disposal agents—either the individual tire owner, the tire distributor, or the local gas station—simply to dump old tires in trash dumpsters, vacant lots, and, ultimately, landfills.[49]

Tire disposal in landfills, however, led to trouble. By the 1980s, the problem had become staggering in terms of both the volume of discarded tires and the decline of the various recycling and reuse or retread approaches used to reclaim waste rubber. Tires in landfills seldom stay covered; they trap landfill gas and "float" to the top, a kind of rubber monster from the lagoon. Due to their shape, tires do not compact and therefore take up a great deal of space in a landfill, invariably shortening its life. They can also be a breeding ground for rodents and mosquitoes, creating vectors for disease. "Spongy" tires, furthermore, can aid the migration of toxic chemicals into underground aquifers, and because they trap used oil as well as volatile gases, they pose a serious fire hazard.[50]

Fires are a special concern at the dangerous, unsightly, and usually unregulated "tire dumps" that have sprung up especially in the last several years. In many areas of the country, landfills have begun to refuse used tires altogether.[51] As a result, speculators established their own collection sites, anticipating a role for used tires in the newly developing interest in waste-to-energy, or in this case, "tires-to-energy." These tire dumps, however, several containing millions of tires, have come under increasing attack by community groups and public health officials for their numerous potential hazards.

On Halloween night of 1983, the worst tire fire in U.S. history occurred at a tire storage site in Winchester, Virginia. The owner of the dump had hoped to profit from the energy crisis by reclaiming the oil content of the tires. But the drop in the price of oil left him sitting with 7 million tires and pressure from the health department to stop accepting more tires and to remove the entire pile due to the increasing threat to public health. Then, one day, a fire was set, by parties unknown. The fire spread quickly, releasing an estimated 300 gallons of oil every minute. The tires continued to burn for nearly *nine months*, creating significant air emissions, contaminating nearby streams, and reducing the tires to an oily "goo." The oil runoff threatened the entire Potomac River basin and forced cleanup crews to erect barriers for the oil while allowing the fire to burn itself out. All told, more than 24 federal, state, and local agencies became involved in dealing with the fire, and the area became an EPA priority list site, with cleanup costs estimated at $12 million. The Winchester fire and several others that subsequently occurred in Hudson, Colorado, Ogden, Utah, and Somerset, Wisconsin, compounded the storage and disposal problem caused by discarded tires.[52]

During the 1970s, with the rise of volunteer recycling and the passage of resource recovery legislation, the EPA and state agencies began to investigate alternative strategies for used tires. Three were explored: improved collection (there were no existing formal mechanisms to develop a steady source of supply for alternative uses); various reuse/retread and recycling or reprocessing approaches, many of which were mature operations requiring little major technological innovation; and disposal via "energy producing systems." Of the three, tire burning was the most sensitive; earlier attempts at open tire burning had become particularly controversial due to air emissions. Most areas of the country outlawed all open burning in response to federal legislation.[53]

Public agencies recognized that recycling and especially retreading of tires—the classic reuse option—offered significant advantages besides the obvious reduction in the trash volume. A new tire, for example, needed five gallons of oil, while a retread required less than two gallons.[54] But the retread market was in sharp decline throughout the 1960s and 1970s, its market share decreasing by as much as 50 percent in one five-year period during the late 1960s. The percentage of all retreaded tires fell from 25 percent in the mid-1970s to as low as 10 percent by the mid-1980s. Even modest efforts to stimulate a retread market through government purchasing guidelines, for example, were undermined when government bodies, such as the Department of the Army, refused to purchase retreads, opting for new tires instead.[55]

The use of recycled or reprocessed rubber, such as reclaimed or devulcanized material, also substantially declined, partly as a result of industry economics favoring a higher proportion of synthetic material in the production of tires. This in turn meant there was less natural rubber to reclaim and more problems for processors. These trends were reinforced by the rise of steel-belted radials that extended the average use life of new tires, but created recycling problems because of the steel used in its production. Even though Congress and the EPA indicated that recycling and retreading should receive priority status in terms of government intervention, programs were limited. Those that were initiated, such as federal government retread tire procurement guidelines mandated by 1978 legislation, were established only after the tire retreading industry sued EPA in 1984. The policy tentatively adopted but still pending in 1988 focused only on government purchases totaling $10,000 or more, not to modest tire purchases, more the norm for smaller public agencies.[56]

Tire recycling or reclamation, already battered by the rise of steel-belted radials and the changing composition of new tires, continued to experience difficulties through the 1970s and 1980s as both federal and local agencies provided only minimal support for new technologies or markets. Most reclaimed rubber went to semipneumatic tire producers, mat manufacturers, and producers of molded products, though during the 1980s, some experimental processes were developed using tire chips for a wide range of products, from cement and asphalt to car bumpers and bridge deck waterproofing.[57]

The most ambitious of these projects was a Babbitt, Minnesota, facility designed to chemically transform ground rubber into a product, "TIRECYCLE," which can substitute for both virgin rubber and plastics in a variety of products, including rubber and plastic mats, bed liners for pickup trucks, solid industrial tires and other thermoplastic molded products. Funding for the plant came from a consortium of state, regional, and local government economic development programs, all with a financial stake in its success. The local economy was further strengthened when one manufacturer moved to the area to be near its raw material supplier. The Babbitt experiment has not yet been duplicated in other areas, a problem compounded by the rubber industry's reluctance to utilize new, recycled materials.[58]

It was already clear by the 1970s that EPA and a number of state and regional public agencies were increasingly attracted to the idea of a comprehensive "solution," rather than the diverse, smaller-scale recycling and reuse possibilities. The cornerstone of this solution was "tires-to-energy," an approach favorably regarded by the rubber industry itself. Tires had a significantly higher Btu content than even coal, upwards of 50 percent more, according to some estimates. When transformed into a derived fuel pellet, "TDF" (tire-derived fuel), for example, was much drier than coal, produced far fewer sulfur emissions, and had an energy content of 15,500 Btus per pound as opposed to coal's 8,000 to 12,000 Btus per pound. The option looked so attractive that initial inquiries into the subject of "abandoned tires" were framed exclusively as studies on "tire incineration."[59]

Experiences with tire burning during the 1970s and early 1980s, however, were similar to those of waste-to-energy. Capital costs were high and the technology relatively unproven. The first whole tire-to-energy plant (as opposed to incinerators that burned tire chips as one of several fuels) was built in Jackson, Michigan, by Goodyear in 1974. It had immediate operational difficulties and discouraged other similar facilities. Other technologies, such as pyrolysis (heating used tires in an oxygen-free environment to produce gases, oils, and other products), and cyrogenics (using extremely low temperatures to process the used tires), also showed significant cost, operational, and product marketing problems. By the early 1980s, tires-to-energy remained the stepchild of the waste-to-energy industry.[60]

With the trash burning in vogue by the middle 1980s, tires-to-energy made a comeback as well. Projects were proposed in Oregon, Texas, Connecticut, and New Hampshire, as new companies, some with financial backing from the rubber industry, attempted to secure a share of this potential new market. Two of the largest were in California, one in the central part of the state near Modesto, the second in the city of Rialto in Southern California's Riverside County.

The Modesto project had its origins in a scheme put together by speculator/land developer Ed Filbin. During the 1960s, Filbin opened Ed's Tire Disposal not far from the future route of I-5, the huge interstate highway between Los Angeles and San Francisco. The mound of tires at "Ed's," which had no storage permit, grew larger and larger over the next two decades as Filbin trucked in upwards of 10,000 to 20,000 tires each day. The county eventually forced him to obtain a permit to shred tires and sell the rubber to a company for kiln-drying lumber, but this idea never came to fruition. Finally, Filbin leased his land—and the tire dump—to Oxford Energy Company in 1985. Soon after, he disappeared from the spotlight, and Oxford bought his business for $2.75 million.[61]

Oxford quickly accumulated many more tires, importing some from Asia via Canada. In just a few years, Ed's Tire Disposal was transformed into a monster tire dump, with more tires—40 million plus—than anywhere else in the world. Oxford decided to stock enough used tires—a two-year supply at all times—to operate a high-tech, economy of scale, tire-burning plant.[62]

A publicly traded firm begun by a New York merchant banking and securities trading firm, Oxford entered the tire-burning business with its purchase of a German technology. The Oxford plant, designed and constructed on a turnkey basis by General Electric, which obtained a 50 percent co-ownership in the company, was its first major project. It hoped to burn upwards of 700 tires an hour, producing 14.4 megawatts of electrical energy.[63]

As Oxford pursued its permits, it warned Modesto public officials and residents concerned about the ash residue as well as the air emissions (which would be substantial) that the immediate alternative—leaving the tire dump as is—would be far more environmentally dangerous. The dense smoke from a major fire at the vastly expanded Modesto site, Oxford officials cautioned, would eliminate all agricultural activities within a 75-mile radius. Hundreds of thousands of gallons of molten rubber would release their witches' brew of toxic chemicals into the ground or the air and ooze from the site, threatening to destroy everything in its path. "It would be a major environmental disaster," one Oxford official warned local residents. "If it catches fire, pack up your bags and go home, because there's nothing you can do about it, and that's the raw truth."[64]

Caught between two potential hazards and a crisis atmosphere engendered by Oxford, Modesto officials felt compelled to proceed with the tire-burning facility, regardless of the consequence. Officials granted the final permits by late 1987 and the plant began operations soon after. With General Electric and the Radian Corporation as partners, and a company linked to one of the largest gas and electric utilities in the country providing additional capital with its purchase of 49 percent of its stock, Oxford, calling its plant "the largest facility of its kind in the world," hoped to replicate its Modesto operation elsewhere.[65]

Oxford's Modesto operation, despite the corporate interest and public agency acquiescence, remained highly controversial and was still bitterly opposed by local residents. The huge tire site, for one, not only continued to be a dangerous fire hazard, but the problem of mosquitoes had become so severe that the site needed to be continually sprayed with insecticides. The project, furthermore, was also the subject of two lawsuits. One involved the California attorney general as well as several environmental groups dealing with the hazards posed by the 300-foot pile of tires. The second suit was brought by a local group concerned about air emissions and the county's lack of regulation over the site. To compound matters, Oxford itself posted a $678,000 loss in 1987.[66]

While the scenario in Modesto unfolded, another new corporate entry, Garb Oil and Power of Salt Lake City, Utah, was staking its bid for a huge tire-burning facility in Rialto, California. This plant was designed to burn up to 10 million tires annually and generate about 35 megawatts of electricity.[67] Garb Oil's Rialto plant needed a number of permits to proceed with the project and initially felt encouraged in part by the attitude of some state regulatory agencies, such as the South Coast Air Quality Management District, which granted some, though not all, of its permits, and particularly the California Waste Management Board, which declined to regulate the plant on the theory that used tires were a source of fuel for an energy plant rather than a waste managed at a disposal facility.[68]

Once again, community activists, concerned especially about air emissions in the dirtiest area within the dirtiest air basin in the country, intervened by pressuring city and county agencies to act. Consequently, two lawsuits were filed by several cities as well as San Bernardino County, successfully challenging the AQMD permits and forcing Garb Oil to undertake an environmental impact report. Garb Oil appealed the decision to file an EIR and halted all work on its project.[69] Without the Rialto plant on line, and its other incineration projects at an early stage of development, the company, set up to exploit the tire-to-energy market, was unable to generate any operating revenues, intensifying its concern over the effective community opposition. In response to criticisms of its plans, company president John C. Brewer told the *Philadelphia Inquirer*, "You've got some people who are just opposed to any kind of progress."[70]

Almost from the moment of its revival in the 1980s, tire burning has been opposed on two fronts: the environmental issues associated with air emissions and ash disposal that also plagued municipal solid waste incinerators; and the extraordinary health and fire hazards associated with huge tire stockpiles. The economics of tire burning, similar to the economics of solid waste incineration, has remained problematic. Companies argue that the extraordinary nature of the problem—fewer tires recycled and reused, but more tires to be disposed of, while landfilling presents serious obstacles—will ultimately create favorable economics even while early losses are absorbed. This argument, however, has only intensified the level of public concern. In the meantime, the reuse option still receives only cursory support, while the burn alternative, proposed by industry and debated by public agencies, threatens to create a firestorm of opposition even greater than the fierce conflicts witnessed by the larger waste-to-energy industry itself.

DIRTY DIAPERS

In an age of convenience marketing, a few one-way products stand out in the way they have supplanted popular reuse products for functional as well as marketing reasons. Among such products, the greatest success story has been the meteoric rise in sales and market share of the single-use, "throwaway" paper diaper.

This product, whose features evolved over time, today consists of an outer layer of waterproof polyethylene plastic, an inner water-repellent liner, and, between the two, a thick layer of absorbent cottonlike material made from wood pulp with traces of dioxin. In recent years, this inner layer has also been treated with superabsorbing crystals made of a plastic compound. The wood pulp or "paper" component comprises about 65 percent of the diaper materials, while the polyethylene cover constitutes another 16 percent. This single-use diaper, now commanding upwards of 85 percent of the entire diaper market, has displaced the multiple-use cotton or linen diaper that prevailed in this country for the first half of the century, and still prevails in most areas outside the United States.

Today, the disposable diaper industry is an $8.4 billion business, producing $3.5 billion in sales annually from nearly 16 billion diapers sold. It is effectively controlled by two giant consumer product companies: Procter & Gamble (whose Pampers and Luvs brand names constitute 46 percent of the market) and Kimberly-Clark (whose Huggies brand alone accounts for another 32 percent market share). These two dominant leaders shape the industry's research, marketing, and product design.[71] The extraordinary growth of the single-use, disposable diaper business, stimulated by large capital investments from the parent companies, has led to continuing attempts to "improve" or differentiate the product. These efforts have addressed such questions as baby rash, product durability (creating, for example, a stronger elastic band), size variation, or gender differences in an increasingly segmented market. Despite the enormous social consequence of the rise of disposable diapers, their environmental hazards remain largely unaddressed.[72]

Up through the 1940s, diapers were, by definition, a reuse product. Cotton diapers were washed and reused hundreds of times. Those that were no longer useful as diapers were recirculated as cloth rags. Most diapers were washed at home, and the handful of diaper services that existed then were either supplemental or luxury services. Disposables made their entry into the U.S. market in the 1960s when Procter & Gamble, which introduced the Pampers brand in 1961, was able to capitalize on the needs of a growing female work force. Sensing an important new market, P&G and later Kimberly-Clark made enormous investments in new product features, such as tape strength, capacity to absorb, and lighter-weight material. Cloth diaper services, meanwhile, which once numbered more than 700 companies, declined precipitously, bottoming out at about 100 companies serving just 700,000 urban customers, when business finally began to pick up in the late 1980s.[73] In this period, new product breakthroughs for cloth diapers have also been minimal, given the absence of any industry-wide R&D or capital investment. Most experimentation has been confined largely to small cottage-industry entrepreneurs producing more "breathable" diaper liners or less unwieldy diaper-pin substitutes such as Velcro.[74]

The rise of disposables and the decline of the cloth diaper are particularly striking in that the substitution occurred as solid waste issues were rapidly emerging. Yet, the issue of disposability arose only after the single-use baby diaper had become, in the public eye, a product of *necessity* rather than simple convenience. The reorganization of the workplace, with its rapidly expanding female work force, the growth of child care services, and the increased level of activity outside the home all reinforced the triumph of the single-use diaper. Few if any programs, on the other hand, were developed by either the public or the private sector to adapt cloth diapers to changing social and economic circumstances.

The solid waste problems posed by disposable diapers are significant, and are distinct from the relatively minor disposal problems of cloth diapers. Without reuse, an extraordinarily large number of diapers enter the waste stream along with fecal matter—the baby's poop—which is more appropriately treated in the wastewater stream. Soiled diapers, in other words, are thrown into the trash can rather than emptied into the toilet. When buried or burned, diapers also contribute, with their chlorinated organics and heavy metals, to problems of leaching, ash toxicity, and air emissions.

The quantity issue is especially dramatic and controversial. Given the lack of detail and specificity in most municipal waste stream analyses, figures on the number and percentage of single-use diapers in the solid waste stream have varied. Industry figures have been typically vague, mostly around .6 percent of the waste stream. The most detailed waste stream analyses, such as the Garbage Project's 1989 study for the city of Phoenix, place those figures much higher. According to the Phoenix analysis, diapers constituted 1.01 percent of *all* wastes buried in noncompacted landfills and 3.6 percent by weight of household waste. Since the noncompacted landfill consists of dirt cover and other similar components, the overall figure of diapers in the waste stream rests somewhere in between, though closer to the household waste estimate. These figures, moreover, correspond with other estimates placing diapers at about 2 to 4 percent of the waste stream.[75]

The waste stream figures for disposable diapers are even more remarkable considering that the product is less than 30 years old. Even as late as 1970, when the Huggies brand was introduced, disposable diapers constituted only 10 to 15 percent of the diaper market. That figure jumped to 64 percent by 1982, and 85 percent just six years later. Though some estimates suggest that paper diaper sales have peaked, the industry has continued to search for new markets, most recently expanding diaper size to account for delays in toilet training as well as creating new uses among older, incontinent adults. As a consequence, the disposable diaper has become the third largest (after potentially reusable/recyclable newspapers and bottles)—and *single fastest-growing*—consumer product in the waste stream.[76]

Health problems associated with soiled diapers compound the environmental issue of the enormous waste generation. When disposable diapers were first produced, a handful of manufacturers attempted to introduce flushable products for convenience and to reduce impact on landfills. But sewer lines often became clogged and the flushable diaper was quickly removed from the market. Manufacturers shifted the burden of feces removal to the consumer, recommending that feces be emptied into the toilet. This advice was impractical in terms of diaper design, contrary to the general marketing strategy emphasizing convenience, and almost universally ignored. Various studies suggest that little if any of the feces are removed before the diaper is discarded.

The presence of fecal matter in the solid waste stream has become a matter of concern and controversy. The most direct statement on the problem was presented by the Kings County Nurses Association of Seattle, Washington, which argued that fecal matter could increase the transmission of disease, including hepatitis and polio, since babies are effective carriers. Such risks would be most pronounced for workers handling the wastes during household collection and landfill disposal. These concerns have been dismissed by other officials, such as a Tacoma, Washington, Health Department supervisor who argued that "of all the landfill issues on the table in the last ten years, disposable diapers have never been raised as an issue."[77]

The paper diaper industry has become increasingly sensitive to arguments about the solid waste problems of disposables, given its high profile and need for consumer acceptance. "We believe that solid waste is a major issue and we want to be part of the solution," commented a Procter & Gamble official in 1988. The solution offered has been a thinner diaper, first introduced in the early 1980s, "which has contributed 50% less in weight and volume [for landfill disposal]," according to P&G.[78] The introduction of these thin, or "supertrim," diapers was not, in fact, based on solid waste considerations, but on marketing strategies, and has clearly become a convenient tangent to the argument about the company's contribution to solid waste problems. Extra-size diapers, another recent innovation, add greater volume and weight per diaper and can serve to postpone toilet training, thus also adding to the overall numbers of diapers in the waste stream. More to the point, the huge R&D resources available to the two big diaper companies have been almost exclusively devoted to market considerations. "The forces that drive developments in disposability are not usually born of altruism," argued a diaper industry trade official. "They come from ideas that money can be made by making new products from scrap and effectively using raw materials."[79]

Indeed, it was a tiny entrepreneurial operation in Greenwood, Colorado, rather than one of the "Big Two," that first attempted to develop and market a "biodegradable" diaper in 1988, a potentially important environmental innovation. This "Chemical-Free TenderCare Biodegradable" diaper contained a starch-free additive designed to allow the diaper material "to be attacked by insects, bacteria, and enzymes in a landfill environment." As a consequence, the company claimed the diaper would take only two years largely to biodegrade with as much as 92 percent decomposed between five and seven years. This compared with 200 to 500 years estimated for the standard disposable. Industry officials dismiss the concern, claiming that degradability is not a clear-cut issue, and that, in any case, the "Big Two" will ultimately produce a more degradable diaper in the near or medium term.[80]

Even more significantly, a Seattle, Washington-based full-service supplier of cloth and disposable diapers began an innovative separation program for used disposables. The fecal matter is removed and washed into the sewer system, while the plastics and pulp are separated and then sold to a plastics recycler and for fertilizer. Though the initial program has broken even in terms of costs, it has the potential to expand successfully, particularly if mechanisms are developed to inexpensively separate the polypropylene and polyethylene in the disposables. Once again, the Big Two have not pursued any research along such lines. Responding to those and other pressures, Procter & Gamble, concerned that environmental criticism could ultimately have an impact on market share, announced its own exploration of a recyclable paper diaper.[81]

Despite these changes, disposable baby diapers threaten to increase their role as the single largest individual manufactured product contributing to mounting solid waste problems. Moreover, the popularity of the product has caused even the industry's strongest critics to retreat from advocating direct public intervention. Even the litter, health, and disposability issues—intensified by the identification of the role of disposable diapers in ocean pollution and in the poisoning of birds and other wildlife who have mistaken diapers for a food source, the presence of dioxin in the paper fiber, and the chlorinated by-products that increase hazardous leachate, ash, or air emissions in landfills and incinerators[82]—have remained more an expression of concern than a cause for action. Meanwhile, the federal government has only exacerbated the problem by creating obstacles for the cloth diaper industry with trade restrictions against Chinese imports of cotton diapers.[83] The resolution of the tension between a high-powered and effective marketing strategy and an intractable solid waste problem, between an effective reuse product and a disposable with a 200- to 400-year decomposition rate, has ultimately turned out to be no contest at all.

The dilemma of diapers and tires and the difficulties involved in the reuse of compostable matter and beverage containers all suggest that the limits of reuse raise larger questions about the structure of production and decision-making in our society. The rise of new products such as steel-belted radial tires and disposable diapers, while meeting certain social needs and improving product performance, presents enormous problems in terms of solid waste disposal and related environmental and health consequences. The solutions raised—incineration of tires, or, in the case of diapers, no real solution at all—only compound the problems by acting as deterrents to successful intensive recycling and a more fully developed reuse strategy. Meanwhile, the vast capital and human resources applied to the development of new products are allocated on the basis of market and profit considerations. Waste issues, and, in a larger context, resource and environmental questions, are excluded from the process. Few, if any, research and development strategies are directed at questions of product recyclability or reuse. And many of the forms of government intervention that do exist tend to reinforce the profit-driven patterns and to sustain the structure of production-related decision-making that exists outside the public domain.

The difficulties of reuse, in contrast to its high place on the second rung of the waste management hierarchy, suggest that the major players in the solid waste arena—the public agencies and waste disposal companies in particular—see themselves as only secondary players in the issue of how wastes—and how much and what kinds of wastes—are generated. Recycling and reuse present a promising but incomplete response to the decline of landfills and eclipse of incineration, but the deeper issue of waste generation is a more difficult area for intervention. And while the prospect of reducing wastes at their *source* occupies the very top of the hierarchy, it has nevertheless become (given the central role of the production system itself) the most elusive and intangible of the alternative strategies to define and carry out.

NOTES

1. For background material on Proposition 11, see "The Life and Death of Prop 11," Alan Cline, *San Francisco Chronicle*, November 7, 1982; "Container Industry Tries to Crush

Prop 11," Gary E. Swan, *San Francisco Chronicle,* September 30, 1982; "Year of the Smear," *San Jose Mercury News,* October 30, 1982; *Los Angeles Times,* September 21, 1982; October 19, 1982; October 21, 1982; October 27, 1982.

2. *San Francisco Chronicle,* November 7, 1982; "Gun Control, Bottle Bill, and Water Law All Defeated," Carl Nolte, *San Francisco Chronicle,* November 3, 1982; *Los Angeles Times,* October 24, 1982.
3. *San Jose Mercury News,* October 30, 1982.
4. See, for example, full-page ad in *San Francisco Chronicle,* November 1, 1982, p. 15; "No on Proposition 11: The Forced Deposit Initiative," Californians for Sensible Laws, n.d.
5. *San Jose Mercury News,* October 30, 1982; *Can and Bottle Bills,* William K. Shireman and others, California Public Interest Group and the Stanford Environmental Law Society, Berkeley and Stanford, 1981.
6. Statement by Nancy Fadeley cited in *Can and Bottle Bills,* William K. Shireman and others, CALPIRG and ELS, 1981.
7. *Petrochemicals: The Rise of an Industry,* Peter H. Spitz, NY, 1988; *Characteristics of the Municipal Solid Waste Stream, 1960–2000,* Franklin Associates, EPA, 1986.
8. *Hazardous Waste in America,* Samuel S. Epstein and others, San Francisco, 1982, pp. 133–151.
9. "Used Oil Recycling Act of 1980," Report No. 96–1415, House Committee on Interstate and Foreign Commerce, September 26, 1980, 96th Congress, 2nd Session; "Oil Recyclers Target Do-It-Yourselfers," Nora Goldstein, *Biocycle,* vol. 29, no. 4, April 1988.
10. *Coming Full Circle,* Environmental Defense Fund, 1988.
11. "Solving the Hazardous Waste Problem: EPA's RCRA Program," Office of Solid Waste, EPA 530/SW-86-037, November 1986, pp. 24–25; *Mining Urban Wastes,* Cynthia Pollock, Worldwatch Paper No. 75, 1987.
12. "Senate Passes Bill on Dumping Infectious Waste," *New York Times,* October 8, 1988; "Used Oil Recyclers Set to Breathe Again," *Waste Age,* January 1987, p. 8; "Oil Recyclers Target Do-It-Yourselfers," Nora Goldstein, *Biocycle,* April 1988; "Future for Used Oil Recyclers Hinges on Key EPA Decision," Lori Leonard, *Recycling Times, Waste Age* and the National Solid Waste Management Association, January 1989, p. 4.
13. "Resource Recovery Project: Evaluation of Compost Marketing Proposals," Portland Metropolitan Service District, n.d.; *The Rodale Guide to Composting,* Jerry Minnich and Marjorie Hunt, Emmaus, Pennsylvania, 1979; "Verdier Process of Garbage Disposal," Raymond A. Williams, *Proceedings of the 1949 Public Works Congress, 55th Annual,* American Public Works Association, Chicago, September 1949.
14. "Resource Recovery: What Recycling Can Do," Daniel Knapp, Berkeley, 1982; *Composting of Municipal Solid Waste in the United States,* Andrew W. Breidenbach et al., U.S. EPA, 530/SW-47r, 1971; *Coming Full Circle,* Environmental Defense Fund, 1988.
15. "Solid Waste Composting Gains New Energy," Nora Goldstein, *Biocycle,* vol. 27, no. 7, August 1986; *Resource Recovery Project,* Portland Metropolitan Service District, p. 7.
16. "25 Years Ago in Solid Waste Composting," Jerome Goldstein, *Biocycle,* October 1987; "Solid Waste Composting Gains New Credence," Eliot Epstein and Todd Williams, *Solid Waste and Power,* June 1988; "Solid Waste Management Status and Disposal Options in Los Angeles County," City of Los Angeles Bureau of Sanitation, Sanitation Districts of Los Angeles County, and County of Los Angeles Department of Public Works, February 1988, Exhibit 4, p. 16; "Solid Waste Composting Picks Up Steam," Nora Goldstein, *Biocycle,* October 1987; "Perfect Answer in New Jersey," Dawn Schauer, *Biocycle,* September 1986.
17. *Resource Recovery,* Daniel Knapp, 1982, p. 19; *Coming Full Circle,* Environmental Defense Fund, p. 35.
18. *Coming Full Circle,* Environmental Defense Fund, p. 37.

19. "Solid Waste Composting Gains New Credence," Eliot Epstein and Todd Williams, *Solid Waste and Power,* June 1988; "Feasibility Evaluation of Municipal Solid Waste Composting for Santa Cruz County," Cal Recovery Systems, December 1983.
20. Composting and Co-Composting Projects, Files presented before Los Angeles City Council, 1985–1989; author's notes, discussion with Rubell Helgeson and Ruth Galanter, Los Angeles City Council, 1989.
21. Epstein and Williams, *Solid Waste and Power,* June 1988.
22. *Annual Report* for 1985, Delaware Solid Waste Management Authority; Interviews with George Savage, 1987; Jeff Haas, Co-compost Plant Manager, Delaware Solid Waste Management Authority, 1987; "Co-Composting for Small Communities," Aga S. Razvi, *Biocycle,* August 1986.
23. "Resource Recovery Project," Portland Metropolitan Service District.
24. In its 1985 Comprehensive Plan, the California Waste Management Board argues that composting is not expected to "rival waste-to-energy or recycling in terms of the percentage of municipal waste that can be diverted into landfills." *A Comprehensive Plan for Management of Nonhazardous Waste in California,* California Waste Management Board, Sacramento, California, 1985.
25. Interview with Ted Heyl, Los Angeles Department of Parks and Recreation, 1987; notes of discussion with Ruth Galanter, Los Angeles City Council, 1989.
26. Interview with Jeff Haas, 1987; Delaware Solid Waste Management Authority *Annual Report,* 1985.
27. "Minnesota Expands Recycling Potential," Dawn Schauer, *Biocycle,* vol. 27, no. 8, September 1986; "Solid Waste Composting Picks up Steam," Nora Goldstein, *Biocycle,* October 1987; Epstein and Williams, *Solid Waste and Power,* June 1988; "Yard Waste Composting—Portland's Success Story," *Resource Recycling,* May/June 1988.
28. "Recycling Plan Mow You Over," *Capital Times,* April 19, 1988.
29. Cited in "Yard Waste Plan Off for '88," Barbara Mulhern, *Capital Times,* May 10, 1988.
30. See U.S. Code of Federal Regulations, vol. 40, part 257, July 1, 1985, the National Archives of the United States; also *Garbage Management in Japan,* Allen Hershkowitz and Eugene Salerni, 1987.
31. "G.E. Completes Construction of Whole Tire-to-Energy Power Plant," News Release, General Electric Corporation, Schenectady, New York, November 10, 1987.
32. Author's notes, discussion with Rubell Helgeson, 1989; see also *Site Visits to Waste Composting Facilities in Austria, West Germany, and the United States,* City of Los Angeles, January, 1988.
33. "Committee Findings and Staff Papers on National Beverage Container Deposits," prepared for the Resource Conservation Committee, EPA, 1978; also author's notes, presentation by Barry Commoner, Northern California Grantmakers, San Francisco, November 4, 1987.
34. *Can and Bottle Bills,* William K. Shireman and others, CALPIRG and ELS, 1981.
35. "The Social Costs of Packaging Competition in the Beer and Soft Drink Industries," Kenneth C. Fraundorf, *Antitrust Bulletin,* 1975; *Mining Urban Wastes,* Cynthia Pollock, Worldwatch Paper No. 76, 1987.
36. "Solid Waste Becomes 'Crisis,'" *Modern Plastics,* January 1988. "Mining Urban Wastes," Cynthia Pollock; also *Wrapped in Plastic,* Jeanne Wirka, Environmental Action Foundation, Washington, D.C., 1988.
37. "Packaging—USA," Eric Outwater, in *Proceedings: First National Conference on Packaging Wastes,* EPA, 1971; "The Rise (and Fall?) of Nonreturnables," John Dernbach, *Environmental Action Bulletin,* November 1, 1974; *Can and Bottle Bills,* William K. Shireman and others, 1981.

38. *Resource Recovery and Waste Reduction,* 4th Report to Congress, EPA, 1977; *Can and Bottle Bills,* William K. Shireman and others, 1981.
39. *New York Times,* August 3, 1974.
40. *Can and Bottle Bills,* William K. Shireman and others; see also *Oregon's Bottle Bill: Two Years Later,* D. Waggoner, Portland, Oregon, May 1974; *Oregon's Bottle Bill: The First Six Months,* Eileen Claussen, 530/SW-109, EPA, Office of Solid Waste, 1973.
41. *Waste Reduction and Resource Recovery,* 4th Report to Congress, EPA, 1977; *The Economic Impact of Oregon's Bottle Bill,* C. M. Gudger and J. C. Bailes, Oregon State University, March 1974; *Some Economic Consequences of the Vermont Beverage Container Deposit Law,* University of Vermont, February, 1975; *Resource and Environmental Profile Analysis of Nine Beverage Container Alternatives, Final Report,* Robert G. Hunt et al., 530/SW-9/c, EPA Office of Solid Waste, 1974.
42. *Can and Bottle Bills,* William K. Shireman and others, 1981.
43. "California's Beverage Container Recycling Act," *Resource Recycling,* November/December 1986; "New Recycling Law Opens Door to Firms Seeing Profit in Trash," Victor F. Zonana, *Los Angeles Times,* September 28, 1987; *San Francisco Chronicle,* June 2, 1988; see also interviews with Tania Lipschutz, 1988 and Christopher Williams, 1989.
44. "Bid to Add Coolers to Bottle Law Fails," Jerry Gillam, *Los Angeles Times,* May 12, 1987; *Legislative Agenda,* Sierra Club, vol. 10, no. 5, March 16, 1988.
45. See, for example "How a Bottle Bill Like Proposition 11 Works in Connecticut," Gary Swan, *San Francisco Chronicle,* November 3, 1982; also *Solid Waste Management Alternatives,* Elliott Zimmerman, Illinois Department of Energy and Natural Resources, 1988, pp. 50–53.
46. *Coming Full Circle,* Environmental Defense Fund, 1988, p. 70; also *Can and Bottle Bills,* William K. Shireman and others, 1981; also Zimmerman, ibid.
47. Presentation to National Solid Waste Management Association, Frank T. Ryan, Vice President, Rubber Manufacturers Association, 1988; also *Waste Age,* April 1987, p. 212.
48. Interview with Frank T. Ryan, Vice President, Rubber Manufacturers Association, 1988; "A Planning Bibliography on Tire Reuse and Disposal," James F. Hudson and Elizabeth E. Lake, Council of Planning Librarians, Exchange Bibliography, no. 1331, September 1977; "Treating Trash as a Valuable Resource," David Morris and Neil Seldman, *Christian Science Monitor,* November 12, 1987; "Avoiding the Trash Heap: the Tire Makers Try All Sorts of Methods to Destroy Old Tires But Not the Environment," Ralph E. Winter, *Wall Street Journal,* April 27, 1972.
49. "Tire Recycling: Proven Solutions and New Ideas," Jerry Powell, *Resource Recycling,* January–February 1987; "Modest Growth Ahead for Rubber," Bruce F. Greek, *Chemical and Engineering News,* vol. 66, no. 12, March 21, 1988; "Tire Recovery and Disposal: A National Problem with New Solutions," Mary Sikora, *Resource Recovery Report,* Washington, D.C., 1986; "Garbage In, Power Out: A Clean Solution to a Heap of Problems," Jay Mathews, *Washington Post,* November 18, 1987.
50. "Americans Treading Softly Around Growing Problem of Scrap Tires," Gary Polakovic, *San Bernardino Sun,* January 30, 1987.
51. "Treating Trash as a Valuable Resource," David Morris and Neil Seldman, *Christian Science Monitor,* November 12, 1987.
52. "All the King's Men," a video documentary, Public Information Assist Team Production, EPA, 1984; also interview with Frank Ryan, 1988.
53. *The Problems and Possible Solutions to Disposing of or Recycling Used Tires,* State of California Department of Transportation, Division of Highways, Transportation Library, January 1975.

54. Background analysis by Voit Rubber Company, cited in "The Problems and Possible Solutions to Disposing of or Recycling Used Tires," California Department of Transportation, 1975; "A Planning Bibliography on Tire Reuse and Disposal," Hudson and Lake, 1977; *Wall Street Journal*, April 27, 1972.
55. Interview with Frank Ryan, 1988; *Government and the Nation's Resources*, Report by the National Commission on Supplies and Shortages, December 1976, Washington, D.C., p. 169.
56. "Tire Recycling Bounces Along," Jerry Powell, *Resource Recovery*, July 1988.
57. "Tire Recycling: Proven Solutions and New Ideas," Jerry Powell, *Resource Recovery*, January–February 1987; Morris and Seldman, *Christian Science Monitor*, November 12, 1987; "Newark Grinds Old Tires into Pavement," *Solid Waste and Power*, vol. 3, no. 1, February 1989, p. 56; Hudson and Lake, 1977.
58. "Give Them a Home on the Range," and "Advancing Science with Innovation," Rubber Research Elastomerics, Inc, Babbitt, Minnesota, n.d.; also "Tire Recycling Bounces Along," Jerry Powell, *Resource Recovery*, July 1988.
59. See the California Department of Transportation Study, 1975; also comments by Michael Rouse, Waste Recovery, Inc., at "Waste Tire Recovery and Disposal: Emerging Solutions," Resource Recovery Report Conference, Crystal City, Virginia, cited in *Resource Recovery*, January–February 1987.
60. *Resource Recovery*, Jerry Powell, January–February 1987.
61. "Built to Last," Hawley Truax, *Environmental Action*, March–April 1988; "Stanislaus Tests Burning for Scrap Tire Disposal," Raymond Clark Simon, Chairman, Stanislaus County Board of Supervisors, *California County*, January–February 1988; *Resource Recovery*, Jerry Powell, June 1988.
62. "Tire Burn Postponed by Protests," Nancy Marrinan, *Modesto Bee*, February 19, 1988; *Modesto Bee*, February 18, 1988; "How to Make 40 Million Tires Disappear," Ann H. Mattheis, *Waste Age*, January 1988, pp. 46–50; Oxford Energy Company, Press Release, November 10, 1987.
63. Oxford Energy Company, 1987 Form 10K, Commission File No. 33-6694; "Oxford Meets Performance Goals Firing Whole Tires," *Power*, October 1988.
64. Cited in "Tire Burner Produces Energy, Generates Controversy," *Refuse News*, January 1988.
65. Oxford Energy Company, 1987 Form 10K, Commission File No. 33-6694; *Annual Report*, 1987, Oxford Energy Company; "To Our Shareholders," Robert D. Colman, Oxford Energy Company, n.d.; "CMS and Oxford Energy Boards Ratify CMS Purchase of Oxford Stock," Oxford Energy Company Press Release, April 14, 1988; "Oxford Meets Performance Goals Firing Whole Tires," *Power*, October 1988; "Whole Tire-Burning Facility Will Help Fight Dealer Problem," Tim Cooney and Jim Davis, *Tire Review*, January 1988.
66. Oxford 1987 Form 10K, pp. 19–20 and F-4; see also *Modesto Bee*, December 10, 1987; February 19, 1988; *Washington Post*, December 7, 1987; "Built to Last," Hawley Truax, *Environmental Action*, March–April 1988.
67. Garb Oil and Power Corporation, 1987 Form 10K, Commission File No. 0-14859; "Massive Pile of Tires Fuels Controversial Energy Plan," Ronald Taylor, *Los Angeles Times*, November 30, 1987; interview with Mary Burns, 1988; Statement by Concerned Citizens of the Inland Empire on the Rialto Power Corporation Resource Recovery Project, Appeals Hearing Board, South Coast Air Quality Management District, November 1986.
68. Interview with Chris Peck, California Waste Management Board, April 14, 1988; "Findings and Decision of the Hearing Board in the Matter of Rialto Power Corporation to Review Denial of Permits to Construct a Resource Recovery Plant in Rialto, California,"

South Coast Air Quality Management District, February 5, 1987; Garb Oil 1987 Form 10K.

69. Garb Oil 1987 Form 10K, p. 2; "Tire Plant Irks Residents," Pat Murkland, *Riverside Press Enterprise,* March 20, 1986; "Brewer Promises to Fight for Rialto Tire-Burning Plant," Pamela Fitzsimmons, *San Bernardino Sun,* June 30, 1986; "Firm Threatens to Sue Fontana over Tire Plant Views," Pat Murkland, *Riverside Press Enterprise,* July 10, 1986.
70. Cited in "Congress Can't Add, So the Taxpayer Pays," Donald L. Barlett and James B. Steele, *Philadelphia Inquirer,* April 14, 1988; Garb Oil 1987 Form 10K, p. 12.
71. Interviews with Mike Jacobsen, Executive Director, Nonwovens Industry Association, 1988; Roger Johnson, Marketing and Development, Kimberly-Clark Corporation, 1988; Kimberly-Clark Corporation, 1987 Form 10K, Commission File No. 1-225.
72. "At Issue: The Disposing of Disposables," parts 1 & 2, Bernard Lichstein, *Nonwovens Industry,* November 1988 and December 1988, pp. 18–23, pp. 20–23; interviews with Tina Berry, Kimberly-Clark, 198; Daniel Briggs, Human and Environmental Safety, Paper Division, Procter & Gamble, 1988; "Diapers in the Waste Stream: A Review of Waste Management and Public Policy Issues," Carl Lehrburger, Albany, New York, August 1988.
73. "Do Disposable Diapers Ever Go Away," Michael deCourcy Hinds, *New York Times,* December 10, 1988; "Diaper Wars: The Struggle to Win the Hearts and Bottoms of a Nation in a Nerve-Racking Battle of the Disposable-Diaper Titans," Brooke Gladstone, *Boston Globe,* October 18, 1987; "Memorable Years in Procter and Gamble History," Procter & Gamble Corporation, Cincinnati, Ohio, n.d.; "Diapers in the Waste Stream," Carl Lehrburger, 1988.
74. "The Disposable Diaper Myth: Out of Sight, Out of Mind," Carl Lehrburger with Rachel Snyder, *Whole Earth Review,* Fall 1988.
75. *The Phoenix Recycling Project: A Characterization of Recyclable Materials in Residential Solid Wastes: Initial Results,* W. L. Rathje et al., the Garbage Project, Bureau of Applied Research in Anthropology, University of Arizona, Prepared for the City of Phoenix, Arizona, Department of Public Works, May 25, 1988; interview with Wilson Hughes, the Garbage Project, 1988; see also *Statewide Resource Recovery Feasibility and Planning Study,* vol. 2: Solid Waste Characterization Report, Environmental Improvement and Energy Resources Authority, State of Missouri, Department of Natural Resources, December 1987.
76. Interviews with Mike Jacobsen, 1988; Roger Johnson, 1988; Franklin Associates, 1986.
77. "Hazards Posed by the Improper Disposal of Disposable Diapers," Patricia K. Greenstreet, Kings County Nurses Association, n.d.; "Soiled Disposable Diapers: A Potential Source of Viruses," Mirdza L. Peterson, *American Journal of Public Health,* September 1974; vol. 64, no. 9, "The Diaper Dilemma," Kathleen Merryman, *Tacoma News Tribute,* January 29, 1988.
78. Interview with Sue Hale, Procter & Gamble, 1988; also interview with Tina Berry, 1988.
79. "At Issue: The Disposing of Disposables," part 1, Bernard Lichstein, *Nonwovens Industry,* November 1988, p. 18; interview with Judith Brunk, Executive Director, Hill an' Dale Family Learning Services.
80. "Disposable Diaper That's Biodegradable," *San Francisco Chronicle,* September 30, 1988; "New Diapers May Be Able to Relieve Environment," Abby Cohn, *Seattle Times,* December 30, 1988; also interview with Mike Jacobsen, 1988; "Do Disposable Diapers Ever Go Away," Michael deCourcy Hinds, *New York Times,* December 10, 1988; "At Issue: The Disposing of Disposables," part 2, Bernard Lichstein, *Nonwovens Industry,* December 1988, pp. 20–23.
81. "What's New in Diapers: Diaper Pulp—Out of the Garbage and into the Garden,"

Barnaby J. Feder, *New York Times,* March 12, 1989; "A Plan to Recycle Diapers," *San Francisco Chronicle,* June 21, 1989.

82. Interview with Rimmon Fay, 1988; "Milk Cartons to be Studied for any Dioxin Traces," Robert A. Rosenblatt, *Los Angeles Times,* December 9, 1988; interviews with Daniel Briggs, 1988, and Mike Jacobsen, 1988.
83. Interview with Kirk Johnson, assistant to Congressman Ron Wyden, 1988; "Fuss Over a Diaper Import Quota," Clyde Farnsworth, *New York Times,* May 31, 1988.

9

THE CHALLENGE OF REDUCTION: GENERATION AT THE SOURCE

A McTOXICS CAMPAIGN

As a target, it seemed both a visible and poignant reminder of the solid waste crisis. The McDonald's foamed polystyrene container, the famous clamshell package surrounding the variety of McBurgers consumed daily, is a highly recognizable and familiar disposable item. McDonald's, in fact, is the single largest consumer of foamed polystyrene containers in the United States, responsible for the use and subsequent discard of 70 million pounds a year, the equivalent of 1.7 billion cubic feet of foam packaging wastes.[1] These containers, furthermore, pose a range of hazards from the wide variety of toxic chemicals used in their production, including chlorofluorocarbons (CFCs), which are used to puff up or to create the foam effect. CFCs, of course, have also been getting a great deal of attention for their contribution to the destruction of the earth's ozone layer, but several of the other, lesser-known chemicals present major problems as well. Moreover, the fast food foam container is yet another quick throwaway that has become a significant factor in the changes and growth of the solid waste stream. For the initiators of this McToxics campaign, it seemed they could not have selected for public action a more richly deserving symbol of an unnecessary and dangerous waste.

For the executives at the McDonald's Corporation, concerned about corporate image and marketing strategies, the McToxics campaign seemed an affront. "That's the way our customers have told us they want their food," one company official complained, arguing that Styrofoam, after all, is just "basically air" and performs a valuable function by "[aerating] the soil."[2] Another executive told protesters that the company had shifted to Styrofoam after the Sierra Club had complained that too many trees were being destroyed to help produce McDonald's packages, of which approximately 30 percent are made of plastic.[3] Once the McToxics campaign, coordinated by the Citizen's Clearinghouse for Hazardous Wastes, and a broader action by the environmental group Friends of the Earth directed at all fast food foam packaging, were initiated, the issues of solid wastes and related hazards including CFCs became joined. Though less in response to the protest actions than to the front page coverage of the ozone depleting/CFC issue, a number of communities, including Berkeley, California, Suffolk County in New York, and the states of Maine and Vermont, began to ban fast food foam containers, especially those made with CFCs.[4]

Stung by the criticism and concerned with the growing number of bans—bills were actually pending in 21 states during 1988—McDonald's initially agreed to phase out its use of CFCs within 18 months. The McDonald's action paralleled the sudden decision of the Du Pont Corporation, the largest single producer of CFCs in the world with a 50 percent market share, to also stop producing CFCs. Both Du Pont and McDonald's had already developed substitute products, though McDonald's also intended to continue to use plastic foam containers, albeit ones made without CFCs.[5]

Despite McDonald's much publicized CFC phase-out, public pressure, including the colorful and persistent McToxics campaign, continued. Then, in the summer of 1988, with a serious-looking Governor Mario Cuomo at their side, McDonald's executives held an Albany, New York, press conference to announce a new recycling program that would accept used Styrofoam containers and convert them into insulation panel.[6]

For the McToxics campaign organizers, however, the core of the problem still remained. "It's the containers themselves that are the root of this McTrash issue," campaign organizer Will Collette said of McDonald's new recycling program, though he called the change in policy of the fast food company a "partial victory" as well. "You still have toxic chemicals used in their manufacture," Collette argued, "you still have the excessive packaging, you still have, in effect, the problem."[7] To make their point, CCHW organized a second phase to their campaign, calling for activists to collect the Styrofoam ("See a burger wrapper, pick it up") and send it by mail to McDonald's chair Joan Kroc or other McDonald's stores or executives. "McDonald's," the McToxics organizers wrote, "needs our help to make their [recycling] plan work."[8]

For the McToxics organizers, the clamshell campaign is a way to challenge the production of a waste generated at its source. The hazards of the production process are one dimension of the problem; so, too, are the kinds and amount of wastes generated. By directly challenging McDonald's, they have brought to the fore an approach largely absent from the discourse over solid wastes. They argue that democratic participation in industrial production decisions is the key to the development of a real "solution" to the waste dilemma, a solution that focuses on reducing, rather than simply disposing of, wastes. And now this solution, embodied in the McToxics campaign as both a symbolic tactic and an issue of substance, seeks its rightful place at the top rung of the hierarchy.

PACKAGING'S ROLE

Waste reduction, as the saying goes, is something that everyone agrees with, but nobody does. The focus of waste management activities has long been on the postmanufacturing, postconsumer end of the waste cycle as a matter of political choice and institutional bias. Intervention in production decisions, including product development, packaging, and marketing strategies, has long been considered a public taboo, better left to the exclusive province of the market. The resulting energy, pollution, and waste disposal costs, among others, have all been considered *external* to the costs of production, and, at best, dealt with infrequently by limited regulations. Waste reduction at the source, on the other hand, has to be considered the most direct form of preventing, avoiding, or reducing the generation of wastes, the increase of pollution, and the added burden of energy and raw materials use. Yet it has remained the most underutilized waste management option, a casualty of the "free market" principle that production decisions are exclusively reserved for the private sector.

Among alternative strategies, recycling, which is a postmanufacturing, postconsumer activity, has been preferred by industry, public agencies, and even a number of environmental and community groups as the best way to stretch landfill capacity. Incineration had primarily been the favored, end-of-the-waste-cycle approach, and continues to be the accepted management strategy advocated by agencies like the EPA, not only for solid waste, but for toxic waste as well. But with the growing challenge of community groups and overall decline in public acceptance, incineration began to give ground to recycling, as public agencies desperately sought to develop greater disposal capacity. The more effective and comprehensive alternatives of reuse and most especially reduction, however, continue to be either minimized or quietly rejected, given their larger, production-related implications. The waste issue ultimately remains a question of disposal and not generation.

In analyzing the overall dimensions of the waste problem, industrial groups, public agencies, and congressional committees have almost exclusively focused policy on where the wastes end up rather than where they come from. The limits of this approach have been underscored by two key factors that emerged in the post–World War II era that changed both the scope and dimension of solid waste problems. On the one hand, the generation of wastes, especially packaging wastes, has grown at a faster rate than population.[9] At the same time, the nature of the waste stream has changed, setting up new environmental, technical, and economic issues at the disposal end of the production cycle. These two factors have been most pronounced in the area of packaging and the particular role of plastics.

Packaging, by all accounts, represents an extraordinarily high percentage of the waste stream, at least one-third of all municipal residential wastes. In some communities, it represents as much as 40 to 50 percent of household waste by volume. By the 1980s, upwards of 43.5 million tons of containers and packaging were discarded into the country's municipal waste stream (see table 9.1).[10]

Table 9–1
PRODUCTS DISCARDED INTO THE MUNICIPAL SOLID WASTE STREAM
(IN MILLIONS OF TONS AND PERCENTS)

	1970		*1984*		*2000*	
PRODUCTS	TONS	%	TONS	%	TONS	%
Durable goods	13.9	12.6	18.6	14.0	22.9	14.4
Nondurable goods	21.6	19.6	34.0	25.6	47.4	29.8
Containers and packaging	39.3	35.6	43.5	32.7	50.1	31.6
Other wastes	35.5	32.1	37.0	27.8	38.3	24.1
Totals	110.3	100.0	133.0	100.0	158.8	100.0

SOURCE: "Characterization of Municipal Solid Waste in the United States, 1960–2000," Franklin Associates, EPA, Office of Solid Waste, PB-178323, 1986.

"Municipal" solid waste is defined in this table to include residential, commercial, and institutional solid waste.

The packaging industry in the United States has developed into more than just a big business in the last 40 years; it has become an essential part of the production and marketing system. Cynthia Pollock of the WorldWatch Institute estimated that packaging expenditures in 1986 amounted to $28 billion, a staggering figure that is, in fact, greater than the net income that year of the nation's farmers.[11] Another estimate based on the net value of all packaging shipped in 1985 placed the number at $55.8 billion, which amounted to 4 percent of the value of all finished goods sold in this country.[12] The U.S. consumer, according to the World-

Watch study, pays as much as nearly $1 in hidden packaging costs for every $10 spent on food and beverages.

Packaging, of course, provides a number of essential services in the movement of goods from producer to distributor and to the consumer. Packages can physically protect a product, maintain its freshness, and insulate it from dust, insects, or inclement weather. Packages, however, are also marketing tools, enhancing the sales value of a particular product and helping to differentiate it from other, similar items. It is this latter activity, especially, that has changed in the contemporary era, linked to developments in marketing, product design, and the material base for production.

The big jump in packaging wastes that took place during the 1950s and 1960s finally caught the attention of public officials by the late 1960s, but only after significant changes in the waste stream had already taken place. Neither the 1965 nor 1970 federal legislation addressing solid wastes had specifically targeted packaging for its responsibility in generating more—and different—kinds of wastes. But public officials at the local level most directly confronted with the issue of disposal had become increasingly unhappy with packaging's role in the growing waste stream. These officials, however, also hesitated when it came to intervention. A 1969 editorial in *Public Works*, a magazine oriented toward local public officials, complained of the problem of excess packaging, where "the carton destined for discard was easily twice the volume of its contents." Yet the editorial adopted a slightly whimsical rather than confrontational tone: this unfortunate generation of new wastes, the editorial lamented, was just "magnificently ridiculous."[13]

Despite the rather tentative concerns of public officials, the packaging industry organizations and their supporters were prepared to battle any form of government regulation of packaging. For example, a number of packaging firms created the Materials Research Council to coordinate the packaging industry's approach to waste disposal problems. This included strong opposition to government intervention, which one industry executive characterized as "overemotional" and a "failure to put the problem in its proper perspective."[14] Similarly, in a speech at the first national conference on packaging wastes in 1969, Arsen Darnay, one of the best-known waste stream analysts, argued forcefully that waste disposal problems from packaging were not the responsibility of the producer and the packager. Any efforts to address the problem by these groups, Darnay argued, were appropriate only in their capacity as private citizens or, as with antilittering campaigns, in order to "anticipate and defuse" any government intervention. More importantly, any "effective regulation of packaging," Darnay asserted, "logically leads to regulation of *all* economic activities." While agreeing that those who "*generate* undesirable wastes" should "bear the costs of cleanup," Darnay concluded, "In the context of packaging wastes, the generator is the *consumer,* not the *producer.*"[15]

A position simply opposing government intervention, however, was not sufficient given the activism of the period. With a large, but relatively diffuse, environmental movement already beginning to take note of the waste issue by the late 1960s, both industry groups and public officials felt increasing pressure to demonstrate a new plan of action rather than relying solely on the landfill strategies of the post–World War II era. Industry groups stressed that the waste management focus had to remain on disposal technologies rather than regulation, taxation, and especially product bans or standards, the most worrisome of the reduction measures proposed during the late 1960s and early 1970s. Even as early as 1969, when waste-to-energy technology in this country was still considered highly experimental, industry officials began to tout it as "the most common and the most practical solution for disposing of solid wastes in the years to come." Burning rather than burying trash, a Dow Chemical executive noted, is "the ultimate solution."[16]

Such a stance, particularly for packagers and plastics manufacturers, was controversial. The plastics industry, including plastics packaging, was already criticized for the problems they caused various disposal methods. There were significant difficulties for landfills due to plastics' inability to decompose. Moreover, few, if any, of the existing 350 or so municipal incinerators of the late 1960s were designed to adequately burn such materials. The burning of polyvinyl chloride (PVC), by then in common use as pipe and for packaging materials, produced hydrochloric acid. In addition to its adverse health effects and contribution to acid rain, PVC tended to corrode incinerator furnace walls and scrubbers, and, in a number of instances, necessitated expensive repairs. A New York City environmental official, in discussing that problem at the 1969 packaging conference, warned, in fact, that the city might resort to either a major tax or even a product ban on this "particular packaging product" in order to prevent further damage of their $250 million investment in incinerator equipment.[17]

The solution, the packagers responded, lay not with banning PVC or with the present technology with all its difficulties, but with the importation of more sophisticated European waste-to-energy systems. The trends in packaging, furthermore, reinforced the shift to incineration because, as Arsen Darnay put it, they would improve "the heating value of the fuel." This was particularly true with the growing use of plastics and paperboard (alone or in combination) in packaging materials, a trend already pronounced by 1970 that would only increase during the next two decades.

"Packaging materials," Darnay concluded in his 1969 speech, "will consequently help to make refuse a better fuel should we decide to recover heat as we dispose of waste."[18] His position was largely adopted by industry interests at the time, and it countered in later years the community protests targeting such symbols of the period as the fast food polystyrene container.

By the 1980s, the brief and tentative interest in packaging tended to fade from policy agendas, even for community opponents of landfills and incinerators. Waste stream analysts, such as Franklin Associates, estimated that packaging, though a huge part of the waste stream and still increasing in actual tonnage, was nevertheless leveling off and even declining slightly as a *percentage by weight* of the total municipal waste stream. Franklin estimated that although packaging wastes increased by more than 4 million tons between 1960 and 1984 (from 39.3 million to 43.5 million), the percentage by weight decreased from 35.6 percent to 32.7 percent. This trend was projected to continue, with an overall increase of more than 6 million tons by the year 2000 (from 43.5 to 50.1 million tons), but a small decline in percentage by weight to 31.6 percent. In effect, the report suggested that packaging wastes, though still considerable, were no longer a growing problem.[19]

The basis for this trend, however, reveals another set of problems for landfills (or incinerators) related to the changing composition of packaging. The fastest-growing component of packaging is plastic, including plastic combined with other materials that are impossible to recycle. These new packaging wastes are thus *lighter by weight but greater by volume,* given the displacement by plastic materials of such heavier and potentially recyclable or reusable materials as paper and glass. Though all waste management planning has been based on measures of weight such as tons per year, for landfills, the key factor determining capacity has been the volume of wastes, especially plastics, which fail to decompose over the life of the landfill and which take up at least three to four times as much landfill space for their weight as other materials.[20] And though tipping fees are assessed by weight, the ash disposal problems are equally volume-based. Packaging problems, especially those centered around plastic-related packages, usually tend to be underrepresented in most current waste stream analyses.[21]

Despite this approach, by the late 1980s, the question of plastics increasingly began to preoccupy policy-makers. While the focus on packaging had receded, new interest focused on the role of plastics and plastic additives, both in terms of packaging and in relation to their growing use in consumer and industrial products. There was new talk again of product bans and some form of industry regulation, though much of this effort was still limited in scope. Nevertheless, one of the centerpieces of the petrochemical revolution had finally achieved recognition as a serious waste issue, more than forty years after it had begun its march through the production process.

THE IMPORTANCE OF PLASTICS

As packaging became an essential part of the system of production, it contributed significantly to a changing set of waste problems. Since World War II, this massive and expanding business dominated by oil and chemical multinationals grew from an industry nurtured by government subsidies into a huge set of corporate activities, with a 1985 payroll of $23 billion and a value of shipments estimated at the phenomenal figure of $138 billion. Plastics demand, according to an industry estimate, reached a level of 48 billion pounds in 1985, up from 19 billion pounds in 1970. It is projected to increase to 76 billion pounds by the year 2000, accounting for about $345 billion in products annually. At current levels, in fact, the volume of plastics used in the United States exceeds that of steel.[22]

The plastics industry is integrated both horizontally and vertically, determined by the technology. There are essentially four stages in the production of plastic products: conversion of raw materials into intermediate compounds or polymers; transformation of those compounds into resins; the conversion of resins into secondary products; and a final stage involving the manufacture of the final product. The first two stages related to the production of resins are dominated by only a handful of large corporations, almost entirely chemical and petrochemical multinational corporations such as Dow, Monsanto, Du Pont, Union Carbide, Mobil, and Allied-Signal, among others. The second set of resin producers are more varied in terms of company size and activity. Overall, there are more than 10,000 processors, with over half employing fewer than 20 people.[23]

The distinctive structure of plastic production translates in effect into two separate industries, each with its own set of production-related characteristics as well as environmental consequences. The production of resins, a complex chemical process involving many techniques, is the foundation of the industry, since all plastics are made from resins. Though there are thousands of plastic products, there are only two main types of resin: thermosets, which are ultimately used in strong and durable products such as car parts, and thermoplastics, which are used for a variety of packaging materials as well as many other industrial and consumer products. Thermoplastics is by far the largest area of resin production, accounting for as much as 80 percent of all the plastic that is produced. In terms of processing operations, by contrast, there are many different techniques used, including blow-molding, foaming, thermofoaming, coating, and finishing. The most widespread methods, responsible for processing nearly two-thirds of all plastic products made in the United States, involve either extrusion or injection-molding.[24]

The plastics industry can be considered essentially an extension of both the chemical and petrochemical industries. These industries supply the necessary raw materials for production, with some chemicals produced almost exclusively for use in plastics production such as vinyl chloride, dimethyl terephthalate (used solely for the production of PET), and styrene (used for foam containers). Since plastic is essentially made by combining petrochemicals with oxygen and various other chemicals, ultimately all types of plastic are derivative of petroleum in one form or another.[25]

Plastic has most often been used as a substitute for other materials, such as glass, wood, various metals, paper, rubber, and porcelain, more than in the development of entirely new products. Plastic production is more capital-intensive (thus involving labor savings for producers but creating fewer overall jobs), while its products, especially when judged in terms of convenience, have proven to be more marketable. As Barry Commoner has argued, decisions to substitute plastics, and, in broader terms, petrochemicals, for other materials in a variety of products are based not so much on their capability to improve the product's function, but are essentially due to these capital, labor, and marketing considerations.[26]

These changes have been most pronounced in the packaging industry itself, in the midst of its own "plastic revolution." Today, plastics make up the third largest—and fastest-growing—segment of the packaging industry, exceeded only by paperboard and metals. The average consumer uses approximately 60 pounds of plastic packaging a year, almost all of which is not recycled, and tends to have an extremely short use life. The shift to plastics for packaging, furthermore, is expected to almost double by the year 2000 from more than 12.7 billion pounds of plastic packaging resins used in 1987 to approximately 23 billion pounds by the turn of the century.[27]

This plastics revolution has spawned a variety of new packaging strategies, such as "nesting." This approach, common in fast-food outlets, involves one or more inner containers placed inside an outer container. When all the containers are discarded, there is a corresponding increase in the volume of wastes generated, while the weight due to packaging remains the same.[28] Plastics packaging can be found on a broad range of products substituting for other more biodegradable and recyclable raw materials. These include containers such as yogurt and other tubs, eggs, and clamshells, where plastics account for 23.8 percent of all materials; bottles (26.7 percent share) such as those for milk, oil, detergent, and

PET-based beverage replacements; and plastic films like Saran Wrap, dry cleaning bags, and candy wrappers (with a 35.6 percent share). All these figures, moreover, are projected to increase.[29] According to the weight-based Franklin Associates estimate (which underrepresents the amount in terms of volume), plastics accounted for 11.5 percent, or about 5 million tons, of all container and packaging waste discards in 1984, but is likely to grow at a far more rapid rate than its two main competitors, paper and glass (see figure 9.2). For example, the plastic grocery bag had about a 25 percent share of the grocery bag business in 1985, which amounted to approximately 22 billion sacks produced that year. But that figure was projected to more than *double* within three years, quickly surpassing the long-standing biodegradable brown paper bag as the dominant package for transporting groceries back to the home. And while plastics maintains a current 17 percent share of the $19 billion food packaging market, projections show that its share could rise to 50 percent of a $44 billion market by the year 2000.[30]

The continual search for new markets and the introduction of new, substitute products by the plastics industry has had an extraordinary impact on the waste stream and the waste management issue. The new products that continue to be introduced at a dizzying clip such as the one-roll, throwaway plastic camera, or the squeezable ketchup bottle with its six different layers of multiresin plastic compounds, present a multitude of waste-related problems. Production of the squeezable ketchup bottle, for example, is estimated to grow from 300 million units in 1985 to 29 *billion* units in 1995, introducing a product impossible to recycle, while displacing a classic reuse item, the glass bottle.[31]

The phenomenal growth of plastic products has created a broad range of environmental hazards and problems, particularly manifest in the area of solid waste disposal. These problems occur at both the front end and back end of the production cycle. At the front end, a number of highly toxic chemicals are used as base materials for production. According to EPA, in fact, five of the six most hazardous chemicals—propylene, phenol, ethylene, polystyrene, and benzene—are commonly used by the plastics industry and, in the case of the first four, account for 44 to 73 percent of the entire use of those chemicals. The production of plastic resins itself generated over 5 million metric tons of hazardous waste in 1984, while the processing end of plastics production was responsible for another 88.7 million metric tons.[32]

The groups most directly at risk at the front end are the workers who handle the chemicals and the residents in the immediate community exposed to emissions. Air emissions of a variety of toxic, noncriteria pollutants are frequently detected at plants that use such chemicals. Polyvinyl chloride (PVC) chemical plants create potentially serious health risks for

Figure 9–2
PLASTICS IN PACKAGING
(percentages by weight)

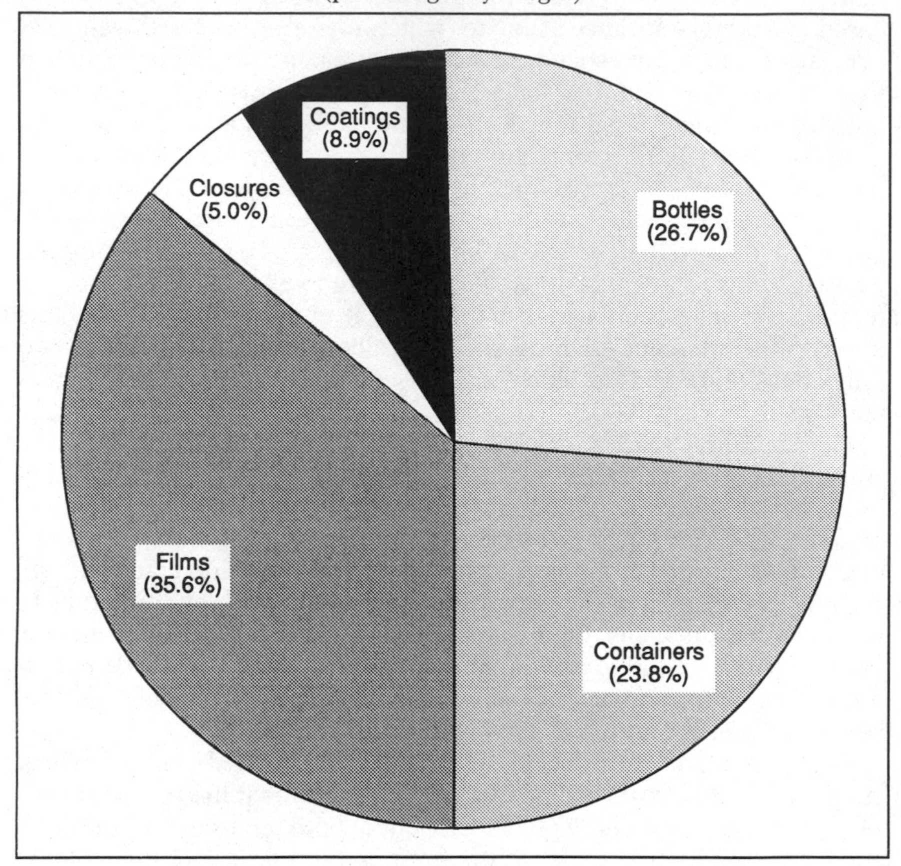

SOURCE: *Modern Plastics*, January 1987.

their work force, including the increased possibility of contracting liver and brain cancer, while workers who manufacture products from PVC have suffered from a respiratory ailment known as "meat wrappers asthma," associated with the PVC film packaging used to wrap meat.[33]

While environmental regulations and recycling pressures have been brought to bear at the front end of plastics production, the back end, where the product is discarded and becomes a solid waste and litter problem, might have longer-term implications.[34] For example, ocean and beach pollution from plastic products such as fishing nets and Styrofoam containers has become a serious eyesore and a hazard for marine life and animals. During one three-hour sweep by volunteers along 157 miles of Texas beach front in 1987, 31,773 plastic bags, 30,295 plastic bottles, 15,631 plastic six-pack rings, 28,540 plastic lids, 1,914 disposable diapers, 1,040 tampon applicators, and 7,640 milk jugs were collected.[35] In the ocean, birds mistake plastic pieces for food, sea turtles get strangled in plastic bags, and seals are found entangled in discarded plastic nets and debris. An estimated 100,000 marine mammals die each year from these problems, causing one marine life expert to suggest that plastics may be as much a problem as oil spills or other forms of discharges into the sea.[36]

But the problem of solid waste disposal seems most inexorable in light of the difficulties associated with the recycling of plastic products. Plastics, for one, represents the fastest-growing category in the waste stream, increasing from just 1 percent (by weight) in 1960 of all municipal solid wastes, to 2.7% or 3 million tons in 1970, to 7.2% or 9.6 million tons in 1984, according to the Franklin Associates estimate. This amounted to a 720 percent increase in plastic wastes in less than 25 years, with future growth predictions—15.5 million tons or 9.8 percent of the waste stream by the year 2000—indicating that the trend will likely continue at approximately the same rate. One industry consultant, moreover, argues that increases by volume are far greater, a claim backed up by more detailed waste stream analyses on the local level.[37]

Those percentages are even more significant given the lack of recycling and biodegradability of nearly all plastic products. Degradability, for one, has direct consequence in terms of landfill capacity since the life of a landfill can be extended by increasing the rate at which the wastes decompose.[38] Since plastic products decompose very slowly, if at all, while inhibiting or preventing other wastes from decomposing, they tend to shorten the useful life of a landfill. Industry officials counter that non-degradibility in a landfill might well be a benefit since it could mean that plastics are not contributing to the problems of leaching. Other industry studies, however, point out that a number of plastic additives, once in contact with water, not only lead to leaching but are likely to contaminate water supplies. "Volatiles and water contamination present potential problems," one trade publication remarked, "when plastic products are burned or dumped in a landfill."[39]

The plastics industry has been forced to deal with the degradability issue in part because of increasing attempts to pass legislation, some successful, that directly address the problem. The technology to manufacture degradable products, such as the plastic loop for six pack containers, is available, while for others it is still in an experimental stage. Although the industry has tended to dismiss the importance of degradability, it has nevertheless begun to invest in research in this area. This is largely in response to the spread of product bans, such as the legislation that outlawed nondegradable six pack loops in sixteen states, a ban extended to the national level when EPA was required by legislation to promulgate similar regulations.[40]

Recycling is an even more sensitive issue for the plastics industry, given the renewed public interest in recycling and plastics' minimal role. In 1984, only about 100,000 tons, or 1 percent of plastic materials, nearly all PET containers, were recycled. Whereas plastics discards were the fastest-growing category of the waste stream, they also had the lowest rate of any category of recyclables. In comparison to the 1 percent figure for plastics, the recycling rate in 1984 for aluminum was 28.6 percent, 20.7 percent for paper and paperboard, glass at 7.2 percent, and rubber, one of the more difficult materials to recycle, still achieving 3 percent, or three times the recycling rate for plastics.[41]

There are several reasons why plastics recycling, despite the recent increase in interest and research and development on the subject, remains more problematic and costly than the recycling of other materials. Plastics degrade when reheated, eliminating any potential reuse strategy for food, drug, and cosmetics containers, including PET bottles, since temperatures have to be high enough to sterilize the container to meet government regulations. Plastic products, furthermore, are likely to be "contaminated" with nonplastic material, thus inhibiting effective recycling, or, at the least, adding substantially to its cost. The opposite of this process is also true, since the inclusion of plastic in other products, such as "paper" diapers, also prevents the recycling of either the product itself or its constituent parts.[42]

The two types of plastics recycling are "primary recycling," which takes place at the front-end of production and involves the use of the waste material in the same production process, and "secondary recycling", which takes place at the back end of the cycle and involves the creation of a new and different product placing fewer physical or chemical demands on the recycled material. Secondary recycling most directly impacts the solid waste issue insofar as it involves products that would otherwise enter the waste stream. One key concern with secondary recycling, however, is that it doesn't necessarily reduce the overall amount of plastic generated for disposal, and, in certain circumstances, *might even increase it*. In nearly all cases, recycled plastic is not used as a substitute for virgin resin, rather, it is often used to replace nonplastic and more naturally degradable materials, thus increasing the total amount of plastic in the waste stream.[43]

The one relatively successful effort at secondary plastics recycling has been the PET bottle of which about 20 percent of the 875 million pounds produced in 1987 was recycled. Though polyethylene terephthalate, or PET, is used for a variety of products such as food packaging "boil-in-bag" pouches and "ovenable" food containers, slightly more than half of the resin ends up in bottle form, especially carbonated soft drink bottles. These have been the primary target of recycling efforts due mainly to their connection with bottle bill legislation. About 75 percent of recycled PET is used as some form of fiberfill stuffing, where it is shredded into pellet form and used in pillows, ski jackets, cushions, and sleeping bags. The rest is chemically reformulated for the eventual manufacture of other products like fence posts, paint brushes, scouring pads, and shower stalls. This latter activity in particular involves both greater energy use and the production of new materials, which could also increase the total amount of plastic generated. And while PET recycling is expected to increase as processing facilities expand in number and size, the amount of PET materials to be disposed of will ultimately still increase significantly, given industry projections that PET use will double between 1985 and the year 2000.[44]

While some modest PET recycling has occurred, recycling of other plastic products, including HDPE or high-density polyethylene bottles, has been far less developed, despite possible use of reprocessed HDPE in products such as garden furniture and lumber boards for boat piers. The best known HDPE product is the plastic milk bottle, though HDPE is also used in bottles for liquid detergent, bleach, antifreeze, and oil. Though PET products have grown astronomically in just a short period of time, sales of HDPE products are still nearly five times that of PET, while recycling rates remain ten times lower, or barely 2 percent. A new technology has been developed, however, to remove nonplastic contaminants (for example, the milk and labels on the milk bottle), a key obstacle in any high-volume recycling effort. Despite this breakthrough, HDPE recycling has not experienced any significant growth, mainly due to the limited amount of source separation involved in most recycling programs. HDPE, like a number of other potentially recyclable plastics, would be best recycled in the context of an intensive recycling program.[45]

The minuscule figures for plastics recycling have become a major embarrassment for the industry. As a result, a number of industry organizations, trade groups, and industry-backed research institutes have recently developed a higher profile around the issue, including the newly formed Plastics Recycling Foundation. These groups have undertaken a large public relations campaign to suggest that plastics are eminently recyclable, partly in the hope of deflecting the pressure for antipackaging legislation and product bans. In 1989, aware of the mounting criticism, the plastics industry launched a major recycling effort.[46]

As with recycling, concern about potential legislation has driven many of these trade groups, along with the industry itself, to actively pursue the development of plastics that better decompose. Some recent successes include the creation of a biodegradable trash bag that is designed to decompose in just a few years compared with the 300 to 500 years it now takes regular plastic bags. But like other plastic degradables, these products cost more, and there are other problems related to product strength and various health-protective regulations. Moreover, some environmentalists fear that the new technology of plastic degradability has not been sufficiently studied and that some toxics may be more readily released during degradation.[47]

The main strategy of the industry groups, however, has been to suggest that recycling of plastics is best accomplished by what industry figures call "waste-to-energy recycling." This play on words (associating recycling with incineration) is in fact the centerpiece of the plastics industry strategy. The preference for incineration, already noticeable among packaging and petrochemical companies in the late 1960s, has become a *cause célèbre*. The combination of the high Btu value of plastics (partly why they are not easily recycled) and the ability of the mass burn system to handle additives, nondegradables, and the wide range of multilayered and mixed-material products (also difficult to recycle) has led the plastics industry to adopt the cause of incineration as its own. Industry leader Du Pont, as an example, has discussed donating company-owned land for use as waste-to-energy sites, an important twist on the merchant plant concept.[48]

The joining of these issues—plastics disposal and waste-to-energy—has also set in relief the two most sharply differentiated approaches to the problem created by the rapid growth of plastics in the waste stream: source reduction and incineration. The plastics industry has repeatedly denied that its products contribute to incineration's environmental problems and has been among the strongest critics of research on the role of plastics in generating hazardous air emissions and ash residue. Their argument, as put forth in one Mobil Corporation ad, "Foam Fast-food Containers: The Scapegoat, Not the Problem," is that "proper incineration" of a product like foam "produces *virtually nothing* but harmless carbon dioxide and water vapor" (our emphasis). "Combustion of plastics in an incinerator," the Mobil ad proclaimed, "contributes no more to pollution than paper, wood, or even leaves."[49]

The defense of incineration contrasts with the hostility toward reduction also paramount in the industry campaigns. The Society of the Plastics Industry (SPI), the main trade group, has adopted its own "integrated" waste management plan that includes incineration, recycling, and landfills, but which categorically rejects source reduction measures, whether in the form of regulations, product bans, or financial mechanisms. Several industry groups were created in the late 1980s to directly lobby against reduction-related initiatives. These included the Plastic Waste Management Council, organized through SPI, whose members commit to a dues structure of $1 million per year for at least three years. The founding companies of this organization are nine of the country's largest plastic manufacturers. Another group, the Polystyrene Packaging Coalition, was organized through SPI and the Food Service and Packaging Institute by over 35 producers and manufacturers of polystyrene products to deal with the movement to ban fast food plastic foam containers. These lobbying efforts, in turn, were strongly supported by major plastic products users, such as McDonald's.[50]

Not in every instance has the industry been able to counteract growing public opposition to the introduction of new products posing significant solid waste problems. In 1985, Coca-Cola launched a test market campaign in Georgia for a plastic can made of clear PET with an aluminum top and a PVC label. This product could potentially seriously undermine existing recycling programs by replacing aluminum (among the easiest and most substantially recycled products, with a well-developed market). But a coalition of opposition groups successfully pressured Coca-Cola, on completing its test run the following year, to withhold introduction of the can commercially until a viable recycling system was developed.

The following year another attempt was made to introduce the plastic can with naturally flavored sodas in four other states. This time, a second and even more extensive opposition campaign led to immediate passage of legislation banning the plastic can in two of the states and the introduction of similar bills in several others. In the face of such rapid-fire opposition, the company behind this new marketing effort, the Original New York Seltzer Company, decided, just one year after launching its campaign, to withdraw its plastic can as well. Despite this opposition, it was likely, given the continuing search by the plastics industry for new markets, that other efforts to introduce the plastic can would reoccur.[51]

The resistance to reduction measures and the continuing drive to expand existing markets and establish new ones, though characteristic to a certain extent of other industries, has become a special feature of the plastics industry. While this industry claims that it has been unfairly singled out as a "symbol" of the waste crisis,[52] it has, in fact, been targeted by community groups precisely for its special role in contributing to the difficulties in managing solid waste today.

Plastics have, in this contemporary era, come to represent a range of both cultural and economic values. Trumpeted as the most direct expression of convenience packaging and product utility, plastics have also been seen as the foremost expression of product wastefulness and social and economic dysfunction. Plastics executives like to argue that the consumer has made a choice, and plastic products are, in this sense, market-driven. Such a choice, however, is never fully presented or explained, and always fails to account for the external economic and social costs imposed by such a product. The plastics industry, in fact, like other major industries, is motivated not by consumer need but by economic performance. It evaluates its products by profit margin and return on investment, and, at times, insofar as it impacts that performance, on company image. It is in this setting, ultimately, that an issue like reduction continues to be addressed, with the realities and constraints of policy-making contrasting sharply with reduction's presumed role. At the top of the hierarchy, reduction remains an elusive strategy, one that the plastics industry has always feared and public agencies have mostly declined to explore.

THE PROGRAMMATIC IMPASSE

Far more than recycling and even more than reuse, reduction initiatives have had an uneven and limited history. The notion of intervening in the production process to accomplish a specific social goal—in this case, reducing or eliminating the generation of waste and the problems of pollution associated with it—has long been resisted by industry forces who have defined such intervention as unwarranted and harmful to the smooth functioning of the market. "Natural market forces in the free enterprise system," one executive with the Glass Container Manufacturing Institute said of source reduction measures, would better serve the society in the long run "than would any government intervention."[53]

Though concerns about "disposability" and "convenience" marketing as well as product redesign and obsolescence developed during the 1950s and 1960s, the first substantial efforts to translate those concerns into policy didn't fully emerge until the late 1960s and early 1970s. Most of the initiatives at that time were local and statewide. Federal activity, as embodied in the 1970 Resource Recovery Act, tended to be limited to research and evaluation of the variety of programs under discussion, or the few already introduced in legislative form. One of the first reduction proposals suggested at the national level, for example, was the "penny a pound proposal" by consultant Leonard Wegman at Senate hearings regarding the Resource Recovery Act. Wegman urged a 1-cent-a-pound disposal charge be levied on the manufacture of consumer products that became waste within ten years after they were made, with the revenues earmarked to help pay disposal costs. Wegman's concept, which eventually evolved into what EPA called a product disposal tax, was part of a series of programmatic ideas that EPA pursued through research studies and public reports during the early and mid-1970s.[54]

In its periodic reports to Congress mandated by the Resource Recovery Act, the reduction strategy was afforded a modest, albeit still significant place in EPA's overview of the solid waste issue. In its third report to

Congress, which represented its most comprehensive analysis of reduction initiatives to date, EPA defined reduction first in terms of three general policy objectives: reduced resource use stemming from product redesign; increased product life or greater durability; and product reuse. These objectives, in turn, were framed by several possible policy strategies, including voluntary efforts to reduce wastes by industry and consumers, incentives for reuse and recycling such as bottle deposits, and various product charges, such as a container tax by weight.[55]

Much of EPA's concept of reduction was framed by concerns over the depletion of nonreplenishable resources, such as various raw or virgin materials used in producing products and packages. In this way, EPA defined reduction as a companion strategy to resource recovery, the main focus of its solid waste strategy, which included both "materials" and "energy" recovery. Reduction was seen as beneficial since it extended the stock of virgin materials and reduced energy usage, complementing the materials and energy savings from recycling and energy recovered from waste-to-energy. By elevating resource depletion to a central role, reduction advocates at the federal level hoped that government intervention would be looked on more favorably "in order to represent the interests of future generations."[56]

At the local level, most reduction efforts were initiated in response to a specific problem or a particular area of concern rather than as a comprehensive or strategic approach to the waste issue. Bottle bills, defined as reduction measures during the 1970s, were by far the most widespread and popularly conceived. Most community efforts narrowly focused on specific recycling programs. Where reduction measures were introduced, they failed to sustain the kind of intense constituency support needed to offset the massive counterattack by industry, as happened repeatedly in the bottle bill fights. Aside from the bottle bills, the few significant reduction initiatives of the 1970s were fiercely resisted by industry and largely abandoned by public agencies.

Two cases are particularly illustrative. During the late 1960s, officials with New York City's Environmental Protection Agency began indicating to plastics and packaging industry executives that they were considering a product ban on plastic containers as well as a federally administered user charge on packaging producers. The revenues from such a program would be used to offset collection, disposal, and R&D expenses, and for possible subsidies for recycling programs. City officials were particularly concerned with the effect of polyvinyl chloride on their existing network of incinerators, a problem that had become particularly acute after 1966, when the city's strong air pollution code forced apartment owners to either upgrade or shut down their own incinerators. By 1970, New York officials, concerned with the stress on their incinerators and equally severe problems with landfilling, began to identify solid waste disposal as the city's most important pollution problem.[57]

These officials increasingly focused on measures aimed at plastics products to reduce the volume of nonfood wastes and a problem discard for their incinerators. Toward that end, the city EPA proposed in 1971 a "deposit bounty system" on packaging, including a special 2-cent tax on most plastic containers. Plastics industry trade groups successfully sued to bar the city from levying this tax, arguing that plastic containers were singled out in the city's efforts to alleviate its waste problem. Industry groups further asserted that plastic containers were in fact far more suitable for disposal by incineration than other types of containers.[58] Having been defeated with its reduction plan, city officials, meanwhile, shifted their efforts to the state level to help underwrite a resource recovery program. These funds, however, were almost entirely earmarked for R&D and other support for waste-to-energy, and eventually locked city officials into a new dependence on incineration as the answer to its waste problems.[59]

The attempt to develop a reduction program in the state of Minnesota was even more striking in what changes its leading public environmental agency went through around the waste issue. Public interest in waste issues and environmental problems was quite substantial in the state during the late 1960s and early 1970s. There were numerous local recycling programs, a strong, locally based environmental movement, and significant public support for a more extensive governmental effort to tackle such issues. In that setting, the Minnesota legislature passed the most far-reaching reduction-related legislation to date, the 1973 Package Review Act. Packaging industry efforts in Minnesota had been most concentrated on blocking a bill to place a deposit on all nonreturnable beverage containers. The bottle bill fight had involved a major confrontation between the industry and its various allies pitted against the state's environmental groups, a confrontation that would repeat itself throughout the decade. Though the Package Review Act was not viewed as a direct alternative to the bottle bill legislation, its passage enabled legislators who acceded to industry lobbying on beverage container deposits to still demonstrate their concern about the waste issue.[60]

The packaging industry, however, quickly mobilized, after realizing the degree of intervention possible under the Package Review Act. Though much of the language in the act was broad and relatively open-ended, it did provide the foundation for an extensive program, depending on the level and specificity of implementation. The act's Statement of Policy cautiously laid out the rationale for a reduction approach, arguing that the "recycling of solid waste materials is one alternative for the conservation of material and energy resources, but it is also in the public interest to reduce the amount of materials requiring recycling or disposal." The legislation stipulated that the Minnesota Pollution Control Agency (MPCA) would be responsible for reviewing which packages "would constitute a solid waste disposal problem or be inconsistent with state environmental policies." After completing its review process, MPCA would have the power to prohibit the sale of particular packages or containers in the state.[61]

MPCA's regulations for implementation of the Package Review Act were specific and detailed. They focused particularly on food, cosmetics, and dry cleaning packages, with their heavy use of plastics. Exemptions were provided for recyclable products and those providing a deposit for reuse purposes. The regulations also allowed industry to withdraw or modify packages that did not meet the adopted guidelines, while leaving the explicit power to the agency to ban a product if the product had not been changed after such a review.[62]

The MPCA regulations were in keeping with the thrust of MPCA activities at the time. MPCA had been founded in 1967, a direct expression of the fusion of progressive and environmental traditions in the state. During the early 1970s, environmental questions, such as the Reserve Mining case involving asbestos contamination of Lake Superior, were hotly debated issues. MPCA functioned as much as environmental advocate as governmental bureaucracy, demonstrating an ability to both mobilize and respond to its constituency.[63]

With the publication of the implementation procedures, the packaging industry, which now saw the Package Review Act as the most explicit challenge to its exclusive control over production and marketing decisions, launched a major challenge to the law. Through a lawsuit as well as a public relations campaign, the industry argued that the law and its companion regulations were unconstitutional because they inhibited interstate commerce, and that the vagueness and imprecision of the statute also constituted the taking of property without due process of law. In 1979, the Minnesota Supreme Court rejected the industry claims about the law's unconstitutionality, but it undercut any effective implementation by finding that the MPCA regulations did not have the force of law and were therefore only guidelines.[64]

Minnesota reduction advocates were not yet completely deterred. While the Package Review case was in the courts, the legislature, in 1977, passed a ban on nonrefillable plastic milk bottles, again in the context of a parallel, but unsuccessful attempt to pass another bottle bill. Once again, the plastics industry intervened through a lawsuit. In 1979, the same year it gutted the Package Review Act, the Minnesota Supreme Court overturned the plastic milk ban, agreeing this time with the interstate commerce argument. Two years later, the U.S. Supreme Court overturned the Minnesota Court's decision, disagreeing that the ban was arbitrary, irrational, and discriminatory. By then, however, the state's dairy industry had substantially converted to the plastic milk bottle and the legislature decided that any implementation of the law would be counterproductive, and thus repealed it that same year.[65]

The MPCA, meanwhile, had already begun to shift from advocate toward a more cooperative stance with industry. "Our goal at the Agency," an MPCA official told the American Paper Institute in 1983, "is to avoid projecting an adversarial image of the 'heavy-handed government regulator.' "[66] Toward that end, MPCA interpreted the 1980 passage of a comprehensive Waste Management Act as strongly favoring large incinerator projects. The legislation established a hierarchy with both reduction *and* resource recovery at the top and landfills at the bottom. New landfills would be allowed only where it could be shown that an alternative to land disposal was not feasible. In implementing the law through its crucial role in permitting, MPCA essentially abandoned its previous reduction-oriented approach and adopted a strong pro-incineration position. Criticized by its own staff for placing incineration at the top of its agenda, MPCA came to be called by its critics the "Minnesota Pollution Collusion Agency."[67]

The evolution of MPCA and the demise of reduction in Minnesota paralleled other developments taking place at both the state and national level during this same period. The California State Waste Management Board, for example, had expressed some interest in the reduction approach during the mid-1970s, when Jerry Brown was governor. Accordingly, the board established a Source Reduction and Packaging Policy Committee to report on possible reduction policies for the state. The committee in turn contracted with a UCLA environmental planning professor who outlined 13 possible policy options for the state board to consider.[68]

These options ran the gamut of reduction measures proposed at the time, though none of them were spelled out to the extent found in the MPCA implementation procedures for the Minnesota Package Review Act. Policies included direct regulation of packages and other individual products including disposable cutlery and plates as well as regulation of durable goods such as household appliances and TV sets; purchasing regulations for state agencies; minimum warranty requirements; disclosure of a product's environmental impact; voluntary waste reduction programs; various tax measures, including those designed by container, by weight, and by unit, as well as a value-added levy on such products as single-use disposable goods; a variable waste collection/disposal fee; a tax or subsidy to help convert from one-way to reusable products; and, of course, the refund on beverage containers, the most controversial of all the proposals under consideration. One committee participant, an executive with the Continental Can Company, argued, in fact, that the report and committee deliberations were "nothing more than one attempting to provide a reason for recommending mandatory deposits on beer and soft drink containers. . . . If this could have been accomplished," the executive concluded, "very little other recommendations would have come from the committee."[69]

As it turned out, the only policy option with full committee support was the recommendation for voluntary efforts, emphasizing public education. A majority of the committee members were industry representatives, and together unanimously rejected nearly all the options, arguing that efforts at intervention would only, as a Crown Zellerbach executive asserted, "disrupt the free market mechanism with resultant higher costs and poorer service to the consumer."[70]

"We found that the subject of source reduction," the committee chair, a member of the League of Women Voters, wrote in transmitting the report, "is intertwined with philosophical values regarding the role of government, the functioning of the free market, and concepts regarding the quality of life as measured by the consumption of material goods."[71] As a result, without any specific policy direction, the report failed to provide a basis to pursue reduction for either the state board or the California legislature.

By the early 1980s, reduction measures had, as in other states, disappeared from the deliberations of California policy-makers. Increasingly during the decade, the CWMB became a strong advocate of waste-to-energy and structured its research and policy recommendations accordingly. Despite the continual placement of reduction at the top of the waste management hierarchy, even by legislative mandate, source reduction programs remained at best "educational," often defined as attempts to convince consumers that the solution—and the problem—were their own.

The experiences in Minnesota and California were duplicated on the national level by a similar transition at EPA and by the lack of sufficient support for nationally mandated reduction programs. The Resource Conservation Committee, for example, reviewed only one program directly related to reduction—the product container tax—with two others—the solid waste disposal fee and beverage container deposits—directed at eliminating discards before they entered the waste stream. Review and regulation of packages and other products were not included in the committee's deliberations nor were warranty requirements or product bans. As it turned out, moreover, the Resource Conservation Committee, similar to the California situation, failed to reach consensus on any of the policy proposals, further demonstrating the political difficulties facing any reduction-oriented approach.[72]

During the early and mid-1980s, research on reduction virtually disappeared from EPA-funded activities. EPA's shift of direction into hazardous waste and its increasing bias in favor of incineration and market initiatives laid the groundwork for this change. Even toward the end of the decade, when EPA began to advocate increased recycling, it continued to favor incineration while relegating reduction to a program of educational activity.[73]

What had also disappeared from EPA's approach, as with other public agencies, was any indication of a truly functioning hierarchy. EPA's advocacy of what it now called an "integrated hierarchy" was based on the *simultaneous* development of all the components of the hierarchy. Priorities were to be determined on a local, case-by-case approach. "Of course," the EPA concluded, "strict adherence to a rigid hierarchy is inappropriate for every community."[74] For EPA and most other public agencies, as well as industry groups who applauded EPA's new emphasis, reduction was marginalized, no longer even a token goal in a postproduction, disposal-driven system.

THE POINT OF PRODUCTION

By the late 1980s, reduction programs appeared increasingly distant and problematic. But then, to the surprise of both industry and waste officials, a new wave of actions by community groups frustrated with the disposal orientation of waste officials and hazards associated with certain forms of production unexpectedly catapulted reduction near the top of policy agendas. The McToxics campaign, the plastic foam container bans, the ban of nondegradable six-pack carriers, and actions aimed at preventing the introduction of the plastic can were all indications of this new level of reduction-related activism. And, similar to the peak period of interest in the early and mid-1970s, a number of states, such as Rhode Island, Massachusetts, and Illinois, began to establish commissions to fund reduction-related research.[75]

Again, much of the new talk about reduction has focused on financial measures, product regulations, and educational activities, quite similar to the framework for reduction policies developed in the earlier period. A new, more comprehensive, graded tax on all packaging materials, with credits and exemptions based on such criteria as reusability, recyclability, and/or recycled content has been proposed. More extensive and encompassing regulatory concepts have also been introduced, including specification of secondary material content, durability and degradability factors, product reusability specifications, mandated reduction of material content per unit-of-product output, more sophisticated labeling requirements revealing environmental impacts, and mandated reduction of a product's potential to create damage at the disposal stage.[76] The most direct form of intervention has remained the product ban, a tactic that may be even more utilized as divisions increase between a technology-based disposal strategy and an intervention-based strategy following the priorities of the waste hierarchy.

Those differing approaches were most strikingly revealed recently in Minnesota. There, the Minneapolis City Council passed in 1989 a stringent plastic packaging ordinance aimed at preventing the sale of styrofoam as well as other plastics such as polyvinyl chloride, low-density polyethylene and polypropylene. That action stood "at odds," as one industry publication put it, with regulations drafted by the Minnesota Pollution Control Agency in conjunction with the plastics industry to identify resins in products in order to help facilitate potential—but still distant and problematic—recycling. MPCA, while promoting incineration, remained reluctant, fifteen years after the Package Review Act, to intervene at the level of production choices.[77]

The hostility toward intervention, while still strong in this country, has appreciably weakened in Europe, where the waste crisis is perceived to be more immediate and the tradition of governmental activity is more secure. Packaging control, especially, has been much more widely employed, such as Denmark's ban on one-way beverage containers. A comprehensive packaging control law in the Netherlands provides the authority to ban "any product that adds unduly to the amount of waste, is difficult to dispose of, frequently littered, or is not sufficiently reusable." Though there are no procedures for implementation, it has nevertheless influenced industry design and marketing decisions. Other countries have developed more specific product regulations and financial disincentives, such as Italy's 13-cent tax on nondegradable plastic bags, adopted in 1988 as part of its overall strategy to eliminate their use by 1991.[78]

In this country, the reduction concept, though revived, has yet to develop into a strategic approach, and, beyond that, to help identify the overall waste issue as essentially a question of production and how it should be controlled. A broad consensus has developed among public officials, industry groups, and even among certain environmental organizations that the reduction option is too politically volatile, too close to sensitive matters of policy and decision-making, to represent a serious alternative. In that context, the "integrated" waste management approach represents a public relations formula, disguising a long-term reliance on incineration. Recycling, meanwhile, simply becomes another way to stretch landfill capacity.[79]

A more systematic and comprehensive approach to reduction, on the other hand, though certain to engender protracted political conflict, nevertheless remains the most environmentally benign and socially equitable option available. The generation and disposal of waste is not simply a matter of dealing with consumption patterns that have somehow gotten out of control and searching for a disposal technology that can contend with escalating economic and environmental costs. It means coming to terms with a system based on privately made production decisions and organizing a management strategy based on the principle that post-manufacturing, postconsumer "discards" are potentially reusable and recyclable.

The debate around reduction, furthermore, mirrors a developing conflict between different advocates of an "alternative" approach. Many of the conventional environmental groups have tended until recently to be involved exclusively in the drafting of national legislation and monitoring of federal agencies, especially EPA. Many of these groups drifted away from the solid waste issue during the early 1980s when the EPA Office of Solid Waste, along with the media, shifted its focus to the hazardous waste arena and Congress largely dropped any attempt to craft new legislative approaches. Instead, organizing around the solid waste issue became the more exclusive province of the grass roots community groups opposing proposals for an unwanted disposal facility, whether a landfill or an incinerator, often without the assistance of the national and regionally based environmental organizations.[80]

By the late 1980s, due partly to the force of community opposition, as well as the more graphic illustrations of the solid waste problem such as the garbage barge, the trash issue forced its way back onto everyone's agenda, including a number of environmental groups. Some of these groups, most particularly the Environmental Defense Fund, challenged the shift to incineration, raising questions about cost-effectiveness compared to recycling and concerns around ash toxicity. EDF especially presented some of the strongest arguments that incineration posed an environmentally hazardous route for disposal.[81] At the same time, however, EDF also became active, seeking out legislative and administrative compromises that would allow incineration to proceed on a more limited and less crisis-oriented basis, a position the industry itself adopted in light of its problems with community opposition.[82]

The community-based movements, however, have been in no mood to compromise. Instead, they have coalesced into larger regional and national networks that have increasingly clashed with many of the more established environmental groups. The position of the community groups on reduction has been slow to evolve as well, with most efforts focused instead on intensive recycling as the primary alternative.[83] Through the product ban actions, however, reduction has *ipso facto* achieved a new prominence among such groups, including the Citizen's Clearinghouse for Hazardous Waste, and a few of the more militant environmental organizations such as Greenpeace and Environmental Action. Environmental Action in particular has been at the forefront of efforts to develop a more strategic focus around reduction with its analysis of the plastics industry and its efforts to ban the plastic can.[84]

The interest in reduction is also likely to grow as issues of hazardous and solid waste are increasingly joined. In one fascinating example of this link, incineration technology provided the connection between these two distinct (in legislative and regulatory terms) waste arenas. Since 1976, a toxic waste controversy has raged over the presence of six Superfund sites containing PCB-laden hazardous wastes in Bloomington, Indiana. In 1985, Westinghouse Electric Corporation, which manufactured PCB-filled capacitators and then disposed of them at local landfills and in sewers, signed a consent decree with EPA to clean up the sites by incinerating the wastes. Westinghouse proposed, with EPA approval, to use municipal solid waste as *fuel* in this first-of-a-kind solid waste-cum-hazardous waste burn plant. The proposal was immediately attacked by community groups who feared that commingling hazardous with solid waste could well generate additional, as of yet unknown, problems.[85]

Incinerator ash represents another particularly noteworthy example of this commingling of wastes, both in terms of its composition and disposal. Some of the ash's more toxic elements can be traced to many of the same industries that produce hazardous wastes, while the process of incineration itself, by concentrating the toxics, transforms municipal solid waste into hazardous waste. The disposal of ash in landfills raises the possibility that a whole new generation of Superfund sites might be in the making. The link with hazardous waste is also instructive in terms of the current congressional and administrative focus on hazardous waste *treatment*, instead of waste *reduction* efforts. There is a parallel between EPA's interest in hazardous waste minimization, which includes both incineration and secondary recycling (that is, recycling that involves the creation of another, potentially hazardous product), and the solid waste concept of "resource recovery," which has also evolved almost exclusively into a disposal-oriented approach.[86]

The current, limited hazardous waste reduction programs, where implemented, have produced modest results, but a comprehensive approach necessarily involves changes at the point of production. This entails more direct participation by the work force at each level of production. These workers, as one congressional study pointed out, are often most familiar with the details of production organization and thus have direct knowledge of how to reduce its hazardous components. That work force, furthermore, is most directly subject to the consequences of hazardous production.[87]

Since the costs of liability and cleanup are so high and, at least in certain circumstances, some of these costs are partially internalized by those industries most responsible for hazardous wastes, there is an incentive for companies to reduce their output of hazardous waste. That process has yet to be applied in the solid waste arena, though the parallels between the two suggest that such incentives might have equivalent results. A focus on reduction, furthermore, would have the effect of broadening the constituent base of support for alternatives, linking community concerns directly to workplace issues.

The political difficulties regarding waste management alternatives such as bottle bills and reduction measures have in part reflected the classic division between "environmentalists" and "labor," which has criticized the failure of environmental groups to address employment, equity, and other social concerns. The conventional environmental movement, furthermore, has functioned primarily as a middle-class movement more concerned with questions of "consumption" (for example, preservation of scenic resources, population growth, or lifestyle changes) than "production" issues (how and why products are produced, and which groups are most "at risk," including the work force at the point of production).[88]

The development of the new community movements in the 1980s has, to a certain extent, challenged those tendencies of environmentalism. These movements tend to be broader in class composition, and have included poor and working-class residents as well as middle-class suburbanites. Some of the community groups are multiracial, led by black or Hispanic residents, as in the LANCER case in Los Angeles. They have also begun to link the issues of hazardous and solid waste in their coalitions and strategies. What these groups lack, however, is a *reduction strategy* that addresses the central role of production and thereby also addresses the concerns of both workers and community residents.

A reduction strategy necessarily requires a reformulation of the waste issue as a matter of democratic intervention in the political process where waste management strategies are selected and the industrial process where the production system is governed. Such a strategy would also need to challenge the marketing presumption that people will always accept and want "convenience" and "disposability." That cultural disposition is a manufactured one, much as the product to be consumed. Whether in terms of waste management, production, or marketing choices, the public remains excluded from the decisions that shape their lives.

In this way, the waste issue can be seen as a metaphor for the broader issues of daily life in our urbanized and industrial society. Our system of production and consumption can be said to a certain extent to meet human and social needs, but this process is only incidental to the goals and structure of the system itself. Many of these needs, in turn, are induced needs, or artificial needs, as Herbert Marcuse wrote 25 years ago, stimulated to allow the system to expand.[89] The solid waste crisis and the dilemmas it has posed in our contemporary period are in part a reflection of how such needs are manufactured and, as a consequence, the system's capacity to create problems that it then fails (and often refuses) to address. The search for alternatives becomes in its own way a search to find the appropriate forms of public intervention to deal with the system's failures. The solid waste debate thus goes beyond the questions of waste management to the question of whether the larger system of production and consumption is indeed an appropriate arena for public decision-making. The outcome of this debate will tell us much about the capacity of people to intervene in the decisions that directly affect their lives.

NOTES

1. "Everyone's Backyard," Citizen's Clearing House for Hazardous Waste, vol. 6, no. 1; Spring 1988; "Action Bulletin," Citizen's Clearing House for Hazardous Waste, issue no. 17, February 1988. See also Jeanne Wirka, *Wrapped in Plastic: the Environmental Case for Reducing Plastics Packaging,* Environmental Action Foundation, August 1988, Washington, D.C. pp. 107; "McDonald's Pullout: The CFC Issue Hits Home," *Modern Plastics,* October 1987, pp. 15–16.
2. "Tarnish on the Golden Arches," Karen Stults, *Business and Society Review,* no. 63, Fall 1987; interview with Karen Stultz, 1988; see also "Everyone's Backyard," Citizen's Clearinghouse for Hazardous Waste, vol. 5, no. 4, Winter 1987.
3. Interview with Will Collette regarding meeting with Shelby Yastrow, April 5, 1988; see also "Action Bulletin," Citizen's Clearing House for Hazardous Waste, issue no. 18, May 1988.
4. "Foam Containers Feeling the Heat," Debra Levi, *San Francisco Chronicle,* June 14, 1988, p. A9. See also "Ban on Wrapping in Plastic Signed," Philip S. Gutis, *New York Times,* April 30, 1988, p. A1.; "Even with Action Today, Ozone Loss Will Increase," James Gleick, *New York Times,* March 20, 1988; "Industry Acts to Curb Peril in Ozone Loss," Philip Shabecoff, *New York Times,* March 21, 1988; "Fighting the Greenhouse Effect," Matthew Wald, *New York Times,* August 28, 1988.
5. "Why Dupont Gave Up $600 Million," Cynthia Pollock Shea, *New York Times,* April 10, 1988; "The Race to Find CFC Substitutes," Philip Shabecoff, *New York Times,* March 31, 1988; "Behind Dupont's Shift on Loss of Ozone Layer," William Glaberson, *New York Times,* March 26, 1988; see also "McDonald's pullout: CFC issue hits home," *Modern Plastics,* October 1987; *New Scientist,* September 3, 1987; "Action Bulletin," Citizen's Clearing House for Hazardous Waste, issue no. 19, July 1988, and issue no. 18, May 1988.
6. Interview with Karen Stoltz, 1988; CCHW Action Bulletin no. 19, July 1988.
7. Interview with Will Collette, 1988.
8. "Send It Back to McDonald's," Citizen's Clearinghouse on Hazardous Wastes *Action Bulletin* no. 20, November 1988.
9. "Bridging the Gap: Packaging Industry and Government," *Waste Age,* November/December, 1971, pp. 12–14. Also, Lanier Hickman, Jr., "The Role of Packaging in Solid Waste Management, 1966–1976," Arsen Darnay and W. E. Franklin, Public Health

Service Publication no. 1855; U.S. Government Printing Office, 1969, 205.; see also Franklin Associates at S-2.

10. Franklin Associates, *Characterization of Municipal Solid Waste in the United States, 1960–2000,* Prairie Village KS, US EPA, Office of Solid Waste, 1986, p. S-2 and table 3-1. See also *Solid Waste Management Alternatives: Review of Policy Options to Encourage Waste Reduction,* Elliott Zimmermann, Illinois Department of Energy and Natural Resources, Springfield IL, February 1988, p. 12.
11. "Mining Urban Wastes: The Potential for Recycling," Cynthia Pollock, Worldwatch Paper No. 76, Washington, D.C., April 1987, pp. 8–10.
12. "Impact of Plastic Packaging on Solid Waste," Christopher Lai, Susan Selke, and David Johnson, paper presented at the Conference on Solid Waste Management and Materials Policy, New York City, New York, February 11–14, 1987, cited in *Wrapped in Plastic,* Jeanne Wirka, Environmental Action Foundation, Washington, D.C., August 1988, p. 9.
13. "Food Packaging and the Solid Wastes Problem," Editorial in *Public Works,* vol. 100, no. 8, August 1969.
14. "Packaging in Perspective," C. Soutter Edgar, in *Proceedings: First National Conference on Packaging Wastes,* EPA, 1971.
15. "The Changing Dimensions of Packaging Wastes," Arsen Darnay and William E. Franklin, presented by Arsen Darnay at the First National Conference on Packaging Wastes, September 22, 1969.
16. "Wastes from Plastic Packages," Thomas Becnel, Environmental Control Systems, Dow Chemical Company, in *Proceedings,* 1971; see also "Incineration of Packaging Wastes with Minimal Air Pollution," Elmer R. Kaiser, in *Proceedings,* 1971.
17. "Packaging Problems from the Perspective of the Public Official," Hugh Marius, in *Proceedings,* 1971; see also "Governmental Organization for Regional Management of Wastes," Richard T. Anderson, in *Proceedings,* 1971.
18. "The Changing Dimensions of Packaging Wastes," Arsen Darnay, in *Proceedings,* 1971.
19. Franklin Associates, supra, 1986, table 3-1.
20. "Is Industry Serious About Solid Waste Recovery?" Robert Leaversuch, *Modern Plastics,* June 1988, especially pp. 67–68; *Plastic Packaging Recycling: The Challenge and Opportunity,* John A. Schlegel and E. E. Fuller, Business Communications Co., Norwalk, Connecticut, February 1988; "Solid Waste Management and the Packaging Industry," Richard D. Vaughan, Bureau of Solid Waste Management, in *Proceedings,* 1971.
21. Franklin Associates, supra, 1986, table 3-1 and p. 3-3.
22. *Plastics AD 2000: Production and Use Through the Turn of the Century,* Chem Systems Inc., a Report Prepared for the Society of the Plastics Industry, Washington, D.C., July 1987; see also Jill W. Tallman, "More Plastics Use Means More Waste, Which Means Plastic Trash Defies Technological Solution," Malcolm W. Browne, *New York Times,* August 6, 1987; "What's New in Plastics," Jennifer Stoffel, *New York Times,* March 5, 1989.
23. See Wirka, 1988, supra, pp. 21–25. See also "Plastics: The Risks and Consequences of Its Production and Use, An Industry Overview with Case Studies," Patricia Lichiello and Lauren Snyder, Graduate School of Architecture and Urban Planning, University of California at Los Angeles, a Client Project Report, Spring 1988; see also the Radian Corporation, *Plastics and Resins Processing: Technology and Health Effects,* New Jersey, Noyes Data Corporation, 1986, p. 3. See also U.S. Department of Commerce, *1982 Census of Manufacturers.*
24. Wirka, 1988, supra, pp. 8, 15; Lichiello and Snyder, 1988, supra, p. 9; Radian Corporation, 1986, supra, p. 1. See also Joel Frados, ed., *Plastics Engineering Handbook of the Society of the Plastics Industry, Inc.,* New York, Van Nostrand Reinhold Company, 1975, pp. 41–42. See also Lai et al. supra, 1987, pp. 11–14.

25. See Wirka, supra, 1988, p. 7.
26. *The Poverty of Power,* Barry Commoner, New York, 1976; also "The Social Use and Misuse of Technology," Barry Commoner, in *Ecology in Theory and Practice,* ed. Jonathan Benthell, New York, 1972.
27. *Plastics AD 2000;* also "Plastics—What Is Their Future?" Frank Sudol and Alvin L. Zach, Department of Engineering, City of Newark, New Jersey, p. 4., cited in Wirka, pp. 9–11; also "SPI on Year 2000: Packaging Will Star," *Modern Plastics,* August 1987, p. 14.
28. Snyder and Lichiello, supra at pp. 11–13.
29. *Modern Plastics,* January 1987.
30. Franklin Associates, supra, 1986, p. I-17. For projected growth, see Franklin at S-2; see also "Forum Reflects Growing Solid Waste Concern," *Modern Plastics,* April 1987, p. 15; Snyder and Lichiello, supra at pp. 58–75.
31. Lai, supra at 13.; cited in Wirka, p. 14.
32. "U.S. EPA, Office of Solid Waste and Emergency Response, *Report to Congress: Minimization of Hazardous Waste,* Washington, D.C., Government Printing Office, 1986, pp. 2–33. See also Congressional Budget Office, *Empirical Analysis of U.S. Hazardous Waste Management,* Staff Working Paper, October 1985. See also Congressional Budget Office, *Hazardous Waste Management: Recent Changes and Policy Alternatives,* Washington, D.C., Government Printing Office, 1985, pp. 17–19. See also Radian Corporation, *Polymer Manufacturing: Technology and Health Effects,* New Jersey, Noyes Data Corporation, 1987, p. 10.
33. See "Pulmonary Function and Respiratory Symptoms in Polyvinyl Chloride Fabrication Workers," M. E. Baser, M. S. Tockman, and T. P. Kennedy, *American Review of Respiratory Disease,* vol. 131, no. 2, pp. 253–58.
34. Snyder and Lichiello, p. 80; see also Michigan Department of Transportation, *Michigan Litter Composition Study,* September 1986; Jill Tallman, *Waste Age,* September 1987.
35. Michael Weisskopf, "Plastic Reaps a Grim Harvest in the Oceans of the World," *Smithsonian,* March 1988, p. 60.; "Beach Cleaners Cite Plastics as Worst Problem," *Los Angeles Times,* October 11, 1988.
36. Frank J. Sudol and Alvin L. Zach, "Plastics—What Is Their Future," unpublished paper, Department of Engineering, City of Newark, New Jersey, 1988, cited in Wirka; for a general discussion of plastic pollution and its impact on the marine environment, also see *Plastics in the Ocean: More than a Litter Problem,* Washington, D.C., Center for Environmental Education, February 1987, pp. 19–29.
37. Franklin Associates, supra, S-2. See also "Is Industry Serious About Solid Waste Recovery?," Robert Leaversuch, *Modern Plastics,* June 1988, pp. 65–76; *Plastic Packaging Recycling: The Challenges and Opportunities,* John A. Schlegel and E. E. Fuller, Business Communications Co., Norwalk, Connecticut, February 1988; "Analyst: Solid Waste Becomes 'Crisis,'" *Modern Plastics,* January 1988, pp. 25–26.
38. "Tufts University Researchers Look to Increase Efficiency of Landfills," Tufts University Office of Communication Press Release, July 5, 1988.
39. "Plastic Additives: Less Performing Better," Bruce F. Greek, *Chemical and Engineering News,* June 13, 1988; "A Wary Eye on Plasticizer Testing," *Chemical Week,* May 27, 1981; "How Safe Are Food Containers," Will Hitchcock and Rose Audette, *Environmental Action,* vol. 16, October 1984, pp. 18–19. See also Radian Corporation, supra, 1987, p. 19.
40. "Is Industry Serious About Solid Waste Recovery?," Robert Leaversuch, *Modern Plastics,* June 1988, p. 73; Wirka, supra, 1988, p. 83; see also *Coming Full Circle,* Environmental Defense Fund, 1988, supra, p. 83 and p. 145 for a list of statute citations by state;

also "Reagan Backs Radon Program, Phase-out of 6-Pack Rings," *San Francisco Chronicle,* October 28, 1988.

41. Franklin Associates, supra, 1986, p. 3-19.
42. "Plastics Recycling Faces Barriers," T. Randall Curlee, *Waste Age,* July 1987, pp. 55–60.; also U.S. EPA, *Incentive for the Recycling and Reuse of Plastics: A Summary Report,* Washington, D.C., Government Printing Office, 1973, p. 8; "Mixed Plastics Recycling: Not a Pipe Dream," Gretchen Brewer, *Waste Age,* November 1987.
43. T. Randall Curlee, ibid. See also Wirka, supra, 1988, pp. 56–60.
44. "PET Reclaim Business Picks Up New Momentum," George R. Smoluk, *Modern Plastics,* February 1988, pp. 87–91; See also "PET Plastic Recovery," *Resource Recovery,* vol. 1, March/April 1982, p. 11.; *Plastic Bottle Recycling Directory and Reference Guide,* Washington, D.C., The Society of the Plastics Industry, 1987, p. 35.; *Plastic Bottle Recycling: Case Histories* (Washington, D.C., The Society of the Plastics Industry, 1986.; *Plastic Bottle Recycling—1986,* Plastic Bottle Institute, New York, n.d.
45. "The Case for HDPE Recycling: Part 1," Andrew Stephens, *Resource Recycling,* vol. 6, September/October 1987, p. 18.
46. Industry groups promoting plastics recycling include Center for Plastics Recycling Research, the Plastics Recycling Applied Research Institute, and the National Association for Plastic Container Recovery, Solid Waste Management Solutions; see also "Activities in the Solid Waste Arena," *The Privatization Report,* May 1988; *Plastic Bottle Recycling Directory and Reference Guide,* 1987.
47. "Degradable Plastics Now on Market," Thomas H. Maugh II, *Los Angeles Times,* January 23, 1988; "At Issue: The Disposing of Disposables, Part 2," Bernard Lichstein, *Nonwovens Industry,* December 1988, pp. 20–23.
48. "Forum Reflects Growing Solid Waste Concern," *Modern Plastics,* April 1987, pp. 15–16; T. Randall Curlee, supra; see also "Incineration of Packaging Wastes with Minimal Air Pollution," Elmer R. Kaiser, in *Proceedings: First National Conference on Packaging Wastes,* EPA, 1971.
49. Ad appearing in *Los Angeles Times,* February 23, 1988, part I, p. 5.; see also Mobil ad "Truth, Fiction, and Solid Waste," in *Columbia Journalism Review,* November–December 1988; see also "Is Industry Serious About Solid Waste Recovery?" Robert Leaversuch, *Modern Plastics,* June 1988; see also Curlee, ibid.
50. "State Legislative Issue Report—Packaging/Solid Waste Related Issues," memo from Mike Donahue, Manager, State Government Relations, to McDonald's Owner/Operators, McDonald's Corporation, Oak Brook, Illinois, n.d. (circa March 25, 1988); see also, "Is Industry Serious about Solid Waste Recovery?" Robert Leaversuch, ibid.; Wirka, supra, 1988, p. 44, who cites The Society of the Plastics Industry, "Policy Statement on Municipal Solid Waste Disposal," as adopted by the SPI Board of Directors, September 18, 1987; also, *Source Reduction Task Force Report,* Rhode Island Waste Management Corporation and Rhode Island Department of Environmental Management, November 1987, p. 15; *Minnesota* v. *Clover Leaf Creamery Co.,* 449 U.S. 456 (1981).
51. See Wirka, supra, 1988, p. 106.; also *Environmental Action,* May/June 1988, p. 2. The anti-plastic-can campaign led to the formation of a new solid waste–environmental group, the Coalition for Recyclable Waste.
52. *Los Angeles Times,* February 23, 1988, part I, p. 5.
53. Statement of Sam Bowman, Glass Containers Manufacturers Institute, in *Proposed Policies for Waste Reduction in California,* California State Solid Waste Management Board, March 1976, p. 87.
54. *New York Times,* July 20, 1969; *Resource Recovery and Waste Reduction,* 4th Report to Congress, 530/SW-600, EPA, 1977.

55. Office of Solid Waste Management Programs, *Resource Recovery and Waste Reduction,* Third Report to Congress, 530/SW-161, EPA, 1975.
56. Cited in *An Evaluation of the Effectiveness and Costs of Regulatory and Fiscal Policy Instruments in Product Packaging: Final Report,* Taylor H. Bingham and others, 530/SW-74c, EPA, 1974.
57. "Garbage Disposal Most Important Pollution Problem," *New York Times,* July 6, 1970; also *New York Times,* July 24, 1969; July 27, 1969; February 27, 1972.
58. *New York Times,* July 16, 1971; also January 31, 1971.
59. *New York Times,* May 15, 1988; *The Burning Question,* Public Information Research Group, New York, 1986.
60. Interview with Bill Dunn, 1988; *Damning the Solid Waste Stream: The Beginning of Source Reduction in Minnesota,* Karen A. Wendt, Annual Report to the Minnesota Legislature, Minnesota Pollution Control Agency, January 1975; also see State of Minnesota, Chapter 116F, "Recycling of Solid Waste," and 116F.06, Minnesota Package Review Act.
61. Minnesota Package Review Act, Chapter 116 F.01 "Statement of Policy," and 116F.06, "Packages and Containers."
62. Regulations for Packaging Review, October 31, 1974, see SR-1, "Applicability and Definition," SR-2, "Criteria," SR-3, "Review Procedure," SR-4, "Exemptions," SR-5, "Information Required for Review."
63. *Reserve Mining Co.* v. *EPA,* 514 F. 2d 492 (8th Cir. 1975); interview with Grant Merritt, 1988; also *The Reserve Mining Controversy: Science, Technology, and Environmental Quality,* Robert Bartlett, Bloomington, Indiana, 1980.
64. *Can Manufacturers Institute Inc.* v. *State of Minnesota,* Minnesota Pollution Control Agency, September 7, 1979.
65. *Minnesota* v. *Clover Leaf Creamery Co.,* 449 U.S. 456 (1981). See also Wirka, supra, 1988, pp. 110–111. See also EDF, supra, 1988, p. 85.
66. Speech by Michael Robertson, Assistant Executive Director, Minnesota Pollution Control Agency, to the American Paper Institute, White Sulpher Springs, West Virginia, September 21, 1983.
67. Cited in "A Tremendous Silence," Brian Lowery, *Minneapolis/St. Paul,* October 1988; also interviews with Grant Merritt; Norm Oppegard, 1988; also *Minneapolis Star-Tribune,* November 4, 1988.
68. *Proposed Policies for Waste Reduction in California,* W. David Conn, California State Solid Waste Management Board, March 1976, especially pp. 15–45.
69. Statement of John E. Gallagher, Manager, Public Relations, Continental Can Company, in *Proposed Policies for Waste Reduction in California,* supra, pp. 78–80.
70. Statement of Michael E. Conway, Crown Zellerbach Inc., in *Proposed Policies for Waste Reduction in California,* supra, p. 74.
71. Statement of Linda Craig, representative of Bay Area League of Women Voters, Letter of Transmittal to *Proposed Policies for Waste Reduction in California,* supra.
72. *Choices for Conservation, Final Report to the President and Congress,* Resource Conservation Committee, Office of Solid Waste Management, U.S. EPA 530/SW-779, July 1979; for a discussion of the policy issues influencing the RCC, see, for example, "An Evaluation of the Effectiveness and Costs of Regulatory and Fiscal Policy Instruments in Product Packaging: Final Report," Taylor H. Bingham, M. Susan Marquis, Philip C. Cooley, Alvin M. Cruze, Edwin W. Hauser, Steve A. Johnston, and Paul F. Mulligan, U.S. 530/SW-74c, EPA, 1974.
73. Municipal Solid Waste Task Force, *The Solid Waste Dilemma: An Agenda for Action,* Draft Report, U.S. EPA, Office of Solid Waste, 530/SW-88-05, September 1988; see also

"EPA and Recycling: An Interview with J. Winston Porter," *Resource Recycling,* vol. 7, no. 3, July 1988.

74. *The Solid Waste Dilemma,* ibid., p. 18.
75. Source Reduction Task Force, State of Rhode Island; Illinois Solid Waste Management Act, P.A. 84-1319, see Section 2 for source reduction policy. See also EDF, 1988, supra, pp. 80–85.; "A Plastics Packaging Primer," *Environmental Action,* July/August 1988; also "War on Disposables", Deborah E. Herz et al., *Biocycle,* vol. 28, no. 10, November/December 1987.
76. See, for example, Elliott Zimmermann, *Solid Waste Management Alternatives: Review of Policy Options to Encourage Waste Reduction,* Illinois Department of Energy and Natural Resources, Springfield IL, February 1988; "Challenging the Disposable Society," Rob Young and Marion Storey, *Biocycle,* vol. 29, no. 4, April 1988.
77. "Minneapolis Council Votes to Restrict Plastic Packaging." *Solid Waste Report,* vol. 20, no. 15, April 10, 1989, p. 114.
78. See Zimmermann, supra, 1988, pp. 48–49. See also Wirka, supra, 1988, p. 85. See also *Los Angeles Times,* January 23, 1988; October 21, 1988; Other countries with less comprehensive packaging controls include Norway, France, and Switzerland.
79. "WTE Development: A Zigzag Course," Harvey W. Gershman and Nancy M. Petersen, *Solid Waste and Power,* August 1988, pp. 14–20. U.S. EPA, *Agenda for Action,* supra, 1988; see also the *New York State Solid Waste Management Plan, 1987–1988 Update,* New York State Department of Environmental Conservation, Albany, New York, December 1987, see, for example, pp. 1–9 of the Executive Summary. See also R. W. Beck and Associates, *The Nation's Public Works Report on Solid Waste, National Council on Public Works Improvement,* May 1987, see, for example, pp. ii–iii.
80. Interview with Will Collette, 1988; also *Hazardous Waste News,* no. 103, Environmental Research Foundation, Princeton, New Jersey, November 14, 1988.
81. See, for example, "Risks of Municipal Solid Waste Incineration: An Environmental Perspective," Richard A. Denison and Ellen K. Silbergeld, *Risk Analysis,* vol. 8, no. 3, 1988; see also Richard Denison, "The Hazards of Ash and Fundamental Objectives of Ash Management," Environmental Defense Fund, January 27, 1988, Washington, D.C.; see also Richard Dennison, "Summary of Available Extraction Procedure Toxicity Test Data for Lead and Cadmium from MSW Incinerator Ash," Environmental Defense Fund, January 7, 1988, Washington, D.C.; see also *To Burn and Not to Burn,* Environmental Defense Fund, New York, 1985.
82. "Recycling Must Be the Equal of Incineration," Frederick Krupp et al., Letter to the Editor, *New York Times,* December 6, 1988; *Hazardous Waste News* no. 103, supra; see also "Current Legislation Could Regulate WTE Ash," Roger D. Feldman and Howard L. Sharfsten, *Solid Waste and Power,* June 1988.
83. Examples of local community groups advocating recycling as an alternative to incineration include the Citizens Against Polluting the Peninsula in Hampton Virginia, Dumpbusters in Spencerville, Ohio, and Residents for a More Beautiful Port Washington. The last group participated in the Coalition to Save Hempstead Harbor, which is a regional group in the Long Island, New York, area. The Alliance for Solid Waste Alternatives in Kings County, Washington, a coalition of seven local community groups and three environmental groups, is another regional group advocating recycling as a viable alternative to incineration. Other regional groups include MassPIRG and NYPIRG. The National Campaign Against Toxic Hazards and Work on Waste USA (formerly the National Coalition Against Mass Burn Incineration) are two nationally focused groups actively proposing intensive recycling while opposing incineration.; see also *Recycling . . . The Answer to Our Garbage Problem,* Citizen's Clearinghouse

for Hazardous Wastes, Arlington, Virginia, May 1987; also *Waste Not*, no. 31, November 22, 1988.

84. See special issue of *Environmental Action*, "Wrapped in Plastic," July/August 1988; also Jeanne Wirka, *Wrapped in Plastic: The Environmental Case for Reducing Plastics Packaging*, Environmental Action Foundation, August 1988, Washington, D.C.
85. See letter from Basil G. Constantelos, Director, Waste Management Division, Region 5, EPA, to Mark R. Kruzan, Indiana House of Representatives, October 3, 1987; also "PCB Plan 'Novel,'" Steven Higgs, *Herald-Times* (Bloomington, Indiana), September 28, 1986; *Roots: A Newsletter of Peace*, People Against the Incinerator, Bloomington, Indiana, July 1988.
86. See, for example, *Cutting Chemical Wastes*, David J. Sarokin, Warren R. Muir, Catherine G. Miller, and Sebastian R. Sperber, INFORM, Inc., New York, 1985. See also *Alternative Technology for Recycling and Treatment of Hazardous Wastes: Third Biennial Report*, Alternative Technology and Policy Development Section, Toxic Substances Control Division, California Department of Health Services, Sacramento, California, July 1986; also *Proven Profits from Pollution Prevention*, Donald Huisingh et al., Institute for Local Self Reliance, Washington, D.C., 1986.
87. *Serious Reduction of Hazardous Waste: For Pollution Prevention and Industrial Efficiency*, U.S. Congress Office of Technology Assessment, OTA-ITE-317, Washington, D.C., September 1986.
88. See "Which Way Environmentalism: Toward A New Democratic Movement," Robert Gottlieb and Helen Ingram in *Winning America: New Ideas and Leadership for the '90s*, ed. Marcus Raskin and Chester Hartman, Boston, 1988; also *Beauty, Health and Permanance*, Samuel Hays, New York, 1987.
89. *One Dimensional Man: Studies in the Ideology of Advanced Industrial Society*, Herbert Marcuse, Boston, 1964; see also *All Consuming Images: The Politics of Style in Contemporary Culture*, Stuart Ewen, New York, 1988.

INDEX

ABOUT THE AUTHORS

Louis Blumberg was a core participant in the UCLA LANCER Project, *The Dilemma of Municipal Solid Waste Management*, which received the 1989 National Planning Award of the American Institute of Certified Planners. He has continued to write, research, and consult regarding problems of waste and other environmental issues. He currently is working with the EPA Region 9 Office in San Francisco, where he lives with his wife and family.*

Robert Gottlieb, a member of the UCLA Urban Planning Program faculty, was project supervisor for the UCLA LANCER Project. His most recent work is *A Life of Its Own: The Politics and Power of Water.* He is also the co-author of *Empires in the Sun: The Rise of the New American West* and *America's Saints: The Rise of Mormon Power* (with Peter Wiley), and *Thinking Big: The Story of the Los Angeles Times, Its Publishers and Their Influence on Southern California* (with Irene Wolt). He lives with his wife and two children in Santa Monica, California.

** This book was co-written by Louis Blumberg in his private capacity. No official support or endorsement by the Environmental Protection Agency is intended or should be inferred.*

Also Available from Island Press

The Challenge of Global Warming
Edited by Dean Edwin Abrahamson
Foreword by Senator Timothy E. Wirth
In cooperation with the Natural Resources Defense Council
1989, 350 pp., tables, graphs, bibliography, index
Cloth: $34.95 ISBN: 0-933280-87-4
Paper: $19.95 ISBN: 0-933280-86-6

The Complete Guide to Environmental Careers
By The CEIP Fund
1989, 300 pp., photographs, case studies, bibliography, index
Cloth: $24.95 ISBN: 0-933280-85-8
Paper: $14.95 ISBN: 0-933280-84-X

Creating Successful Communities: A Guidebook to Growth Management Strategies
By Michael A. Mantell, Stephen F. Harper, and Luther Propst
In cooperation with The Conservation Foundation
1989, 350 pp., appendixes, index
Cloth: $39.95 ISBN: 1-55963-030-2
Paper: $24.95 ISBN: 1-55963-014-0

Resource Guide for Creating Successful Communities
By Michael A. Mantell, Stephen F. Harper, and Luther Propst
In cooperation with The Conservation Foundation
1989, approx. 275 pp., charts, graphs, illustrations
Cloth: $39.95 ISBN: 1-55963-031-0
Paper: $24.95 ISBN: 1-55963-015-9

Natural Resources for the 21st Century
Edited by R. Neil Sampson and Dwight Hair
In cooperation with the American Forestry Association
1989, 350 pp., illustrations, index
Cloth: $39.95 ISBN: 1-55963-003-5
Paper: $24.95 ISBN: 1-55963-002-7

The Poisoned Well: New Strategies for Groundwater Protection
By the Sierra Club Legal Defense Fund
1989, 400 pp., glossary, charts, appendixes, bibliography, index
Cloth: $31.95 ISBN: 0-933280-56-4
Paper: $19.95 ISBN: 0-933280-55-6

Reopening the Western Frontier
Edited by Ed Marston
From *High Country News*
1989, 350 pp., illustrations, photographs, maps, index
Cloth: $24.95 ISBN: 1-55963-011-6
Paper: $15.95 ISBN: 1-55963-010-8

Research Priorities for Conservation Biology
Island Press Critical Issues Series
Edited by Michael E. Soulé and Kathryn A. Kohm
In cooperation with the Society for Conservation Biology
1989, 110 pp., photographs, charts, graphs
Paper: $9.95 ISBN: 0-933280-99-8

Rivers at Risk: The Concerned Citizen's Guide to Hydropower
By John D. Echeverria, Pope Barrow, and Richard Roos-Collins
Foreword by Stewart L. Udall
In cooperation with American Rivers
1989, 220 pp., photographs, appendixes, indexes
Cloth: $29.95 ISBN: 0-933280-83-1
Paper: $17.95 ISBN: 0-933280-82-3

Rush to Burn
From *Newsday*
Winner of the Worth Bingham Award
1989, 276 pp., illustrations, photographs, graphs, index
Cloth: $29.95 ISBN: 1-55963-001-9
Paper: $14.95 ISBN: 1-55963-000-0

Shading Our Cities: Resource Guide for Urban and Community Forests
Edited by Gary Moll and Sara Ebenreck
In cooperation with the American Forestry Association
1989, 350 pp., illustrations, photographs, appendixes, index
Cloth: $34.95 ISBN: 0-933280-96-3
Paper: $19.95 ISBN: 0-933280-95-5

Wildlife of the Florida Keys: A Natural History
By James D. Lazell, Jr.
1989, 254 pp., illustrations, photographs, line drawings, maps, index
Cloth: $31.95 ISBN: 0-933280-98-X
Paper: $19.95 ISBN: 0-933280-97-1

These titles are available from Island Press, Box 7, Covelo, CA 95428. Please enclose $2.00 shipping and handling for the first book and $1.00 for each additional book. California and Washington, DC, residents add 6% sales tax. A catalog of current and forthcoming titles is available free of charge.